伦敦开题
TRANSFERS

15 International Young Architects in London
15位国际新生代建筑师的伦敦建筑教育体验及实践

With Introduction by Sir Peter Cook **彼德·库克爵士撰写导言**

主编：潘岩　协作编辑：丹尼尔·丹卓　祖贝尔·苏提　卡塔琳娜·迪昂斯伯鲁　翻译：李真

Chief Editor: Yan Pan　Associate Editors: Daniel Dendra　Zubair Surty　Katerina Dionysopoulou Translator: Zhen Li

中国建筑工业出版社

丽
修
Lis
Silv

迈克
米切尔
Michael
Mitchell

阿曼多
赫南德兹
Armando
Hernandez

里奥纳多
拉塔瓦
Leonardo
Lattavor

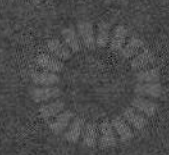

洛乌
昆资
Raoul
Kunz

娜奈德
亚科斯基
Nannete
Jackowski

卡提娅
阿柯曼
Katja
Ackermann

提奥
降托格鲁
Theo
arantoglou

玛若
卡利马尼
Maro
Kallimani

蒂诺斯
帕帕蒂米卓
Dinos
Papadimitriou

大卫
蓉基曼
David
ajchman

沃肯
奥卡纳格鲁
Volkan
Alkanoglu

潘岩
Yan
Pan

丹尼尔
丹卓
Daniel
Dendra

祖贝尔
苏提
Zubair
Surty

目录

第一部分 Part1

P002-017 卡提娅·阿柯曼 Katja Ackermann
P018-033 沃肯·奥卡纳格鲁 Volkan Alkanoglu
P034-051 丹尼尔·丹卓 Daniel Dendra
P052-065 阿曼多·赫南德兹 Armando Hernandez
P066-085 娜奈德·亚科斯基 Nannete Jackowski
P086-101 玛若·卡利马尼 Maro Kallimani
P102-121 洛乌·昆资 Raoul Kunz
P122-137 里奥纳多·拉塔瓦 Leonardo Lattavor
P138-151 迈克·米切尔 Michael Mitchell
P152-165 潘岩 Yan Pan
P166-179 蒂诺斯·帕帕蒂米卓 Dinos Papadimitriou
P180-193 提奥·萨隆托格鲁 Theo Sarantoglou
P194-209 丽莎·修娃 Lisa Silver
P210-225 祖贝尔·苏提 Zubair Surty
P226-239 大卫·塔基曼 David Tajchman

P240-241 导言 彼德·库克 *Peter Cook* Introduction
P242-243 本书建筑师分布图 Transfers Atlars
P244-245 编者按 潘岩 *Yan Pan* Words From Editor

第二部分 Part2

P248-257 卡提娅·阿柯曼 娜奈德·亚科斯基 洛乌·昆资 里奥纳多·拉塔瓦 里卡多·奥斯图斯 Katja Ackermann Nannete Jackowski Raoul Kunz Leonardo Lattavor Ricardo Ostos
P258-261 丽莎·修娃 Lisa Silver
P262-263 迈克·米切尔 Michael Mitchell
P264-265 玛若·卡利马尼 Maro Kallimani
P266-267 阿曼多·赫南德兹 Armando Hernandez
P268-269 沃肯·奥卡纳格鲁 Volkan Alkanoglu

卡提娅
阿柯曼
Katja
Ackermann
沃肯
奥卡纳格鲁
Volkan
Alkanoglu
丹尼尔
丹卓
Daniel
Dendra
阿曼多
赫南德兹
Armando
Hernandez
娜奈德
亚科斯基
Nannete
Jackowski
Part1
第一部分
玛若
卡利马尼
Maro
Kallimani
洛乌
昆资
Raoul
Kunz
里奥纳多
拉塔瓦
Leonardo
Lattavor
迈克
米切尔
Michael
Mitchell
潘岩
YanPan
蒂诺斯
帕帕蒂米卓
Dinos
Papadimitriou
提奥
萨隆托格鲁
Theo
Sarantoglou
丽莎
修娃
Lisa
Silver
祖贝尔
苏提
Zubair
Surty
大卫
塔基曼
David
Tajchman

N Circular
A 406 (S)
(A 12, A 13)
N. Circular
A 406(S)&(W)
(A 12, A 13)
(A 503, A 10, A 1)

卡提娅
阿柯曼
Katja
Ackermann

Self

2002/03 University College London, The Bartlett School of Architecture, London / UK
-Master of Architectural Design 2003

2002 Büro für Urbane Projekte, Leipzig / Germany
-Consultant Architect for urban regeneration and rehabilitation projects

2001 Norwich University, School of Architecture, Northfield VT / USA
-Guest Juror and Adjunct Professor - Twentieth Century Architectural Theory

John Anderson Studio, Burlington VT / USA
-Consultant Architect for local high-end residential design projects

Birdseye Design and Building Company, Richmond VT / USA
-Assistent Project Architect for local high-end residential design-build projects

2000/01 Freeman - French - Freeman Architects Inc., Burlington VT / USA
-Architect and Assistant Project Architect for commercial and institutional projects 1999

1999 Lohan Associates Inc., Chicago, IL / USA
-2 month internship

1998 VRA Architects, Chicago, IL / USA
-2 month internship VRA

1997/2000 Illinois Institute of Technology, Chicago, IL / USA
-Bachelor of Architecture - Honors

1996 Kraus & Krusenbaum Architects, Leipzig / Germany
-2 month intermship 2

1994/97 Hochschule für Technik, Wirtschaft & Kultur (HTWK) Leipzig, Germany
-Intermediate Diploma in Architecture 1997

自述

伦敦大学，巴特雷特建筑学院，伦敦/英国
-建筑设计硕士 2003

城市项目办公室，莱比锡/德国
-城市再生项目顾问建筑师

诺威奇大学建筑学院，诺斯菲尔德 维蒙特/美国
-客座初级助理教授—20世纪建筑理论

约翰•安德森工作室 伯林顿， 佛蒙特州 /美
-本地高端住宅项目顾问建筑师

博兹艾瑞设计建造公司，里士满，佛蒙特州 /美
-本地高端住宅项目助理项目建筑师

自由人—法国—自由人建筑师公司，
伯林顿， 佛蒙特州 /美国
—建筑师，商业和助理项目建筑师1999

洛寒公司，芝加哥，伊利诺伊州/美国
-2个月实习

建筑师事务所，芝加哥，伊利诺伊州/美国
-2个月实习

伊利诺斯技术学院，芝加哥，伊利诺伊州/美国
-建筑学学士—荣誉毕业

考斯与库森巴姆建筑师事务所，莱比锡/德国
-2个月实习

经济文化技术大学(HTWK) 莱比锡/德国
-建筑学中级文凭 1997

If one wants to discover new continents, one has to be willing to loose sight of the coastline. - Andre Gide

假如你想发现新大陆，那你必须甘愿失去海岸线的视界。——安德鲁 •纪德

Fleetingness has penetrated all aspects and levels of daily life. Contemporary temporality is the status quo. Production, working and living habits are the grounds on which this development towards a more and more ephemeral and incon-stant culture is taking place. Patchwork biographies characterized by job hopping, partner and address changes are the sign of a transforming society. Statistics prove the trend of increased mobility - a global migration. We constantly move between the mobile office and the Ikea living room and change places with increasing speed. Speed blurs con-tours, borders. Old rigid relationships are being lost. Things are not meant to last forever anymore. Everything is in constant motion and we always have to be on guard and ready to replace facts we have just accepted with new informa-tion. Locations change or have never really existed in the first place - if one considers the digital vertigo. "Here I was born and here I shall die...." does not exist anymore. Time and space overlap thus creating a perceived and physical effect of singularity. Virilio understands man as a permanent passenger, as a fleeting inhabitant of transportation devices (com-pare Virilio 1978).

In this evolving society decentralization and networking are fundamental concepts of economical, political and social relationships. Creating small swarms like interconnected clusters promises more efficiency, less maintenance and there-fore less waste of energy. New business structures, as an example, are based on a loosely connected network of highly specialized teams which stay connected through an umbrella organization, where they define their goals and organize their work. As every project differs from others, the composition of a particular project team is unique in accordance with the assignment it has been given.

In ancient times places of worship were built to last and eternity was a symbol of status. The construction of such places took decades, sometimes centuries. It was not only planned and re-planned repeatedly while work was in pro-gress, the work was also interrupted for planning purposes. In the eyes of human perception eternity is valued higher than temporality. This phenomenon has to be re-considered and re-evaluated. The design of the fleeting calls for spon-taneity and a philosophy of doing. Under this aspect, architecture has to become a transforming entity, the value of which will lie in its ability to adapt, to change, to transform and to be reborn all the same. Contemporary planning has not reached an appropriate level of mobility to keep up with the existing social changes and is overwhelmed by the task and its complexity.

The future? Recycling is the future. Transformability is the future. Adaptability is the future. We can already build on countless projects that deal with multifaceted solutions and aspects enshrined in the 'architecture of the fleeting'. Her-man Hertzberger's swimming pontoon house boat offers an individual lifestyle, Archigram's 'Cushicle', a full living module which can be carried on one's back, is one of the most mobile solutions, Robert Winkel's Smart House allows a maximum of variation, Shigeru Ban's Log House is based on a recycling concept which incorporates empty beer crates used as foundations as well as card board tubes and tarp as roof structure. On a larger scale, Kaas Osterhuis' pavilion allows motion and transformation through pneumatic 'muscles'.

Modular container structures with maximum assembly flexibility have been presented and discussed by a number of creative people. The mass prefabrication culture of the 1960s has now graduated and achieved a state of mutilated re-petitiveness in the early 2000s. More will follow. This should merely be our inspiration, because on the whole we still need to let go of the old principles and patterns. Fleetingness - this phenomenon is an active way of life, a natural state of culture from which many people still have to learn to use new options. In order to investigate, new forms of living innovation have to become even more innovative. Seemingly without orientation - one encounters a situation where one can move more freely since there is no wrong answer, no wrong option. The loss of the old principles and guide-lines for those who can manage means more freedom.

暂息性已经刺透了日常生活的各个层面。当代的暂息性即是现状。生产、工作和居住的习惯是更趋向于稍纵即逝与变化无常的文化产生发展的土壤。由跳槽、换偶、搬家所描绘的拼贴式传记正是这个转型中的社会的表征。统计数据证实了移动性增长的趋势——这是一种全球化的迁移。我们以不断增长的速度，频繁地穿梭于移动化办公与宜家式的住居空间之中，变换着我们的落脚点。速度模糊了轮廓、边界。旧有的刚性关系失落了。事物也不再追求持久。万物总是处在不断的变化之中。我们不得不保持警惕，并且随时准备用新的信息取代我们刚刚接受的事实。如果论及数码眩晕，你会发现场所不断变换着，或从未真实存在过。“生于此，死于斯……”再也不复存在。时间与空间重合因而创生出对奇点的知觉与生理的影响。维瑞罗把人类理解为永久的乘客，交通工具上短暂的居民。（比一较，维瑞罗，1978）

在这个进化社会中，分散化和网络化是经济、政治和社会关系的基本概念。创造小的组团比如相互联系的簇群确保了更高的效率，更少的维护和更少的能源浪费。举例来说，新型的商业结构是基于高度专业化的团队松散联系的网络关系，这些团队则通过定义共同目标和组织各项工作的伞状组织得以相互联系。每一个项目都不同于其他的，一个特定项目小组根据其所分配的任务而具有独特的构成。

在古代，修建崇拜性的场所是为了持久，不朽是地位的象征。在这些场所，建造过程会持续几十年，甚至几百年。他们不仅在修建过程中被反复地设计与再设计，建设工作也时常因为规划的意图发生变化而被打断。在人类感性的眼中，永恒的价值高于无常。这一现象必须予以再考虑与再评估。短暂性设计倡导自发性和一种行动的哲学。在这个层面下，建筑不得不成为一种可以变化的实体，其价值仍基于适应、改造、变形与再生能力。当代的设计还达不到适当的灵活性标准，以跟上现有的社会变化，就已被这一任务与它的复杂性所颠覆。

什么是未来？循环使用是未来，可变性是未来，适应性是未来。我们已经建造了难以计数的项目提供多侧面的解决方案和暂息性建筑的神圣。赫尔曼•赫尔佐格的游水浮动船屋提供了一种个性化的生活方式，建筑电信派的“Cushicle”，一种可以背在背上的完整居住模式，是最具有机动性的解决方式之一。罗伯特•温克的智能房屋允许最大限度的变化。坂茂的木屋是基于再生的观念，集合空的啤酒箱作为基础，用纸筒和防水布作为屋顶结构。在大的尺度上，卡斯•乌斯特休斯的亭子则是通过充气“肌肉”来运动和变形的。

一些富有创造力的人已经提出与讨论过拥有最大装配灵活性的箱式模块结构。20世纪60年代的大规模预制文化现已完成，并在本世纪初期得到非完全性的重复。更多模式将会出现。这应该只是我们的启示，因为总体来说，我们需要运用旧的法则和模式。暂息性——这个现象是一种活跃的生活方式，一种人们还得学习如何使用新的选择的自然状态的文化。为了研究的需要，新的居住革新方式必须变得更富有革命性。表面看来似乎没有方向——因为没有错误的解答与选择，人们遭遇到拥有更多自由度的情境。对于善于驾驭的人来说，旧有法则与指导的失落意味着更多的自由。

Fleetingness
暂息性

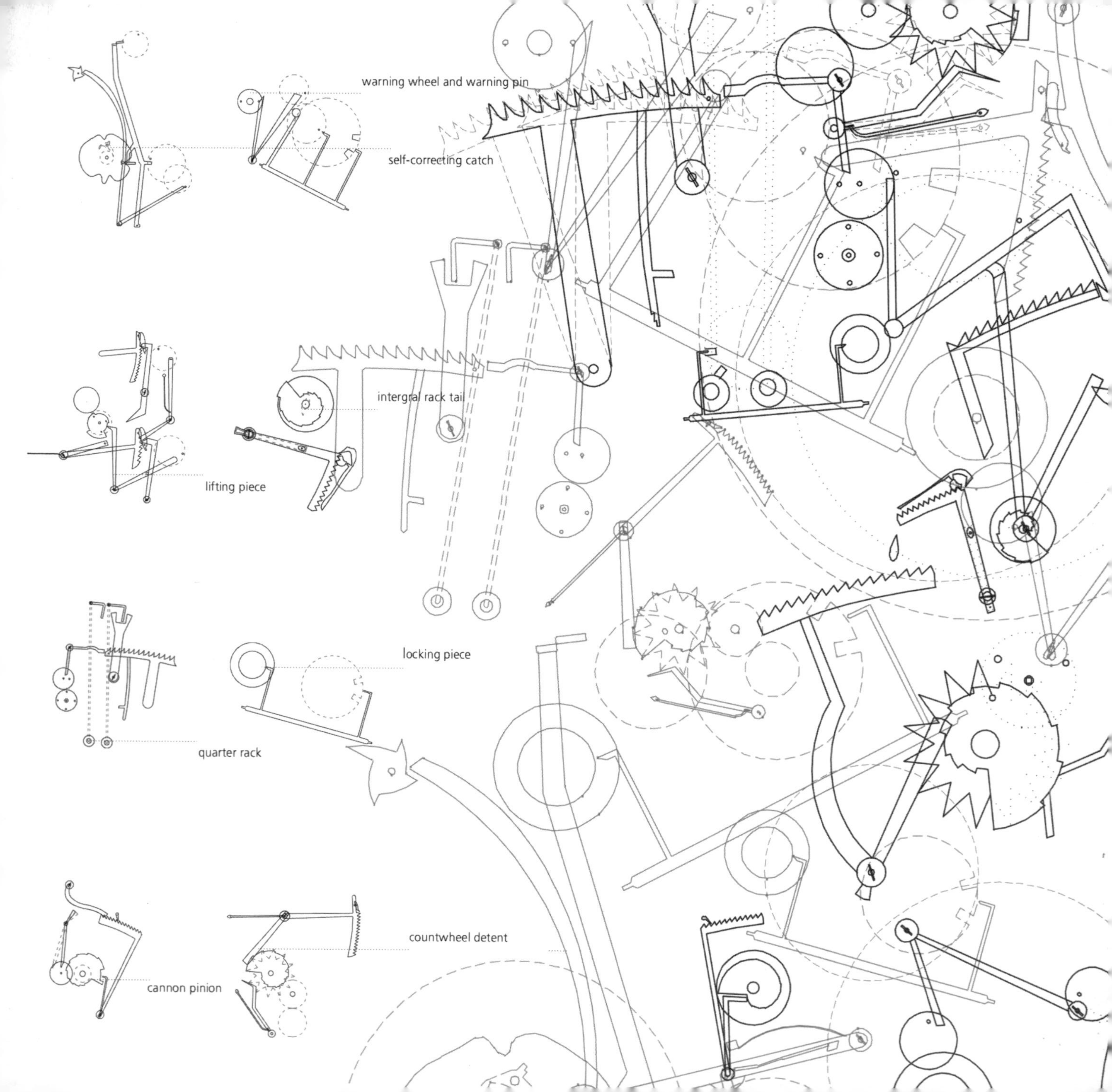

warning wheel and warning pin
self-correcting catch
intergral rack tail
lifting piece
locking piece
quarter rack
countwheel detent
cannon pinion

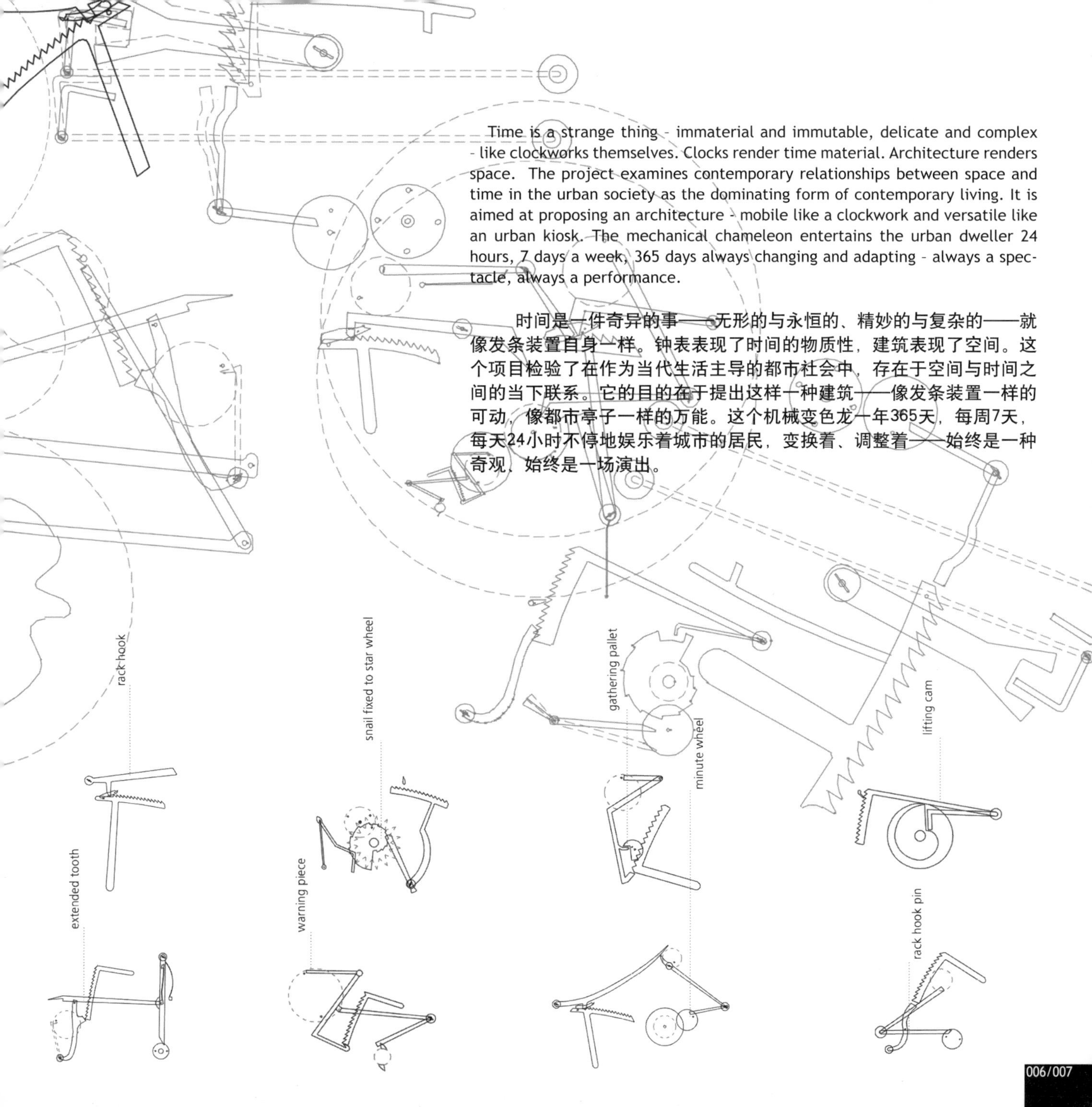

Time is a strange thing - immaterial and immutable, delicate and complex - like clockworks themselves. Clocks render time material. Architecture renders space. The project examines contemporary relationships between space and time in the urban society as the dominating form of contemporary living. It is aimed at proposing an architecture - mobile like a clockwork and versatile like an urban kiosk. The mechanical chameleon entertains the urban dweller 24 hours, 7 days a week, 365 days always changing and adapting - always a spectacle, always a performance.

时间是一件奇异的事——无形的与永恒的、精妙的与复杂的——就像发条装置自身一样。钟表表现了时间的物质性，建筑表现了空间。这个项目检验了在作为当代生活主导的都市社会中，存在于空间与时间之间的当下联系。它的目的在于提出这样一种建筑——像发条装置一样的可动，像都市亭子一样的万能。这个机械变色龙一年365天，每周7天，每天24小时不停地娱乐着城市的居民，变换着、调整着——始终是一种奇观、始终是一场演出。

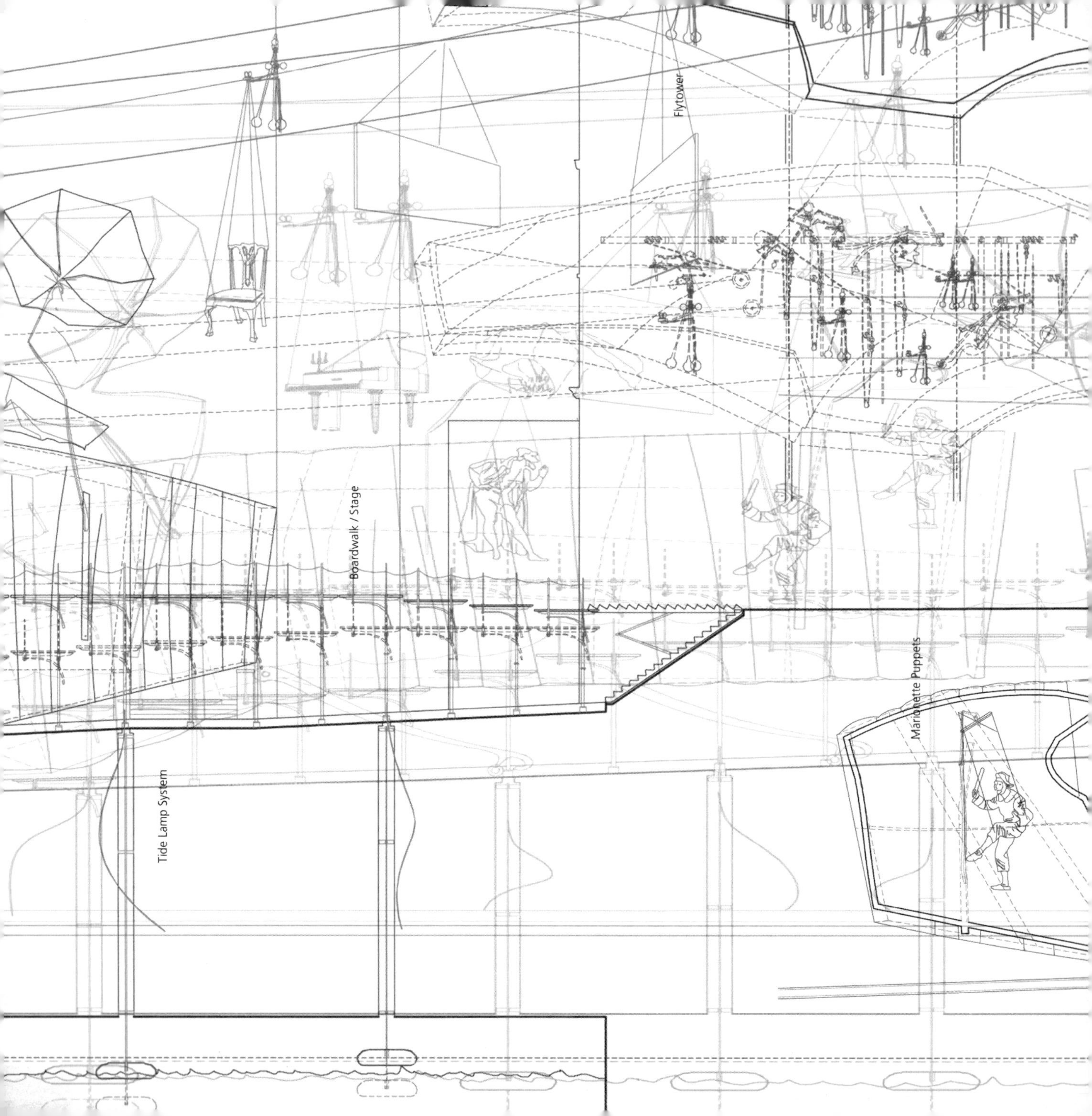
Flytower
Boardwalk / Stage
Marionette Puppets
Tide Lamp System

The city is the stage, the urban fabric the curtain, and everybody an actor and spectator at the same time. The play: a unique mix of routine and improvisation.The mechanical theatre makes change visible so that its physicality can be observed. One is able to experience trans-formation, speed and motion, unlike in a digital culture where performances and operations are oblique, hidden and coded. The intention is to reveal its operations and involve the spectator. The theatre becomes an event, a structure with a temporary set up which can be experienced anew at any given day or minute. Being both urban and public, it involves the spectator, the passerby, the pedestrian. Back drops and props, scenes and pictures, acts and pieces change over time.

以城市为舞台，都市肌理为幕布，每个人同时既是演员又是观众。表演：一种例行公事与即兴创作的独特混合体。机械式的剧场使变化成为可见的，其物质性是可以被观察的。有别于数码文化中间段、隐秘和编码式的展示与操作，人们可以在其中体验交通、速度与运动。设计的意图是展示剧场的运转，并让观众参与进来。剧场由此成为一个事件，一个在任何时间都可以体验如新的临时装置。作为城市的与公共的场所，它把观众、过客和行人统统卷入。背景与道具、布景与图象、表演与作品随着时间的流逝不断变换。

Counterweight System

Timepiece Narr(o)wtives
时间片段叙事

- Clockwork theater of time and space - Thesis project 2003

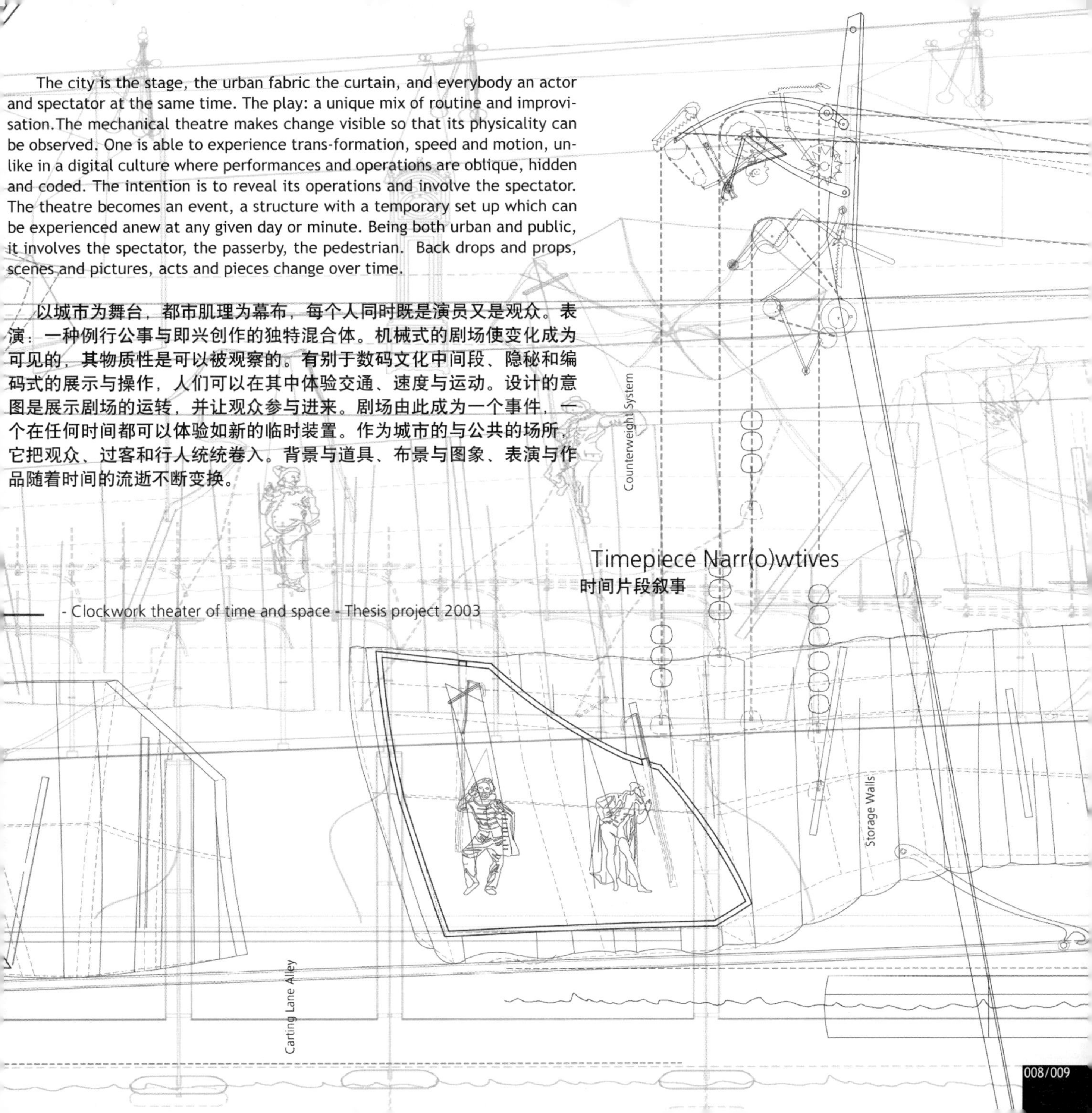

Storage Walls

Carting Lane Alley

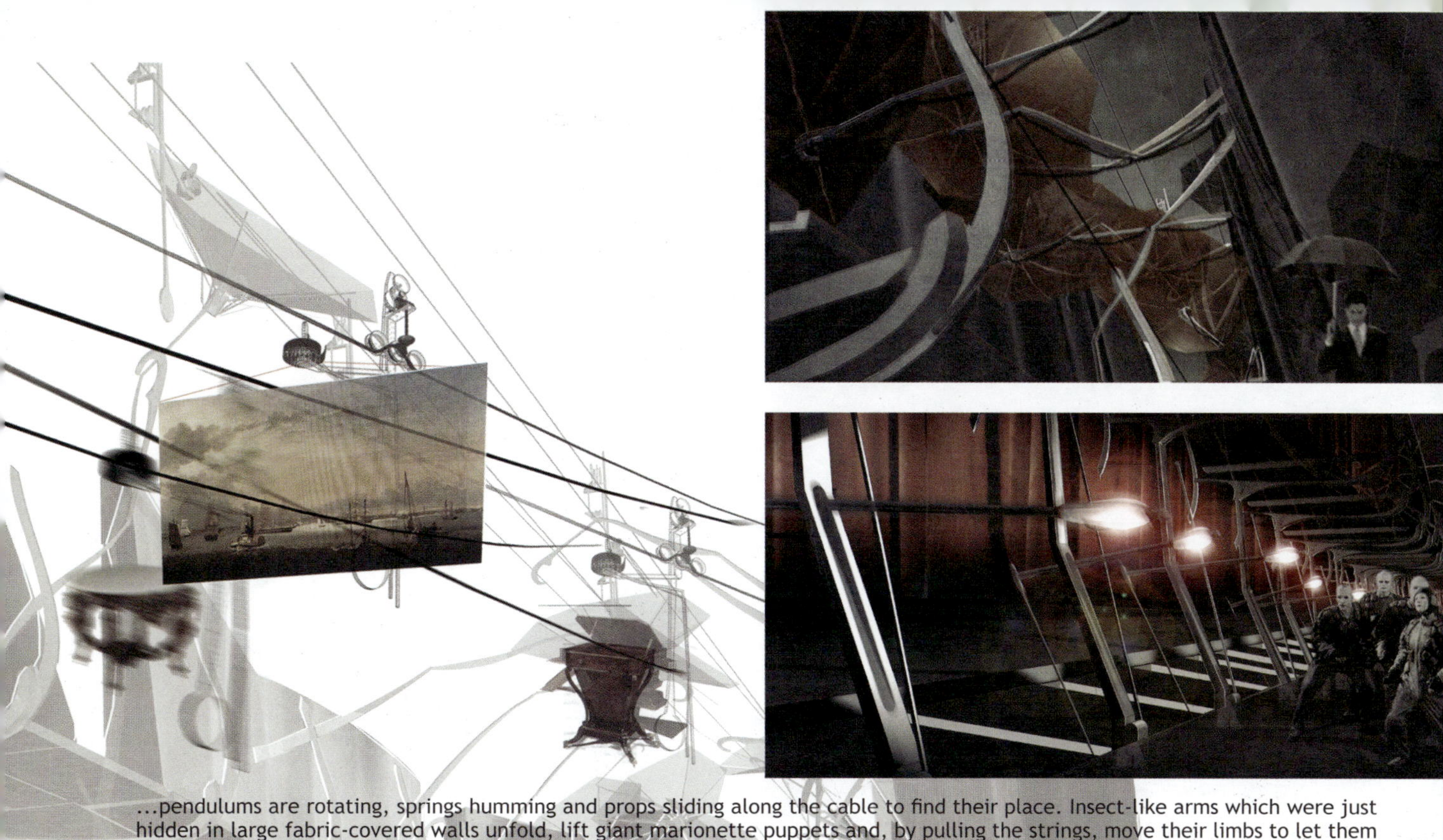

...pendulums are rotating, springs humming and props sliding along the cable to find their place. Insect-like arms which were just hidden in large fabric-covered walls unfold, lift giant marionette puppets and, by pulling the strings, move their limbs to let them gesture; dance if you like. The piano plays from above and the grandfather clock announces high noon. Later it is quiet again the boardwalk lifts slowly at night and sways with the rhythm of the tide waters to guide romantic hearts to the river side. until

……钟摆往复运动，弹簧嗡鸣，道具顺着缆索滑动寻找自身的位置。关节式的手臂藏在宽大织物覆盖的墙后，展示、提升起巨型的提线木偶，通过拉动绳子使木偶活动起来，做出各种姿势。如果你喜欢的话，跳舞也没问题。钢琴声从上面传来，落地式的大摆钟宣布着正午的到来。接着一切又陷入沉寂，直到夜晚木板铺就的步行道缓慢地升起，随着潮水的节奏摆动，引导着浪漫的心来到河边。

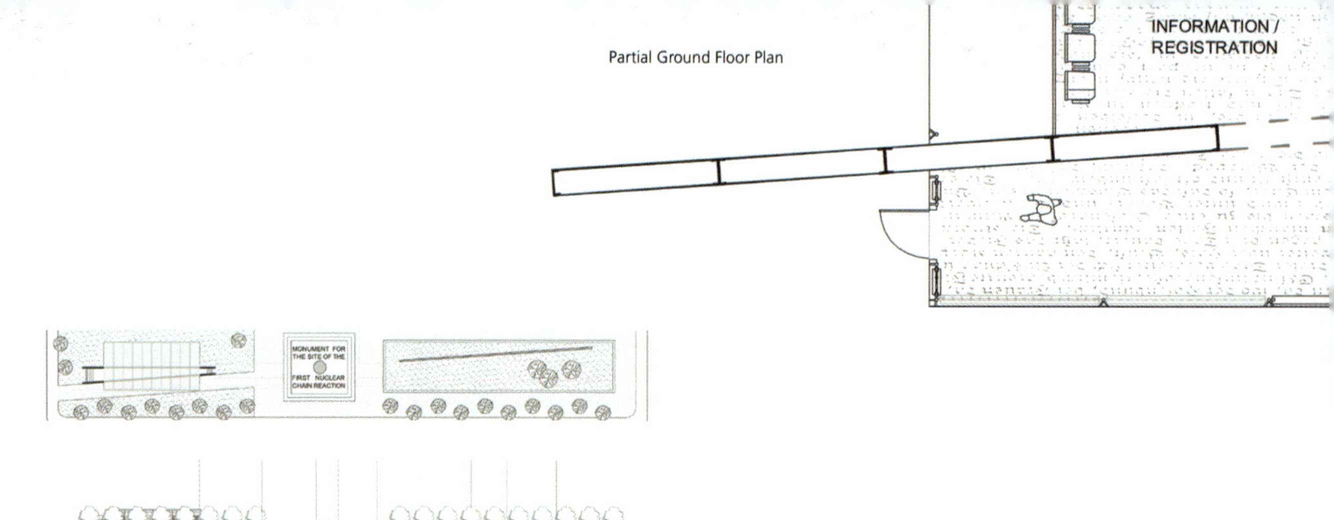

Partial Ground Floor Plan

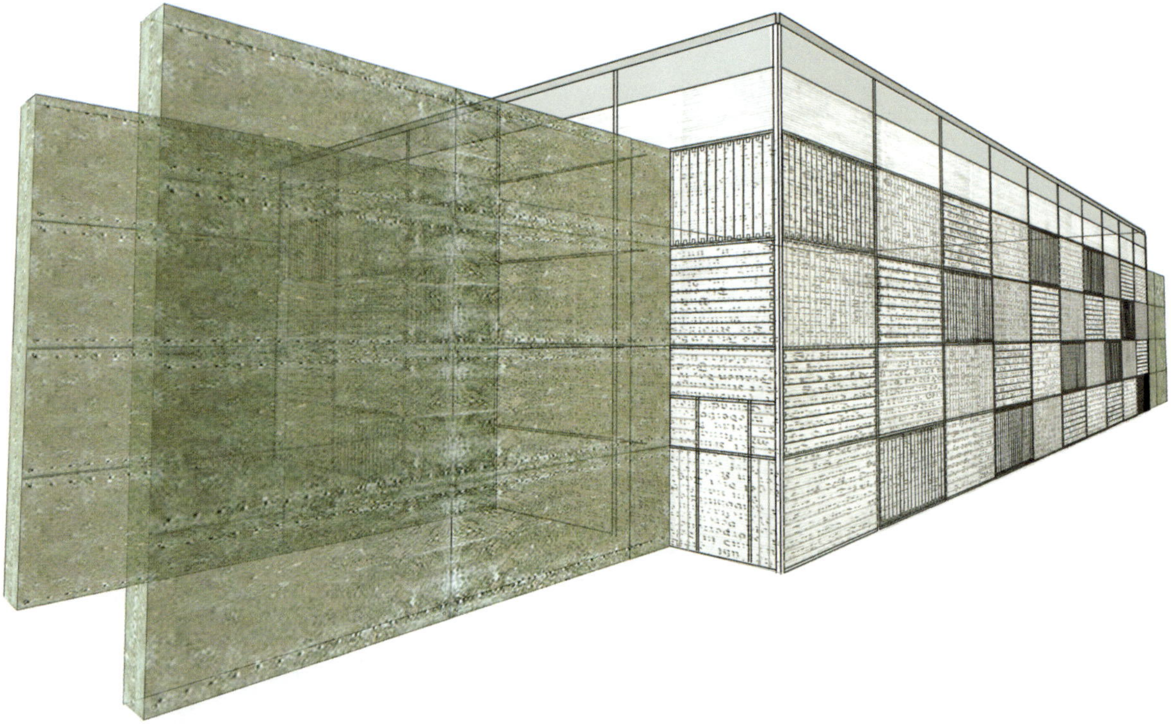

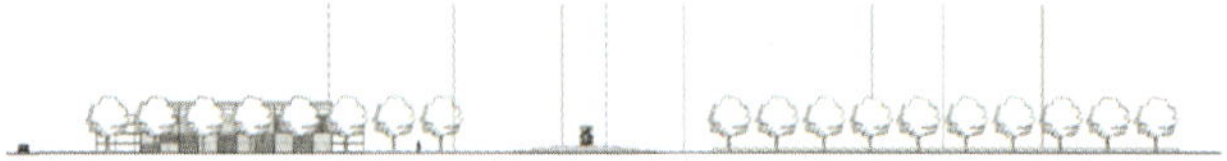

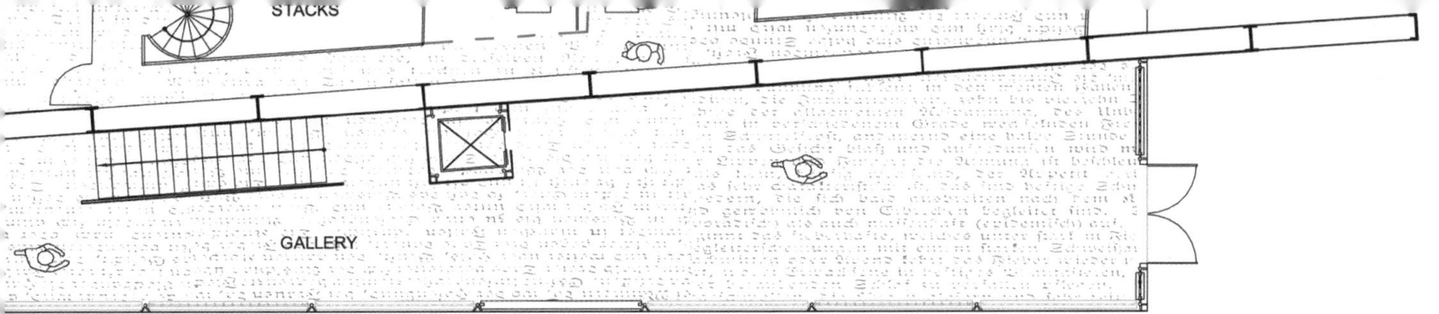

The building as a whole emphasizes the cross path of past and future - the **intersection** of old and new, opaque and transparent, continuity and interruption.

- Rare Book Library at the University of Chicago Campus, Chicago / Illinois - Project 1998

建筑作为整体强调由过去和现在构成的十字路径——一个旧与新、不透明与透明、连续与中断的**节点**。

——芝加哥大学珍本图书馆，芝加哥／伊利诺伊—项目1998

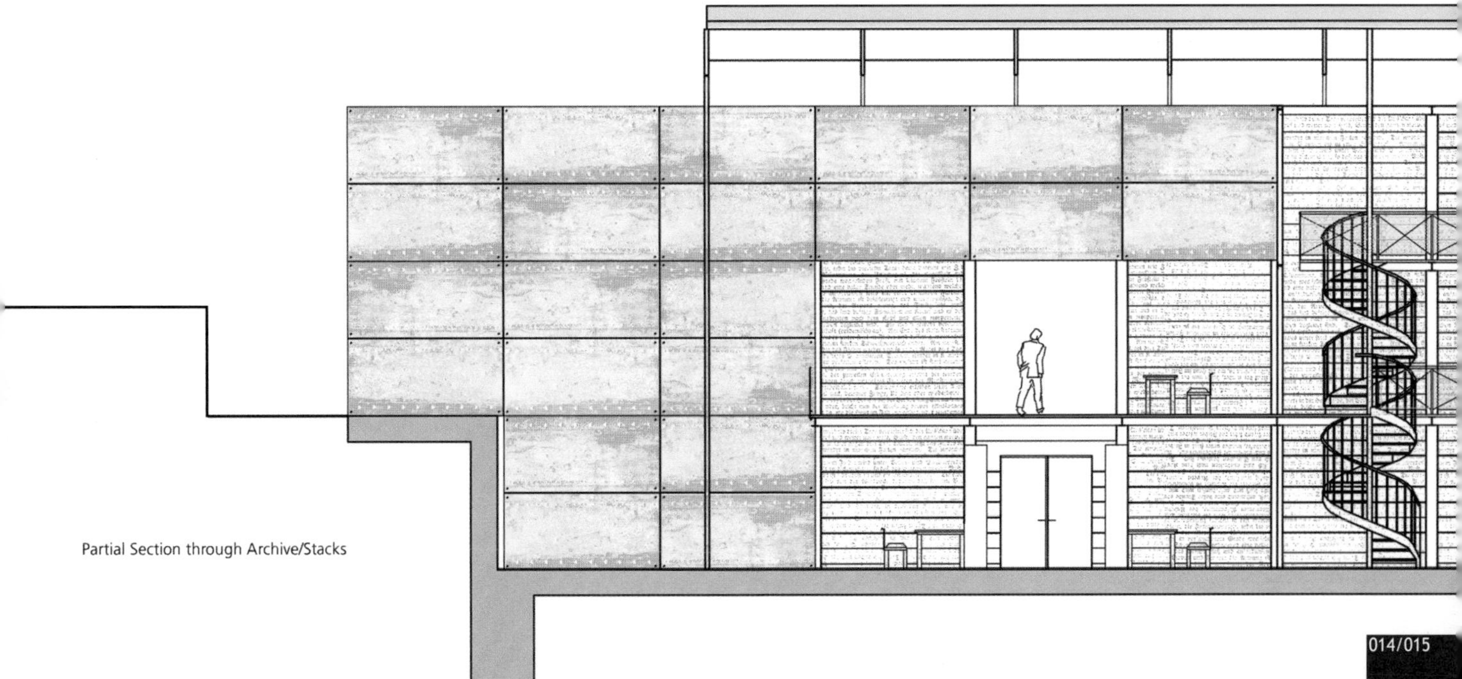

Partial Section through Archive/Stacks

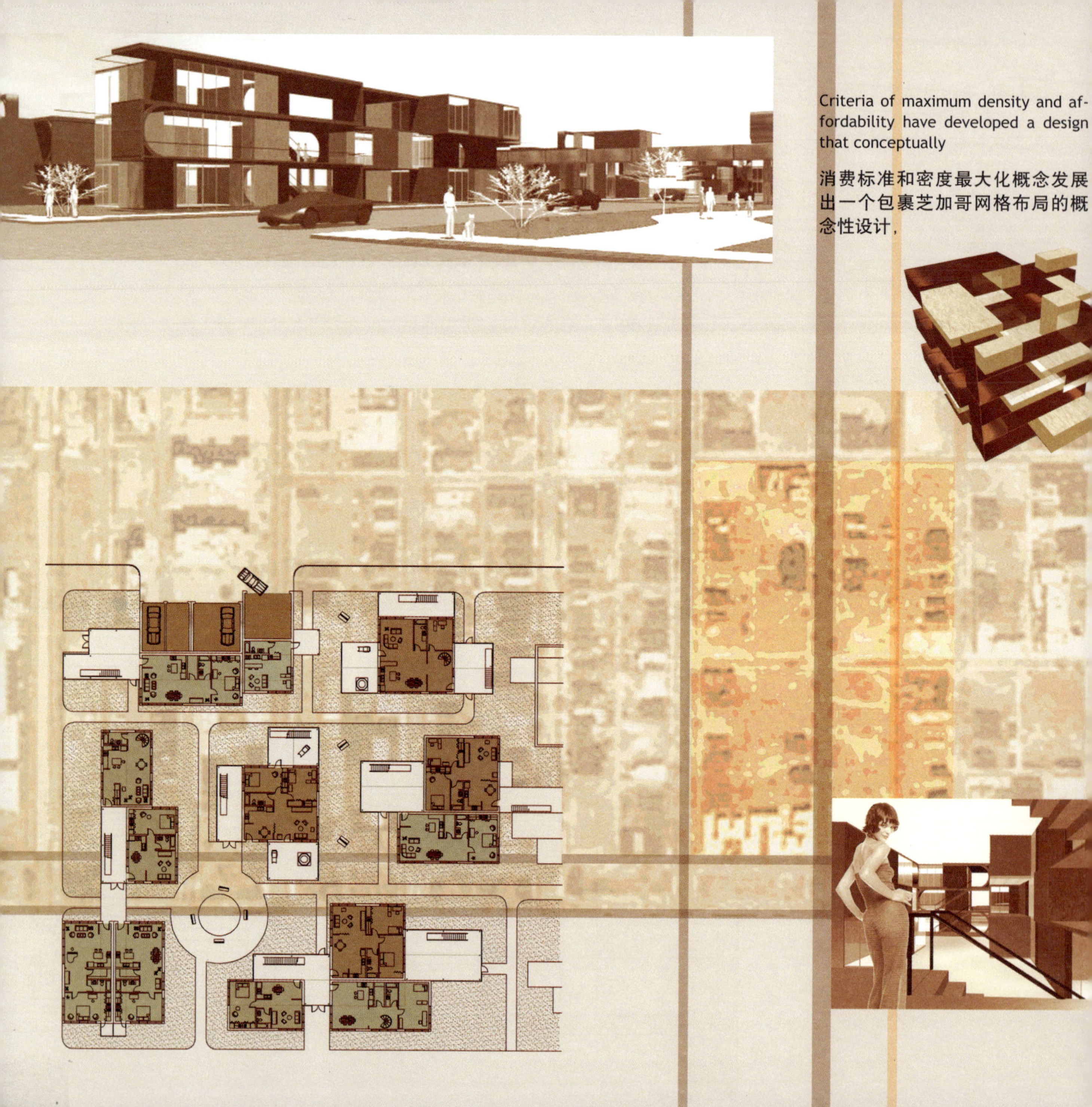

Criteria of maximum density and affordability have developed a design that conceptually

消费标准和密度最大化概念发展出一个包裹芝加哥网格布局的概念性设计,

Wrapped up the Chicago Grid layout, so as to create a mix of commercial & residential spaces.

从而创造出一个商住混合空间。

- Washington Park Housing, Chicago Southside - Project 2000

— 华盛顿公园住宅，芝加哥南部—项目2000

Central London, Stratford
Chelmsford, Romford
A12
N. Circular A406
Ilford, Barking
Docklands (A13)
Huntleigh
HEALTHCARE
Healthcare...it's part of our name
Huntleigh Healthcare
Sales Tel: 01582 745700
24hr Rental Helpline
Lo-call 0345 342000
V737 GVS

沃肯
奥卡纳格鲁
Volkan
Alkanoglu

DESIGN PROJECT: black_the robot
VOLKAN ALKANOGLU
MASTER OF ARCHITECTURAL DESIGN PROGRAM 2003
THE BARTLETT SCHOOL OF ARCHITECTURE, UC LONDON

curriculum vitae
+ volkan alkanoglu
- born 29.10.1976 in izmir/turkey
- since 1976 permanent resident of germany
- 1997-2002, studies of architecture, university of applied sciences duesseldorf, germany
- 2000, exchange studies, university of fine arts berlin, hdk berlin, germany
- 2001, diploma thesis in architecture, university of applied sciences duesseldorf
- 2002, full-time scholarship for post-graduate studies by german academic exchange service (daad)
- 2002, "master of architectural design", the bartlett school of architecture, uc london, uk
- sept. 2003, m.arch degree with distinction
- 2005, Registered Architect at AKNW,
- 2005, Registered Architect at ARB,
- 2006, Annual 10th British " Young Architect of the Year Award" Nomination
+ work experience, freelance, projects, teaching position
- july-sept. 2001, arx architects, nyc, usa
- okt.-dez. 2001, rkw architects, duesseldorf
- jan. 2002, webdesign for "euroshop-messe duesseldorf 02"
- feb. 2002, gruppe mdk architects, cologne
- feb.-june 2002, teaching position at the university of applied sciences duesseldorf
- april 2002, organisation of the architecture exhibition "kick-off", ballhaus nordpark, duesseldorf
- may 2002, kalhoefer-korschildgen architects, cologne
- june 2002, schusterarchitekten, duesseldorf
- 2004-2005, foster and partners, London
- 2005-2007, Future Systems, London
- 2007, Asymptote Architecture, New York
- 2007, Unit Master AA School Summerschool
- 2008, Teaching Associate, Option Studio, GSD Harvard University
+ selected exhibitions
- 2003, Slade Gallery, "Black - The Robot", 2003
- 2004, Architecture Foundation Athens, "Hellenikon Metropolitan Park"
- 2005, Royal Academy of Arts, "Istanbul Metamorphosis"
- 2006, New London Architecture, "Black Series"
- 2007, Architectural Association, "Rising Suprematism",

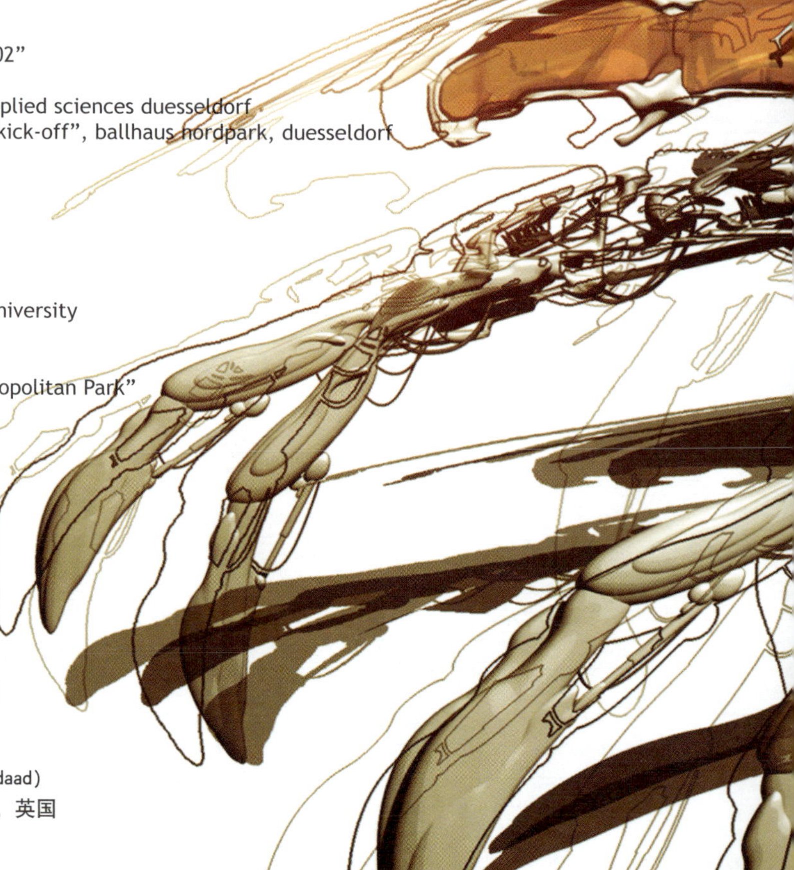

设计项目：黑_机器人
沃肯•奥卡纳格鲁
简历
+ 沃肯•奥卡纳格鲁
– 生于1976年10月29日 伊兹密尔／土耳其
– 1976，获得德国永久居住权
– 1997–2002，学习建筑，杜塞尔多夫实用科技大学，德国
– 2000，交换学生，柏林艺术大学，hdk柏林，德国
– 2001，执业文凭论文，杜塞尔多夫实用科技大学
– 2002，德国学术交换机构提供全日制研究生学习奖学金（daad）
– 2002，"建筑设计硕士"，巴特雷特建筑学院，伦敦大学，英国
– 2003，九月， 建筑学硕士优秀毕业
– 2005，AKNW，注册建筑师
– 2005，ARB，注册建筑师

- 2006，第十届英国年度青年建筑师提名

+ 工作经验，自主工作，项目，教学职位

- 2001,7月–9月．arx建筑师事务所，纽约，美国
- 2001,10月–12月．rkw建筑师事务所，杜塞尔多夫
- 2002，1月．“欧洲商店–措施，杜塞尔多夫02” 网页设计
- 2002，2月．mdk建筑设计集团，科隆
- 2002，2月–6月，任教于杜塞尔多夫实用科技大学
- 2002，4月，组织建筑展“开球”，包豪斯北公园，杜塞尔多夫
- 2002，5月，卡霍夫–寇申根建筑师事务所，科隆
- 2002，6月．鞋匠建筑师事务所，杜塞尔多夫
- 2004–2005，福斯特建筑师事务所，伦敦
- 2005–2007，未来系统建筑师事务所，伦敦
- 2007，渐近线建筑师事务所，纽约
- 2007，AA建筑学院暑期学校任教
- 2008，GSD哈佛大学，助教

+ 展览

- 2003，斯莱得画廊，“黑–机器人”
- 2004，雅典建筑基金会，“海棱尼孔公园都市”
- 2005，皇家美术学院，“伊斯坦布尔变形”
- 2006，新伦敦建筑，“黑系列”
- 2007，建筑师联盟(AA)，“上升的至上主义”

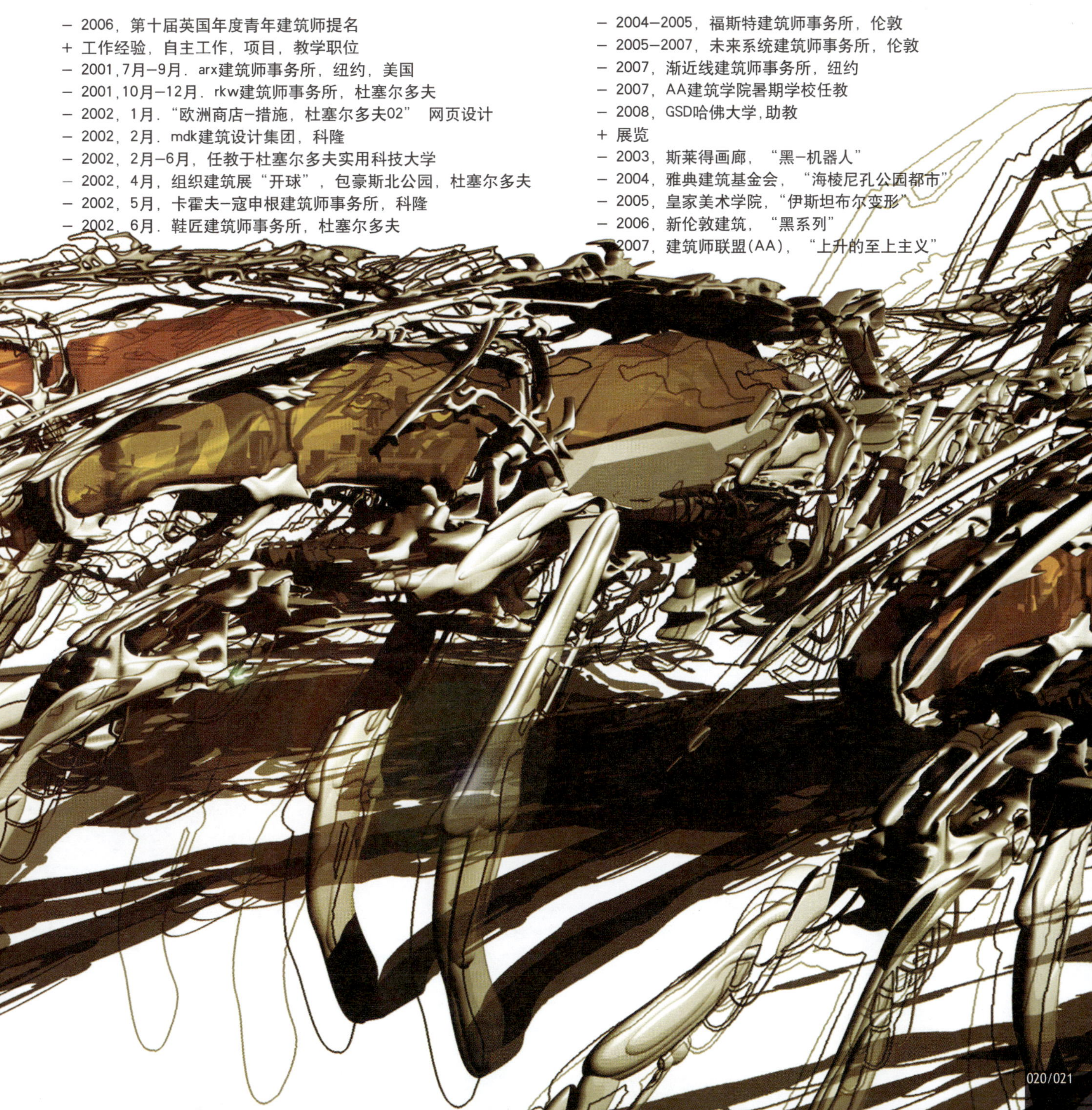

ARCHITECTURAL ETHOS:

THIS PROJECT WAS DEVELOPED AS A MASTER OF ARCHITECTURAL DESIGN SCHEME. IT IS SUPPOSED TO BE ABSOLUTELY FICTIONAL. THE INTENTION OF THIS IDEOLOGY IS THE FREEDOM OF LEARNING AND CREATING A PERSONAL DESIGN LANGUAGE. FICTION DEVELOPS CREATIVITY AND CREATIVITY SOMETIMES BECOMES REALITY. SO START CRYING YOUR HEART OUT...

>建筑气质：

本项目是硕士阶段建筑设计，是一个完全虚构的项目。这种思维方式的意图是自由地学习、创造出一种个性化的设计语汇。虚构可以发展出创造性，创造性有时又会变成现实。所以，从现在起喊出你的心声……

>SURVEY OF THE DESIGN PROJECT- BLACK_THE ROBOT

MY WORK IS BASED ON THE DEMAND OF THE PART THREE TOPIC OF THE MASTER OF ARCHITECTURAL DESIGN PROGRAM. SUBJECT MATTER OF THIS „100+ PROJECT“ IS A HOTEL IN COMBINATION WITH LEISURE AND ENTERTAINMENT. THE HOTEL SHOULD OFFER ACCOMODATION FOR 100 PEOPLE, WHO ARE INTERESTED IN CERTAIN LEISURE ACTIVITIES. THE COMBINATION BETWEEN ACCOMODATION AND FREETIME ACTIVITY SHOULD BE EXPRESSED IN TECTONIC ARCHITECTURE. THE INTENTION WITH MY PROPOSAL, WAS NOT TO DESIGN A SOLID COMPLEX OF BUILDINGS, BUT RATHER TO CREATE 100 INDIVIDUAL SPACES, SPREAD OVER THE CITY OF LONDON.ONE OF THE REASONS FOR THIS SEGMENTATION, IS THE POSSIBILITY TO BE MORE FLEXIBLE AND NOT TO BE DEPENDING ON ONE URBAN CONTEXT. IT WAS IMPORTANT FOR ME TO INTEGRATE DIFFERENT URBAN SITUATIONS IN THE CITY WITHIN THIS PROJECT. THE APPROACH LED TO THE IDEA, TO DEVELOP A DEVICE, THAT IS ABLE TO MOVE AND PROVIDE SEVERAL ACTIVITIES. THE VISITOR IS ACCOMODATED IN A WALKING ROBOT. HE IS ABLE TO NAVIGATE THE MACHINE TO HIS „PLACE OF DESIRE“, TO EXPERIENCE THE CITY FROM A UNIQUE ANGLE AND TO ENJOY THE COMFORT OF TAKING HIS APPLICATIONS WITH HIM. THE NUMBER OF 100 ROBOTS IN THE CITY IS ALSO A POSSIBILITY TO CONFIGURE CERTAIN ABSTRACT SITUATIONS.THE LARGE NUMBER OF MACHINES AND THEIR NEED FOR SUPPLY INVENTS AUTOMATICALLY THE ORIGIN OF A BASE, OR EVEN SEVERAL BASES IN DIFFERENT LOCATIONS COVERING THE CITY LIKE A GRID. THE COMMUNICATION AND INFRASTRUCTURE OF THE ROBOTS CAN TURN TO A NEW FUNCTIONAL MATRIX. THE DENSE BRANCHING OF THIS SYSTEM WILL ENFORCE EVERYONE TO LOVE THE ROBOT. A NEW SUCCESFUL world wide BRANCH IS BORN! viva la globalisation...

>设计项目的调研　黑色_机器人

我的作品是基于硕士设计课程第三阶段的主题的要求。“100+项目”是一个结合休闲与娱乐的综合性旅馆。这个旅馆将为100位对特定的休闲活动感兴趣的客人提供住宿。居住和休闲活动的结合应该以建构性的建筑加以表达。我的提案的意图不是设计一个固定的建筑综合体，而是创造100个单独的空间，散布在伦敦各处。这种分割的一个理由，是一种通往更加灵活的可能性，并且不依赖于单一的城市文脉。在这个项目中，整合城市里不同的都市状况对我来说非常重要。

发展概念的方法是发展一种装置，它能够移动并且提供多种活动。游客被安置在一个行走的机器人里。他驾驶着这个机器到达他所“渴望的宫殿”，从一个独特的角度来体验这个城市，并享受这种随身携带的舒适。城市中100个机器人的数量也是设置特定抽象情景的一种可能性。大量的机器和他们对供给的需要会自动地造成一个基地的起源，甚或是网格般覆盖城市的不同地点的若干基地。这些机器人之间的交流和其基础组织可以转变成新的功能性矩阵。系统密集的分支将会加强人们对这种机器人的喜爱。一种新的世界范围的分支诞生了！万岁，全球化！

>DESIGN

THE DESIGN OF THE ROBOT WAS A PROCESS THAT WENT FROM MODELMAKING OVER DRAW-INGS ILLUSTRATIONS, 3D MODELLING AND ANIMATIONS. TO REACH A BETTER ADAPTABILITY TO THE NARROWNESS OF THE CAPSULE, I DEVELOPED A MORPHOLOGICAL SERIES OF THE INTERIOR. MINIMIZING THE SPACE, BUT STILL TRYING TO KEEP THE QUALITY OF FORM WAS A SUPERIOR TASK. THE ILLUSTRATED SHAPES ARE FLEXIBLE AND RELATE TO AN ASSIGNED DESIGN IDEA. IT IS BASED ON BLACK SHAPE OBJECTS IN SPACE. DUE TO BE A MOVABLE OBJECT, PARAMETERS LIKE DYNAMIC AND SPEEDINESS ARE IMPORTANT AND HAVE INFLUENCED THE CHARACTERISTICS OF THE PROJECT.

>设计

机器人的设计是一个来自于通过手绘图、3D模型和动画形成的模型建造过程。针对于舱体的狭小，为获得一种更好的适应性，我发展了一个系列形态学的室内空间。把空间最小化，但仍试图把保持形态的品质作为首要任务。插图所描绘的形态是灵活的，并与指定的设计概念相关。这是基于黑暗是如何塑造物体的。由于这是一个可动物体，诸如动力和速度这些参量是很重要的，并且对项目的特征也产生了影响。

conceptual section of the **base**
基地的概念部分

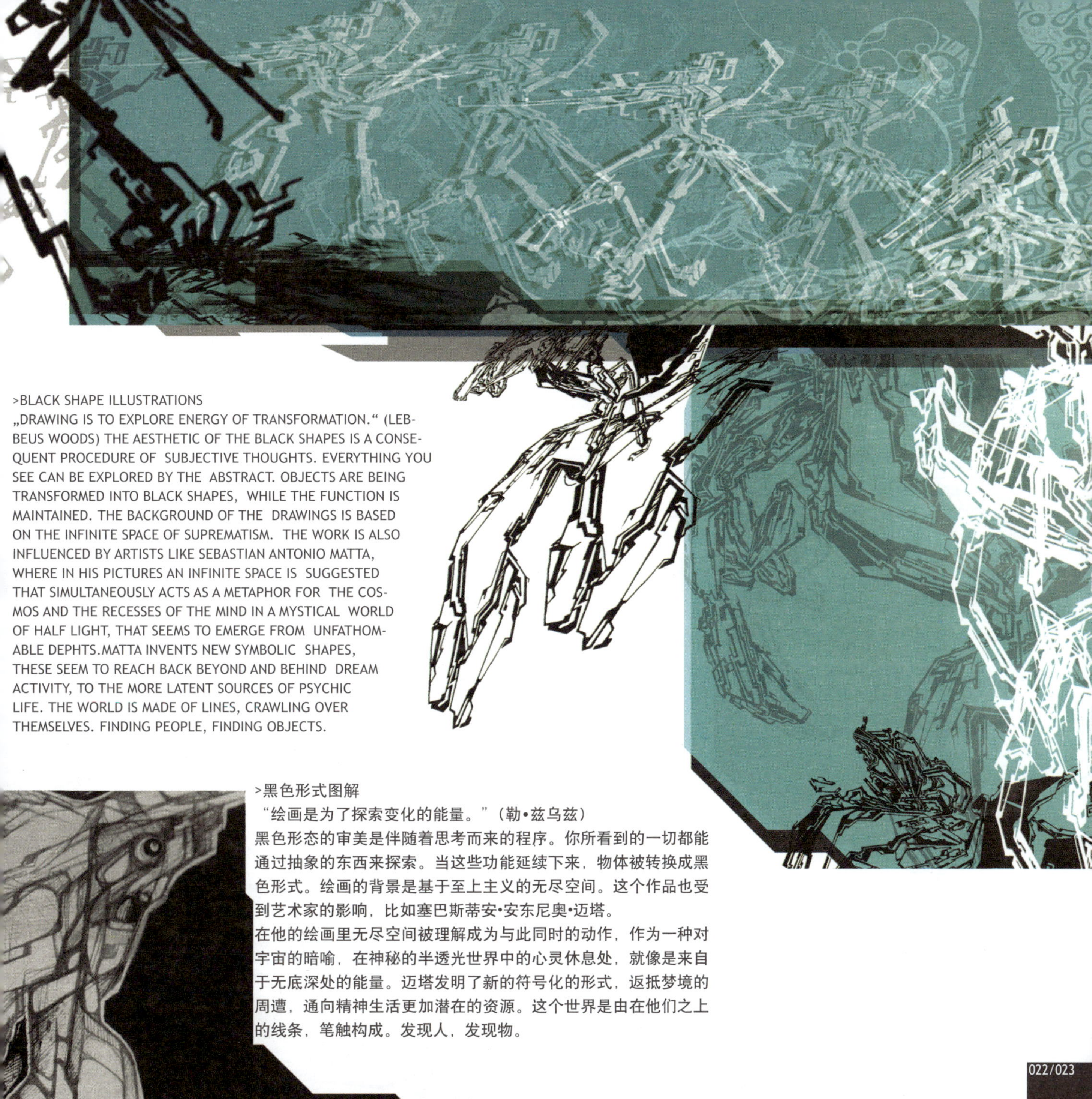

>BLACK SHAPE ILLUSTRATIONS

„DRAWING IS TO EXPLORE ENERGY OF TRANSFORMATION.“ (LEBBEUS WOODS) THE AESTHETIC OF THE BLACK SHAPES IS A CONSEQUENT PROCEDURE OF SUBJECTIVE THOUGHTS. EVERYTHING YOU SEE CAN BE EXPLORED BY THE ABSTRACT. OBJECTS ARE BEING TRANSFORMED INTO BLACK SHAPES, WHILE THE FUNCTION IS MAINTAINED. THE BACKGROUND OF THE DRAWINGS IS BASED ON THE INFINITE SPACE OF SUPREMATISM. THE WORK IS ALSO INFLUENCED BY ARTISTS LIKE SEBASTIAN ANTONIO MATTA, WHERE IN HIS PICTURES AN INFINITE SPACE IS SUGGESTED THAT SIMULTANEOUSLY ACTS AS A METAPHOR FOR THE COSMOS AND THE RECESSES OF THE MIND IN A MYSTICAL WORLD OF HALF LIGHT, THAT SEEMS TO EMERGE FROM UNFATHOMABLE DEPHTS.MATTA INVENTS NEW SYMBOLIC SHAPES, THESE SEEM TO REACH BACK BEYOND AND BEHIND DREAM ACTIVITY, TO THE MORE LATENT SOURCES OF PSYCHIC LIFE. THE WORLD IS MADE OF LINES, CRAWLING OVER THEMSELVES. FINDING PEOPLE, FINDING OBJECTS.

>黑色形式图解

“绘画是为了探索变化的能量。”（勒•兹乌兹）

黑色形态的审美是伴随着思考而来的程序。你所看到的一切都能通过抽象的东西来探索。当这些功能延续下来，物体被转换成黑色形式。绘画的背景是基于至上主义的无尽空间。这个作品也受到艺术家的影响，比如塞巴斯蒂安•安东尼奥•迈塔。

在他的绘画里无尽空间被理解成为与此同时的动作，作为一种对宇宙的暗喻，在神秘的半透光世界中的心灵休息处，就像是来自于无底深处的能量。迈塔发明了新的符号化的形式，返抵梦境的周遭，通向精神生活更加潜在的资源。这个世界是由在他们之上的线条，笔触构成。发现人，发现物。

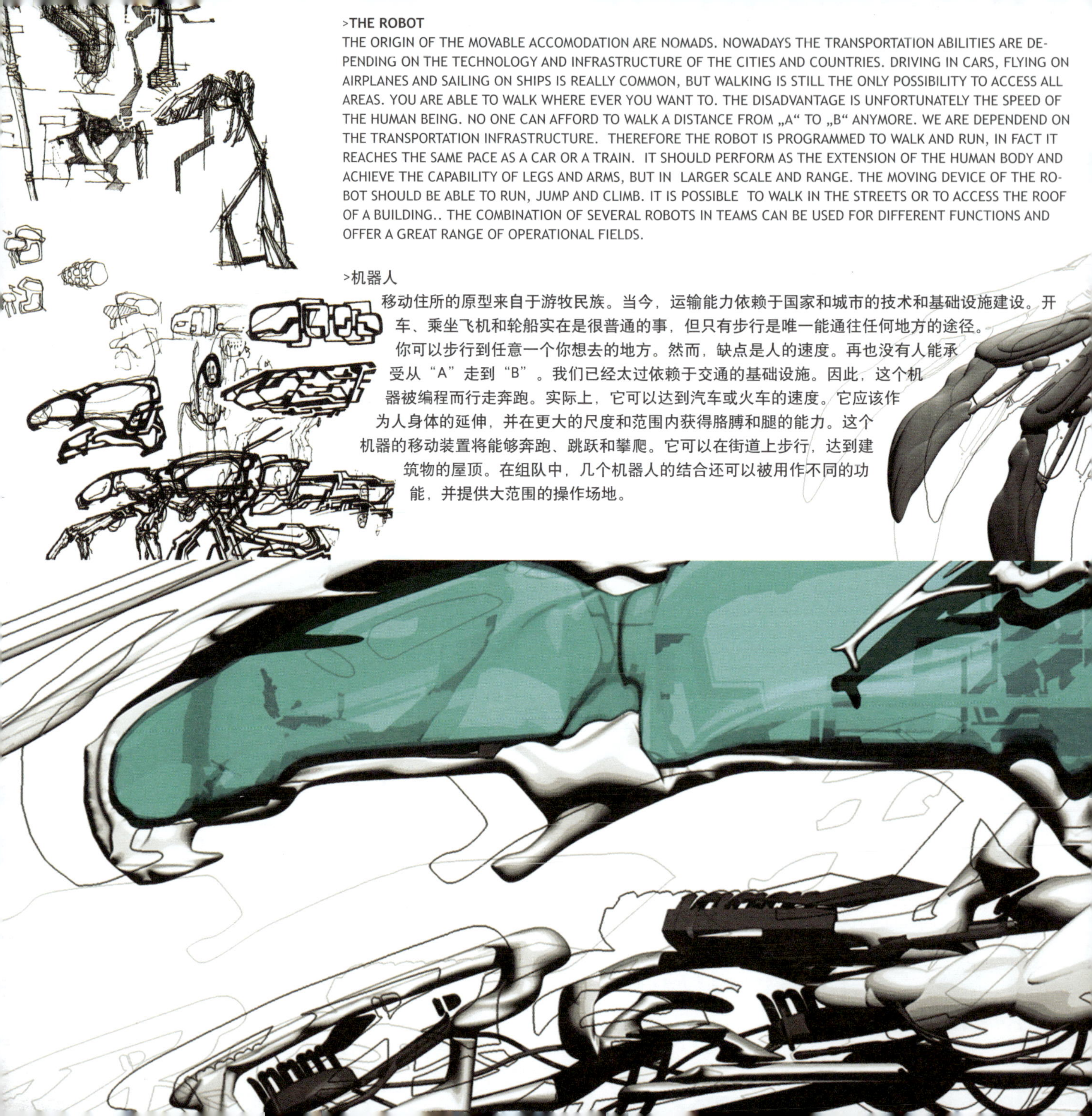

>THE ROBOT

THE ORIGIN OF THE MOVABLE ACCOMODATION ARE NOMADS. NOWADAYS THE TRANSPORTATION ABILITIES ARE DEPENDING ON THE TECHNOLOGY AND INFRASTRUCTURE OF THE CITIES AND COUNTRIES. DRIVING IN CARS, FLYING ON AIRPLANES AND SAILING ON SHIPS IS REALLY COMMON, BUT WALKING IS STILL THE ONLY POSSIBILITY TO ACCESS ALL AREAS. YOU ARE ABLE TO WALK WHERE EVER YOU WANT TO. THE DISADVANTAGE IS UNFORTUNATELY THE SPEED OF THE HUMAN BEING. NO ONE CAN AFFORD TO WALK A DISTANCE FROM „A“ TO „B“ ANYMORE. WE ARE DEPENDEND ON THE TRANSPORTATION INFRASTRUCTURE. THEREFORE THE ROBOT IS PROGRAMMED TO WALK AND RUN, IN FACT IT REACHES THE SAME PACE AS A CAR OR A TRAIN. IT SHOULD PERFORM AS THE EXTENSION OF THE HUMAN BODY AND ACHIEVE THE CAPABILITY OF LEGS AND ARMS, BUT IN LARGER SCALE AND RANGE. THE MOVING DEVICE OF THE ROBOT SHOULD BE ABLE TO RUN, JUMP AND CLIMB. IT IS POSSIBLE TO WALK IN THE STREETS OR TO ACCESS THE ROOF OF A BUILDING.. THE COMBINATION OF SEVERAL ROBOTS IN TEAMS CAN BE USED FOR DIFFERENT FUNCTIONS AND OFFER A GREAT RANGE OF OPERATIONAL FIELDS.

>机器人

移动住所的原型来自于游牧民族。当今，运输能力依赖于国家和城市的技术和基础设施建设。开车、乘坐飞机和轮船实在是很普通的事，但只有步行是唯一能通往任何地方的途径。你可以步行到任意一个你想去的地方。然而，缺点是人的速度。再也没有人能承受从“A”走到“B”。我们已经太过依赖于交通的基础设施。因此，这个机器被编程而行走奔跑。实际上，它可以达到汽车或火车的速度。它应该作为人身体的延伸，并在更大的尺度和范围内获得胳膊和腿的能力。这个机器的移动装置将能够奔跑、跳跃和攀爬。它可以在街道上步行，达到建筑物的屋顶。在组队中，几个机器人的结合还可以被用作不同的功能，并提供大范围的操作场地。

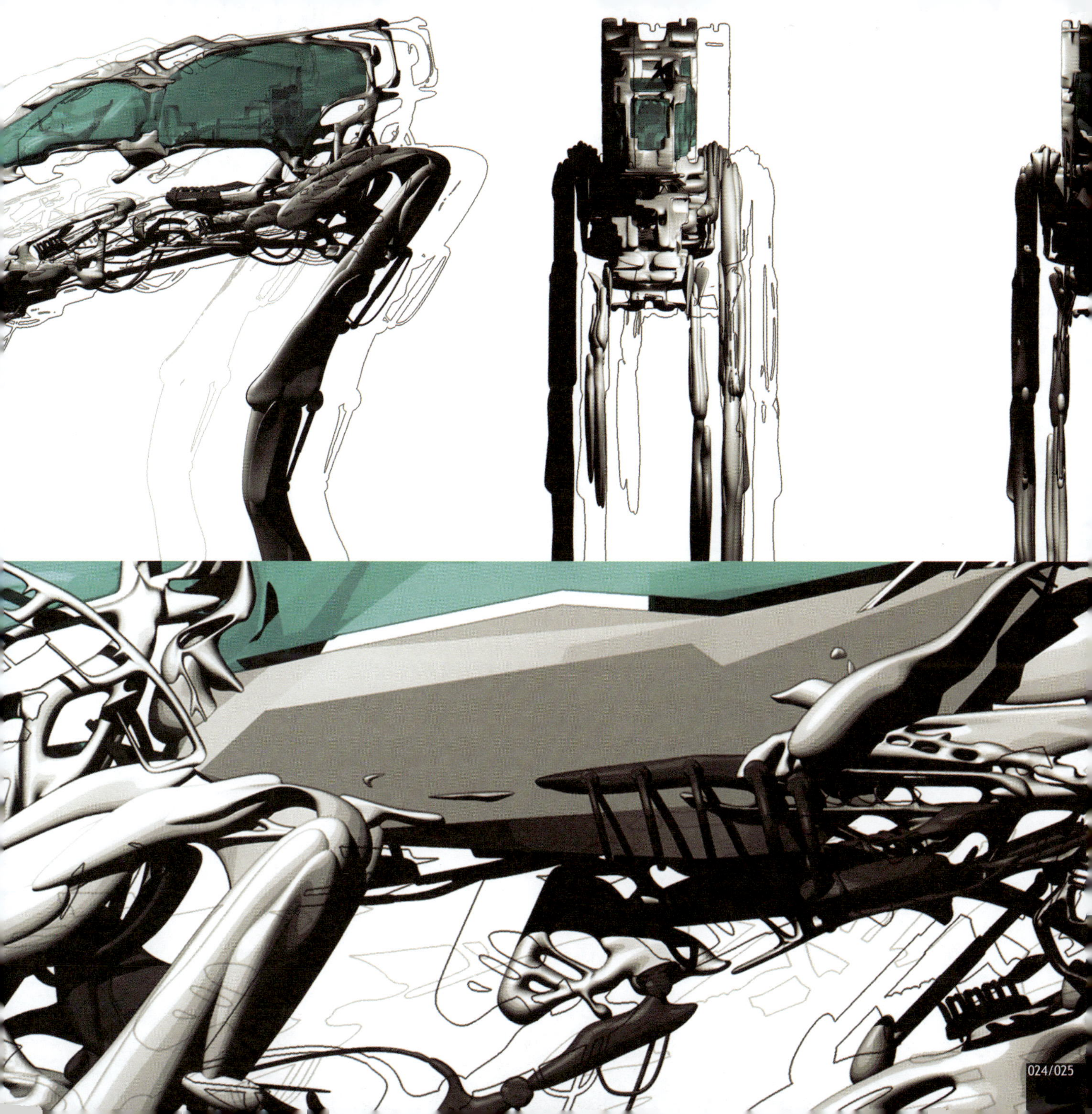

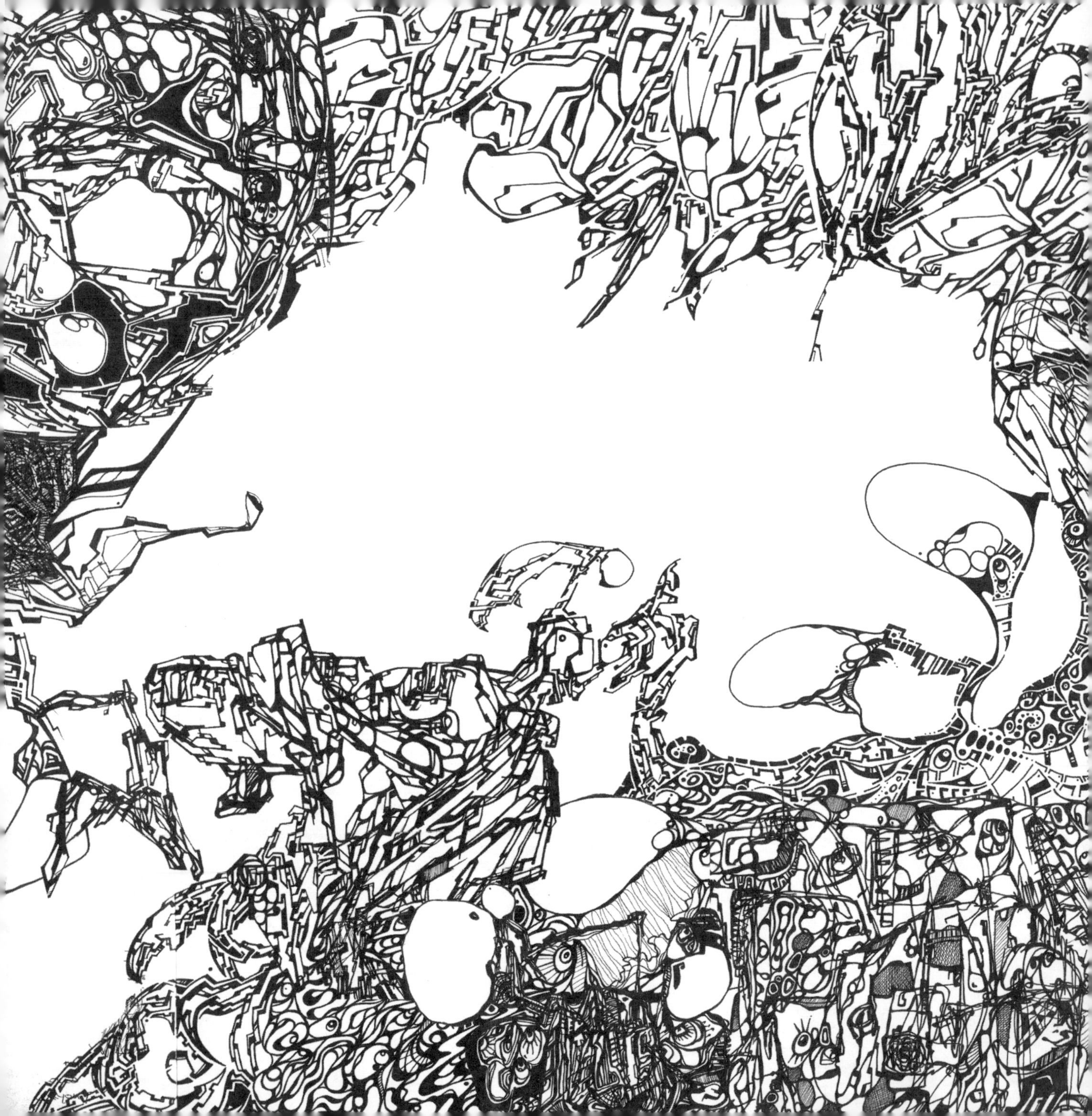

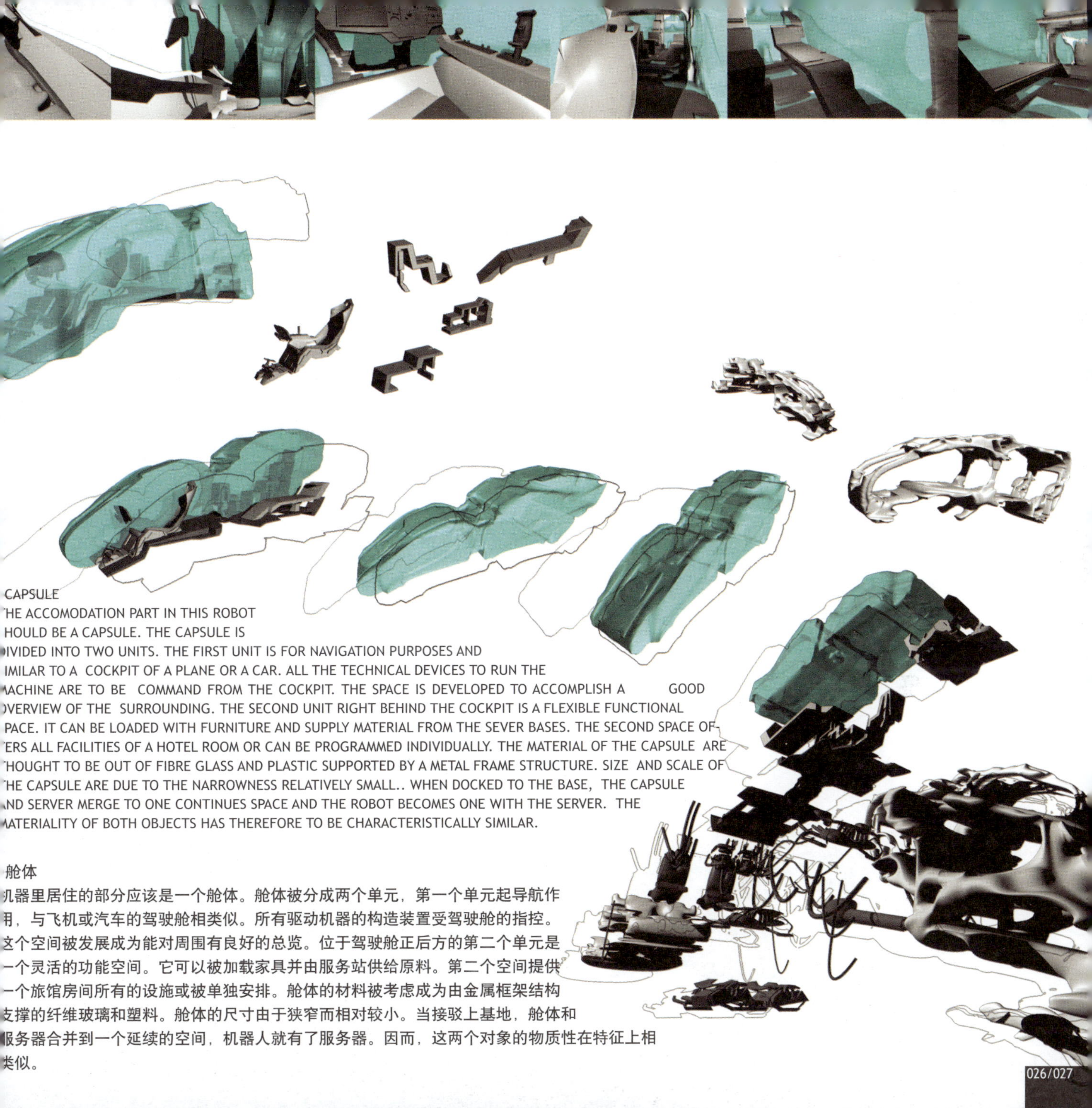

CAPSULE

'HE ACCOMODATION PART IN THIS ROBOT HOULD BE A CAPSULE. THE CAPSULE IS IVIDED INTO TWO UNITS. THE FIRST UNIT IS FOR NAVIGATION PURPOSES AND IMILAR TO A COCKPIT OF A PLANE OR A CAR. ALL THE TECHNICAL DEVICES TO RUN THE ACHINE ARE TO BE COMMAND FROM THE COCKPIT. THE SPACE IS DEVELOPED TO ACCOMPLISH A GOOD VERVIEW OF THE SURROUNDING. THE SECOND UNIT RIGHT BEHIND THE COCKPIT IS A FLEXIBLE FUNCTIONAL PACE. IT CAN BE LOADED WITH FURNITURE AND SUPPLY MATERIAL FROM THE SEVER BASES. THE SECOND SPACE OF-'ERS ALL FACILITIES OF A HOTEL ROOM OR CAN BE PROGRAMMED INDIVIDUALLY. THE MATERIAL OF THE CAPSULE ARE 'HOUGHT TO BE OUT OF FIBRE GLASS AND PLASTIC SUPPORTED BY A METAL FRAME STRUCTURE. SIZE AND SCALE OF 'HE CAPSULE ARE DUE TO THE NARROWNESS RELATIVELY SMALL.. WHEN DOCKED TO THE BASE, THE CAPSULE ND SERVER MERGE TO ONE CONTINUES SPACE AND THE ROBOT BECOMES ONE WITH THE SERVER. THE ATERIALITY OF BOTH OBJECTS HAS THEREFORE TO BE CHARACTERISTICALLY SIMILAR.

舱体

几器里居住的部分应该是一个舱体。舱体被分成两个单元，第一个单元起导航作用，与飞机或汽车的驾驶舱相类似。所有驱动机器的构造装置受驾驶舱的指控。这个空间被发展成为能对周围有良好的总览。位于驾驶舱正后方的第二个单元是一个灵活的功能空间。它可以被加载家具并由服务站供给原料。第二个空间提供一个旅馆房间所有的设施或被单独安排。舱体的材料被考虑成为由金属框架结构支撑的纤维玻璃和塑料。舱体的尺寸由于狭窄而相对较小。当接驳上基地，舱体和服务器合并到一个延续的空间，机器人就有了服务器。因而，这两个对象的物质性在特征上相类似。

sequence of robot docking to server
机器人泊入服务器的步骤

>CONCEPTUAL IDEA

THE SUPERIOR IDEA IS THE ROBOT TO BE AN INTERPRETATION OF THE HUMAN BEING. THE ROBOT IS AN EXTENSION OF THE HUMAN BODY. IT SHOULD TRY TO COPY THE FUNCTION OF THE BODY AND BE AS SIMILAR AS POSSIBLE TO THE HUMAN. THE MOVEMENT OF THE HUMAN BEING IS SUPPORTED AND ENLARGED BY TECHNICAL DEVICES. NAVIGATION AND MOVEMENT OF THE BODY FOLLOWS INSTRUCTONS OF THE BRAIN. SIMILAR TO THE CENTRAL NERVOUS SYSTEM, THE CAPSULE HAS AS WELL A CONTROL CENTER. THE NAVIGATION AREA IS THE COCKPIT OF THE ROBOT AND RESPONSIBLE FOR THE MOVEMENT. COVER AND CLOTHES OF THE HUMAN BEING IS INDIVIDUAL DESIGNATED AND IS DEPENDENT ON THE TASK. IF HE FEELS COLD, HE PUTS WARM CLOTHES ON. IF HE IS GOING TO EXERCISE, HE WILL WEAR CLOTHES FOR THAT SPECIAL ACTIVITY. HUMAN BEINGS LEAVE THEIR ACCOMODATION AND PREPARE THEMSELVES FOR THE SPECIFIC MOMENT THEY GOT TO FACE AT THIS DAY. DUE TO THE FUNCTION HE PREPARES HIMSELF. THE ROBOT SHOULD ACT IN THE SAME WAY AND JUST LOAD THE FUNCTIONS HE WILL NEED. NOTHING WITHOUT FUNCTION IS GOING TO BE PLUGGED INTO THE CAPSULE.

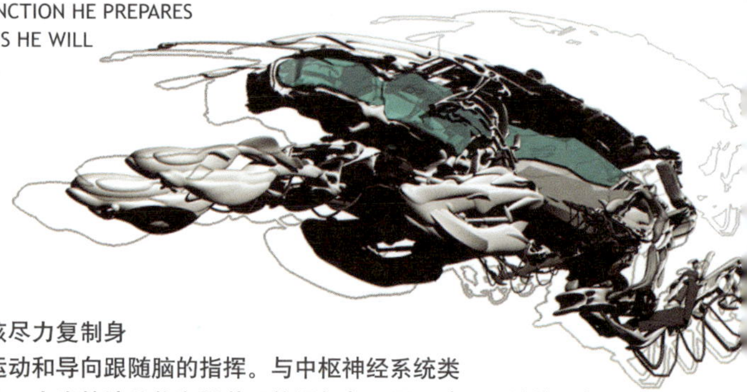

LIVE IN THE SAME WORLD,
LIVE IN THE ROBOTIZED WORLD,
GET EVERYTHING YOU DESIRE,
WHAT ARE YOU DOING?
WHAT DO YOU NEED?
SAY, SELECT AND GO!

概念意图

这个超级想法是把机器人作为人的一种转译。机器人是人体的延续。它应该尽力复制身体的功能，尽力与人相似。人类的活动被技术设施所支持和放大。身体的运动和导向跟随脑的指挥。与中枢神经系统类似，这个舱体也有一个控制中心。导航区域是机器人的驾驶舱并对运动负责。人类的遮盖物和服装是依据任务不同而独立设计的。如果觉得冷，他就穿上暖和的衣服。如果要进行体育运动，他会穿为这项活动的特殊服装。人们离开住所，为这一天要面对的活动做好准备。依据功能他为自己做好准备。机器人应该以同样的方式行为并且只装载他会需要的功能。没用的东西不会被装载到舱体上。

生活在同一个世界，
生活在一个机器人化的世界，
获得你渴望的一切，
你在做什么？
你需要什么？
说吧、选择吧、行动吧！

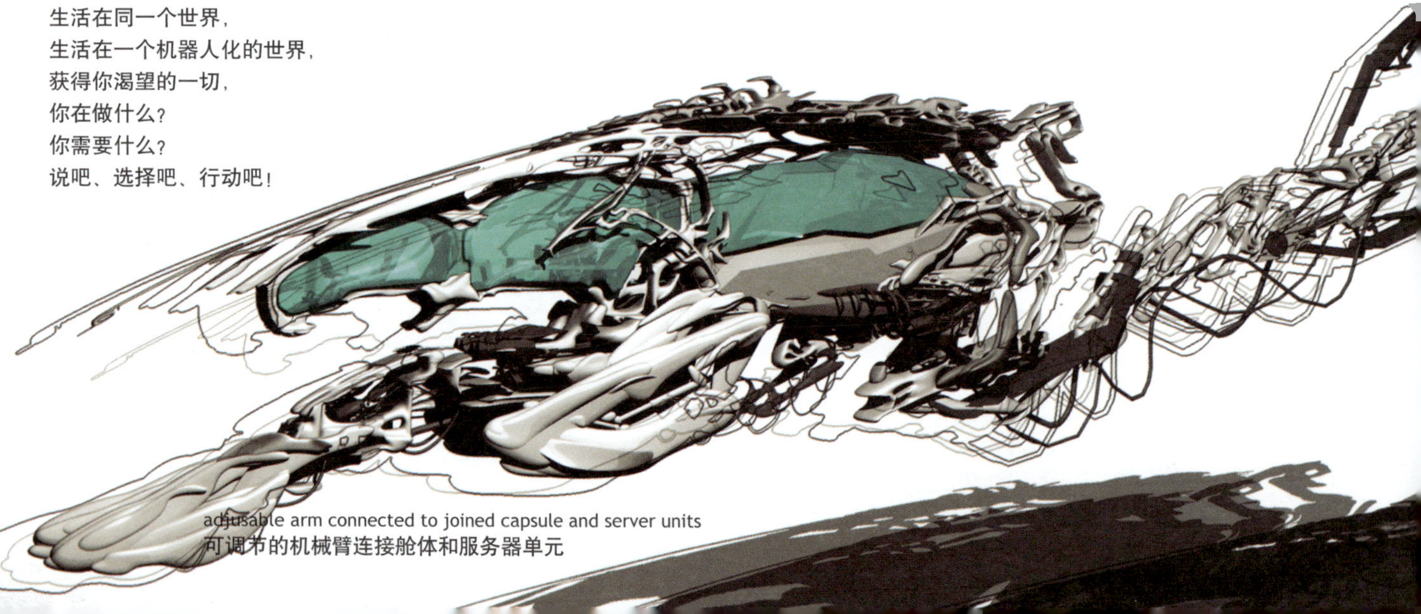

adjusable arm connected to joined capsule and server units
可调节的机械臂连接舱体和服务器单元

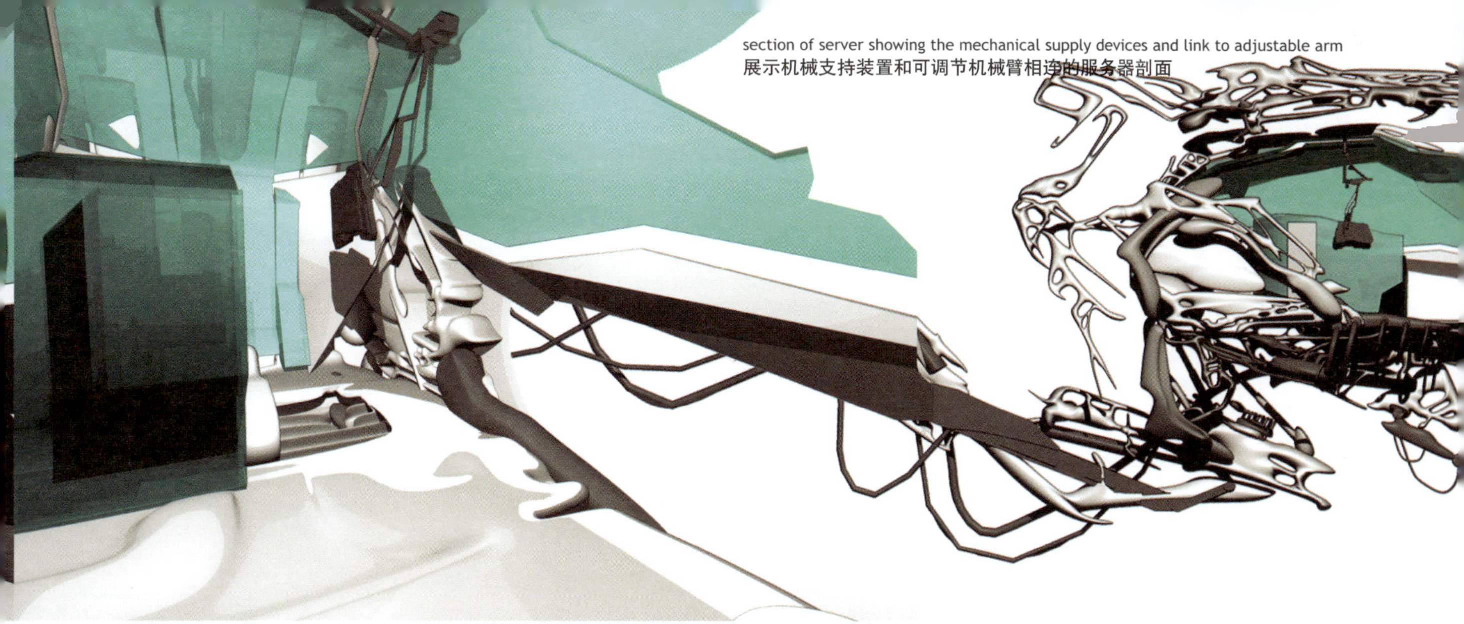

section of server showing the mechanical supply devices and link to adjustable arm
展示机械支持装置和可调节机械臂相连的服务器剖面

>THE SERVER

ADJUSTABLE SERVER MONITOR TO RUN AND PROGRAM THE UNIT. THE SOFTWARE OFFERS A RANGE OF PRODUCTS AND IS CONNECTED TO THE MAIN BASE COMPUTER. THE INTERIOR AND SUPPLY CAN BE ORDERED FROM THE SUPPLY CENTRE AND WILL AUTOMATICALLY BE DELIVERED BY AIR MAIL TUBES. FURNITURE HAS TO BE INFLATED WITH THE INSTALLED INLETS BEFORE USE AND CAN BE DEFLATED AFTER NECESSITY. OTHER PRODUCTS WILL BE DELIVERED IN AIR CUSHION PROTECTED PLASTIC VESSELS.THE FACILITIES OF "DESIRE" CAN THEN BE PLUGGED AND LOADED TO THE ROBOT UNIT. WHILE THE ROBOT IS OFF THE SERVER IS STILL CAPABLE TO OFFER ALL DUTIES. WHILST PLUGGED TO THE ROBOT, THE ADJUSTABLE ARM CAN TRANSPORT THE UNION TO CERTAIN INHABITATIONS IN THE AIR FOR A GREAT VIEW OF THE CITY.

服务器

可调节的服务监控器是用来驱动和编程这个单元。软件提供一系列的产物，并与主基地计算机相连。室内所需供给可以通过服务中心订购并通过航空邮件管道送达。家具在使用之前必须被展开安装，在不需要的时候被压扁。其他产品由气枕保护的塑料包裹递送。“期望”的设施可以被插入和载入到机器人单元中。当机器人被关闭，而服务器仍有能力提供所有的职责。当插入机器人时，可调整的手臂可以移动至空中的特定器具位置，展示城市伟大景观。

breakdown of server elements
服务元件的解体

>**BASE**

AS THE DESIGN PROPOSAL IS DEALING WITH 100 ROBOTS SPREAD OVER THE CITY AND WALKING THROUGH LONDON, IT IS NECESSARY TO HAVE A BASE STATION. IN EARLIER SCHEMES THE BASE WAS JUST USED AS A SUPPLY CENTER. DURING THE TERM THE DEFINITION AND FUNCTION OF THE BASE WAS EXTENDED. THE BASES WILL BE A LARGE STRUCTURE HOSTING THE FUNCTIONAL CAPSULES, SUPPLY, LOGISTIC AND SPACE FOR TECHNIQUE. THE PROGRAM IN EACH OF THE BASES WILL BE DIFFERENT AND VARY. AFTER DEVELOPING THE ROBOT, THE BASE STRUCTURE IS BECOMING MORE IMPORTANT AND THE SCALE IS CHANGING FROM SMALLER TO LARGER ELEMENTS. THE BASE IS THOUGHT TO BE A PART OF THE HOTEL AND HOST THE SPACES THAT ARE NECESSARY FOR THE DEMAND OF A HOTEL FACILITY. FURTHERMORE EACH ROBOT HAS A PERSONAL DOCKING AREA (SERVER) AND OFFERS SPACE FOR THE COSTUMER. WITHIN THIS INDIVIDUAL SPACE, IT WILL BE POSSIBLE TO LOAD THE FUNCTIONAL ELEMENTS/INTERIOR INTO THE CAPSULE. THIS PROCESS WILL BE OBSERVED BY A CENTRAL NAVIGATION SYSTEM. THE ROBOT BECOMES PART OF THE ROOM AND IS A EXTENSION OF THE SERVER. WHEN THE ROBOT TAKES OFF, IT TAKES A PART OF THE ROOM WITH IT. THE SERVER REMAINS IN PLACE BUT OFFERS STILL EVERY FUNCTION. THE HOTEL HAS CUSTOMERS WITH THE INTENTION OF A SHORT TERM STAY. BUT THE COLLECTION OF ACCOMODATION SPACE REMAINS TO BE SIMILAR TO HOUSING FACILITIES. ACTUALLY THE MOST INTERESTING PART OF THIS SCHEME IS THE JUXTAPOSITION OF THE DESIGN LANGUAGE USED TO INTERPRET THE BASE AND THE URBAN CITY CONTEXT. HOW CAN THE TECTONIC ASSIMILATE OR DIFFERENTIATE?

基地

正如设计提案是让100个机器人分散、行走在伦敦城里，一个基站是必要的。在早期的计划中，基地只是一个供应中心。在学期当中，基地的定义和功能扩展了。这些基地将会是一个大型的结构，集合功能舱体，供应，后勤和技术空间。在每一个基地的程序都将是不同和多样的。在机器人的发展之后，基地结构变得更加重要，并且尺度从较小的元素向较大的改变。基地被考虑成为旅馆的一部分，并包容这些对于一个旅馆设施要求是一种必需的空间。进而，每一个机器人有一个私人的入坞区（服务器），并为客人提供空间。在这个个体空间中，将在舱内载入功能性的元素/室内。这个过程将会被中心导航系统加以观察。机器人变成这个房间的一部分和服务器的一种扩展。当机器人离开时，它带走这个房间的一部分。但服务器保持不变，并提供所有的功能。旅馆有一些客人是短期逗留。但是住宿空间仍保持与住宅设施相类似。事实上，这个计划最有趣的部分是介入基地的与城市文脉的设计语汇的毗邻性。这种构造怎么同化或区分?

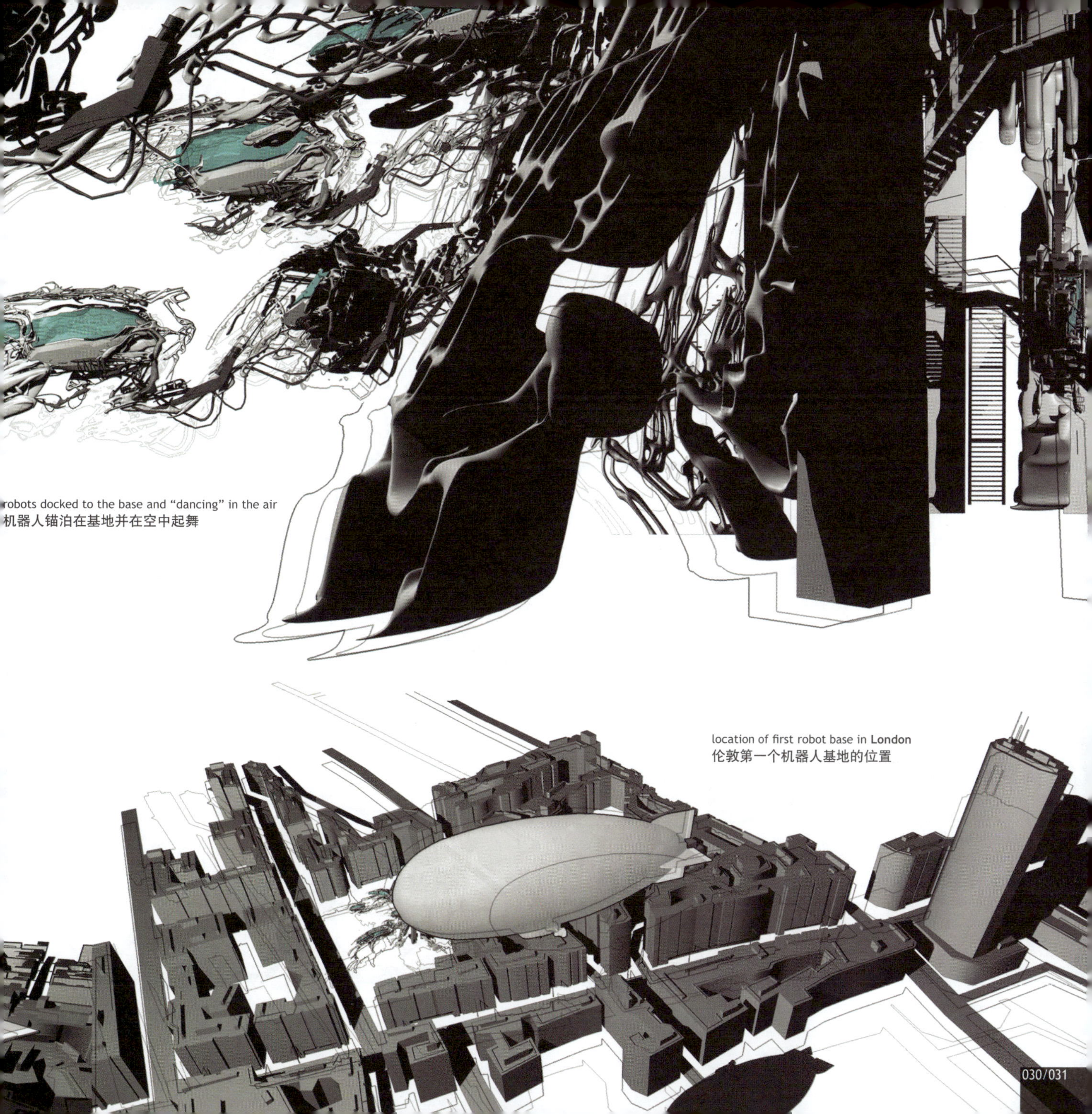

robots docked to the base and "dancing" in the air
机器人锚泊在基地并在空中起舞

location of first robot base in **London**
伦敦第一个机器人基地的位置

LEA INTERCHANGE
Stratford, Dalston, Hackney
A106 Stratford International
Freight Depots
New Spitalfields Market
Central London, Docklands
Blackwall Tunnel
A12
BURGER KING
L004 DHF

丹尼尔
丹卓
Daniel
Dendra

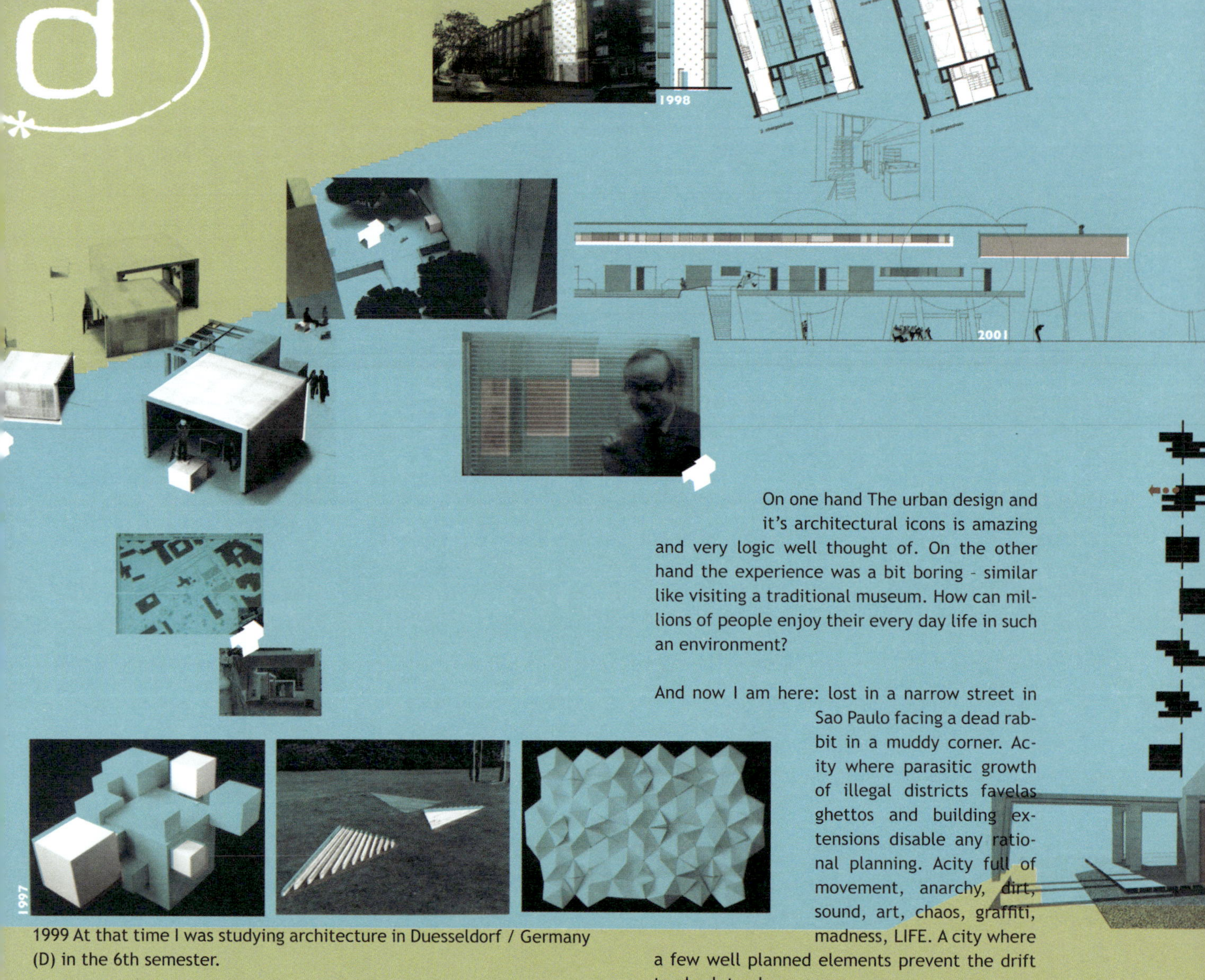

On one hand The urban design and it's architectural icons is amazing and very logic well thought of. On the other hand the experience was a bit boring - similar like visiting a traditional museum. How can millions of people enjoy their every day life in such an environment?

And now I am here: lost in a narrow street in Sao Paulo facing a dead rabbit in a muddy corner. Acity where parasitic growth of illegal districts favelas ghettos and building extensions disable any rational planning. Acity full of movement, anarchy, dirt, sound, art, chaos, graffiti, madness, LIFE. A city where a few well planned elements prevent the drift to absolute chaos.

1999 At that time I was studying architecture in Duesseldorf / Germany (D) in the 6th semester.

Everything I was told was to be very rational and pragmatic when designing a building / or an urban area. This minimalist approach where form should follow the function and the design is inspired by the genius loci (spirit of the) locus seemed to be a holy relict from the Bauhaus modernist movement.

Just yesterday I was inside one of the masterpieces manifestations of modernism: BRASILIA.

Would architecture not be much more lively and interesting if it could adopt some of these anarchistic qualities?

Like Alice I decided to follow the rabbit.

But how is it possible to make architectural design dirty?

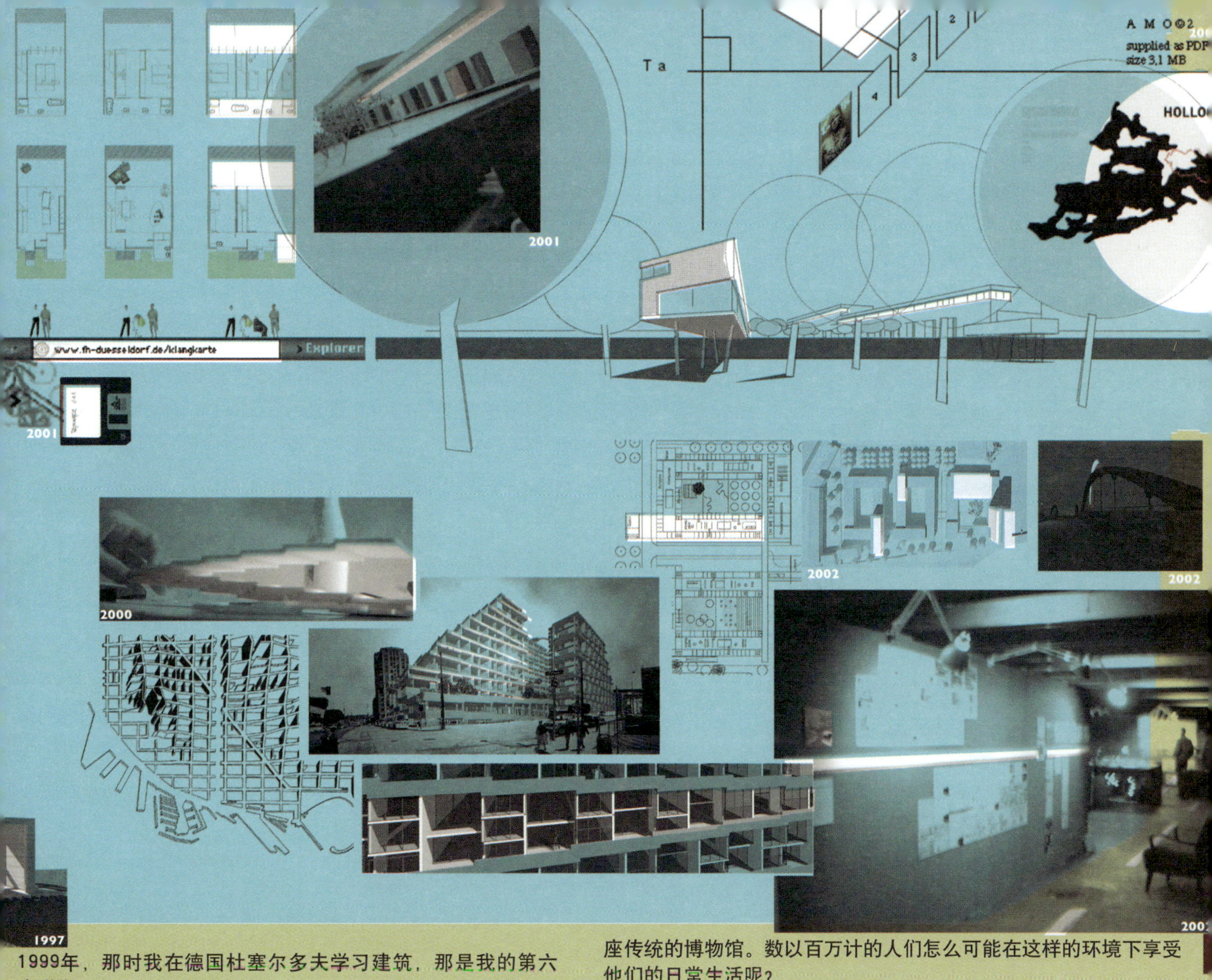

1999年，那时我在德国杜塞尔多夫学习建筑，那是我的第六个学期。

我被告知的每件事都教我在设计一座建筑或城市地段时要理性和实际。形式要追随功能，设计要从地方风格(场所的精神)中获得灵感，这样的极简主义方式看起来就像包豪斯现代主义运动的残留圣物。就在昨天，我还在现代主义的杰出范例之内：巴西利亚。

一方面，城市的设计及其建筑标志物是令人惊叹的，极富逻辑地让人深思；另一方面，这种体验又有些乏味，就像参观一座传统的博物馆。数以百万计的人们怎么可能在这样的环境下享受他们的日常生活呢？

现在，我在这里，在圣保罗的狭窄街道上面对着泥泞角落里的一只死兔子。一个非法犹太人棚户区寄生式生长着的城市。这个城市充满了运动、无政府状态、肮脏、噪声、艺术、混乱、涂鸦、疯狂、还有生活。仅有少数规划元素能防止其彻底滑向混乱。

接受这样一些无政府主义的品质难道不会让建筑更加生动有趣吗？和爱丽丝一样，我决定跟随这只兔子。

但是怎么样才能让建筑设计变脏呢？

overlay of movie and site structures – time / ab

STEP 1

All this happened just shortly before my diploma final thesis design.

The task: create an urban bridge between highway intersections in Basel / Switzerland (CH).
Time: 3 months.
I started with a traditional analysis of the site and the surroundings. The next step was; to forget everything I learned about design; not to create a consistent and rational design; but to create an abnormal state by overlaying the found urban topology with an unattached structure or random forms to create the further design. The site situation of constant flux by a node of highway and rail structures inspired me to look at movie structures, e.g. road movies.

The movie: Two Lane Blacktop (1970): Director: Monte Hellmann.
I tried to adopt cinematographic techniques like {montage},{hidden face} or {depth of focus}.

„Suddenly, in the same image, there exist two spaces that belong in different periods. [...] Two times, in all the relativness, interpenetrate, the whole surface fragments so that another space starts to become visible."

(GODARD)

The structures for the urban layout evolved from a sequential overlaying of urban and movie structures. The character of the design is based on the specific history of artificial adaptation of the topography of site. The use of secondary characters in the movie to describe one of the leading actors is then translated into further artificial modifications of the topography.
Some other parts of the form generation: random. The result: not dirty enough; too superficial. Three months were definitely too short to change my personal way of designing.

graphical movie plot
b1
b2
b3
b4
(max)
¿+
++¿
ent of topography
historical condition < a
natural condition
rail banks < b
artificial adaptation
(inter-)national interest
ghway topography < c
movie topography > d
artificial adaptation
local interest
pment
Aussichtspunkt
Tribüne
Bühne
Weg
Sport + Freizeit
Aussichtspunkt
Nutzungen auf der Aussenhaut.

步骤一

所有这些都发生在我本科毕业设计之前的很短的时间里。任务：设计一座位于瑞士巴塞尔公路交叉口之间的城市桥梁。时间：三个月。

我从对基地和周边环境的传统分析开始。接下来：忘记我学过的关于设计的一切；不做一个协调而理性的设计，而是以独立结构或随意形式覆盖寻找出的城市拓扑学来创造进一步的设计，来创造出反常状态。由公路和铁路结构形成的不断流涌的基地状况激发我开始关注电影的结构，比如，公路片。

影片：《双车道柏油路》（1970）：导演：蒙特·豪尔曼

我试图采用电影摄影术的技巧，比如“蒙太奇”，“画外音”或者“景深”。“突然，在同一个画面中，存在两个属于不同时代的空间．［……］两次，以他们的关联、渗透，整个表面的碎片以至另一个空间变得可见。”（戈达尔）

城市设计的结构由城市与电影结构的有序层叠发展而来。设计的特质基于对基地地形的人工顺应的特定历史。电影中用来烘托领衔主演的次要角色则被转译成对地形更进一步的人工塑造。

其他部分的形式生成：随意。结论：不够脏；太肤浅。三个月显然太短不足以改变我个人的设计方法。

步骤二

我需要得到一些建议。我得继续我的研究。我记起了伦敦，一个先前造访时并不喜欢的城市。这座城市太脏太吵闹，我不能把这座城市和任何美丽的图像联系起来。它看起来像是能让我深化我的研究的完美场地。巴特雷特学院，城里最具试验性的学院之一。我申请了一份奖学金。中奖。我要在七个月之内开始课程了。

STEP 2

I had to get more advice. I had to continue my studies. I remembered London to be a city I didn't like when I was visited it in former times. This city was too dirty and loud; I couldn't associate any nice pictures with this city. It seemed to be the perfect place for my further studies. The Bartlett: one of the most experimental schools in town. I applied for a scholarship. Bingo. I will start the course in seven months.

STEP 4

Learning more about the Situationists movement during my stay in the Netherlands I wanted to start my own derivation to possibly find a solution to my task.

I started with an operation to distinguish random points in the city by overlaying a Japanese recipe with the taxonomy of a supermarket. At these random points I placed notices looking for a stranger to invite for a Japanese dinner.

As a present the random/unknown visitor gave me a cooking-glass. From now on the process was non-linear. Several things happened parallel and some paths were dead-ends, for instance dealing with lost+found objects and trying to convert them into architectural forms.
To make a long story short the cooking-glass enabled me to discover hidden phenomena and events along an as well hidden section through the city: the underground river Fleet.

This whole operation led me to my actual site: Smithfield Meat-Market. At daytime, when the rest of the city is passing through its peak of activity, this space is (closed) not in use. By this it is hidden and lost in a different way.

The final tectonic proposition is to bring back activity to this place during the lost time-period by using a rail system that is overlaid with the urban fabric of the meat market and the surrounding context. This systems allows a swarm of floating objects to create an indeterminate space by reconfiguring it according to diverse needs during altered periods of a day, week, year etc.

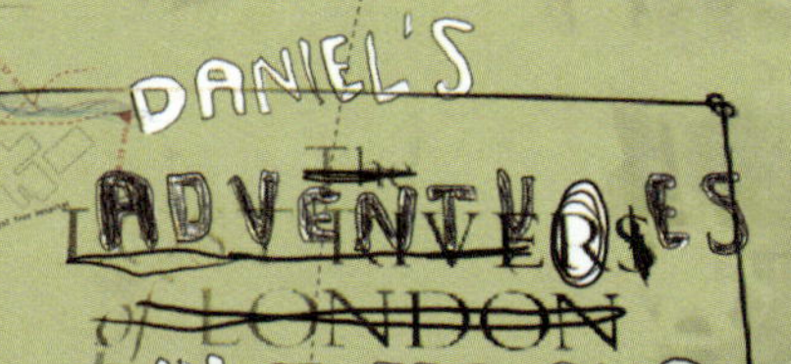

步骤四

在荷兰对于情境画派运动了解更多以后，我想开始我自己的语源，找到一个可能完成我的目标的解答。

我以覆盖日本菜谱(a2)的方式开始分辨城市中的任意点(a1)的操作，菜谱是超市目录式(a3)的。在任意点上我张贴布告(a4)寻找陌生人，邀请他来吃日本晚餐。

作为一份礼物，随机的/不相识的访客(a5)需要给我一个烧杯(a6)。从现在开始，进程是非线性的了。几件事情平行发生，有些路径是死胡同(20)。比如处理失落与找寻(20a)的事物并把它们变成建筑的语言。

长话短说，烧杯使我能够发现沿着同样隐藏着的穿过城市的剖面(a–a')的隐藏的现象(a7)和事件：地下河(a8)。

这整个的操作将我带到了实际的基地：史密斯费尔德肉市场(B)。在白天，当城市的其他部分正在活动的高峰时，这个空间被关闭不用。它以这样不同的方式隐藏并且失落。

最后的建构的命题是在失落的时间用铁路系统(b8)给这里带回生机，铁路系统被肉市场和周边文脉的城市肌理所覆盖。这个系统允许一大堆漂浮的物体(b7)，根据一天、周、年等的变化的周期的需要重组来创造不确定性的空间。

用以描绘空间随意用途的表述程序将是被发现的沿着伦敦河剖面的想像和事件。这样命题转化为从烧杯中看城市的个人导游册。

步骤三

毕业之后，我更愿意不去做设计，因为我还是对我的设计方式不满意，这使我不得不去办公室实践。我去了鹿特丹为雷姆・库哈斯都市建筑事务所即O.M.A/A.M.O工作，我在他们那里做一个不需要设计的更概念性的设计任务。

我做了一个关于德国Ruhrgebiet地区的研究，这地方有650万人口，以弗兰克・劳埃德・赖特设计的洛杉矶式的广亩城密度居住着。我们的处理方法非常无政府主义，也很概念，但没有使用任何建筑学上的设计。我对这种情形感觉非常好。在我的研究中我力图探究自动生成的设计模式。这项试验花费了我绝大多数的时间和精力，最后是一个死胡同，因为我不具备发展出更深一步结果将研究结果发展之更深一层的知识。等到巴特雷特的气氛帮我找回设计的乐趣已经太晚了。

The program that is used to picture the possibly random uses of the space are the phenomena and events found along the Fleet-section of London. By this the proposition transforms into a kind of personal guidebook of a city seen through
the cooking-glass.

STEP 3

After my diploma I preferred to stop designing because I was still not satisfied with my approach and the approach I had to practice in offices.

I went to Rotterdam to work at Rem Koolhaas the Office for Metropolitan Architecture O.M.A / A.M.O on more conceptional tasks without the need to design.

I was working on a study for a region in Duitsland-Germany called Ruhrgebiet, a place of 6.5 million inhabitants with the density of Los Angeles Frank Llyod Wright's design of Broadacre city. Our approach was very anarchistic and conceptual but without any use of architectural design.

I felt very comfortable with this situation.

During my studies I wanted to explore auto generative design methods. This experiments took most of my time and energy and were dead ends because I didn't have the knowledge to develop the outcome to a further stage. Too late the environment in the Bartlett helped me to find the joy of designing again.

THROUGH THE COOKING GLASS

DANIEL'S ADVENTURES
13
the french elvis
JEAN DANIEL FRANCOISE is
playing from thursday to
sunday in a hidden club in
chalk farm - to see french
elvis playing one has to
enter a kebab-shop called
marathon, pass the counter
and enter the hidden club
through a small passage.
there are a lot of unique
places like this in london
that are hidden somewhere
- upstairs - in the courtyard - behind the bakery -
15
the next day i went to the
supermarket to buy some food.
in my pocket i found the
notice with the recipe of the
CHICKEN ITAMERO meal. on the
spur of the moment i decided
to buy all the ingredients.
since i am an architect i
made a plan of the supermarket and mapped where i found
the ingredients for the
chicken-dish.
hungry 腹減った
Do you fancy a dinner?
Do you like chicken?
For free?
a phenomena that is strongly
related to free-time in london ....
many people .... spend a lot ......
of their time .... queuing at the bank
in front of the club .... or
time at the pond in hempstead heath
Security - be alert

STEP 5

Maybe I started too late with the actual tectonic design or the design of situations at a specific site (B). Maybe I didn't use the opportunities of the experimental graphic environment at the Bartlett in the most efficient way. Maybe one year is just too short for this very open-minded kind of Master course. But the stay at the school and especially in London (A) definitely helped me to get closer to the path of the rabbit. And this step is still not finished. I shouldn't forget about all the mentioned dead ends and experiments (20) but use parts of the outcome and process in my further architectural design practice.

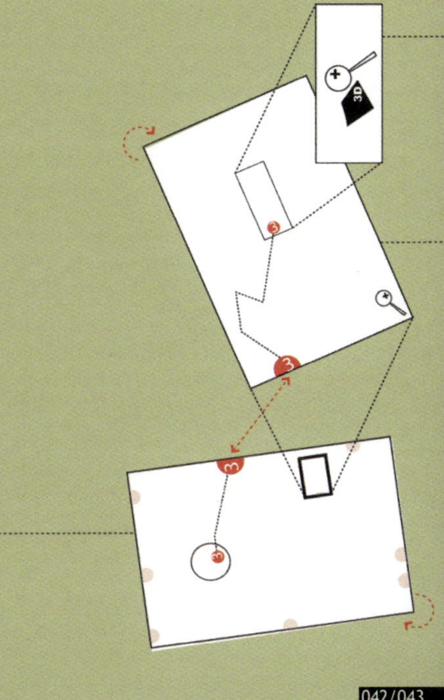

步骤五

可能我实际建构性的设计和针对特定场所状况的设计开始得太晚。可能我没能以最有效的方式利用巴特雷特的试验性图形设计氛围。可能一年对于一个非常开放性的硕士课程来说还是太短。但是在这所学校，特别是在伦敦的驻留，毫无疑问地帮我接近了兔子的路径。并且这一步还没有结束。我不该忘记所有这些提到的死胡同和试验，而应该将部分成果和过程运用到我进一步的实践中去。

NEXT STEP
Some decisions and collaborations helped me to get closer to my aim >>> to be continued.
Find the next chapter at
www.mobileresponseunit.com/followtherabbit

接下来的步骤
一些决定和合作帮助我靠近我的目标〉〉〉待续
请在www.mobileresponseunit.com/followtherabbit
寻找下章。

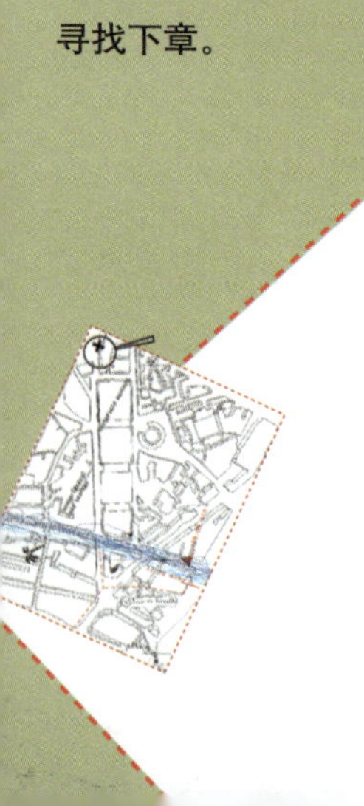

personal modification
<<< london underground
herb market
rails at deck height
smell-scape
washing
clothes
wholesale butcher
hidden-tunnel
<<< docking station / entrance
<< 24h bar
uv-light
connection to the existing rail-system
additional elements
capable of docks
<dropping>
Caution. Wet Floor: is written in bold letters on a bright yellow sign. And indeed it is dropping at frequent intervals from the ceiling. The water drops are creating an amazing shimmery structure on the floor. Each new drop is changing this surprising configuration. This combination of liquid and gravity was very common
for different aleatoric art techniques. The first ever used chance art was called blot drawing and consists of nothing more then ink blots that where dropped on a
piece of paper. This procedure was first mentioned in the eighteenth century my the English watercolorist Alexander Cozens as a tool to create original composi-
tions of landscapes by Leonardo da Vinci as an adv
technique developed over time, the most famous one is the Surrealists' Decalcomania, where ink or
Jackson Pollock
TANDOORI JERK CHICKEN
SANDWICH SCAPE
Ready to
044/045

16
06
THE OLD MAN AND THE MACHINE
THROUGH
THE
COOKING
-GLASS
3D
LTD.
KODAK
35
GUARD
HUNTER

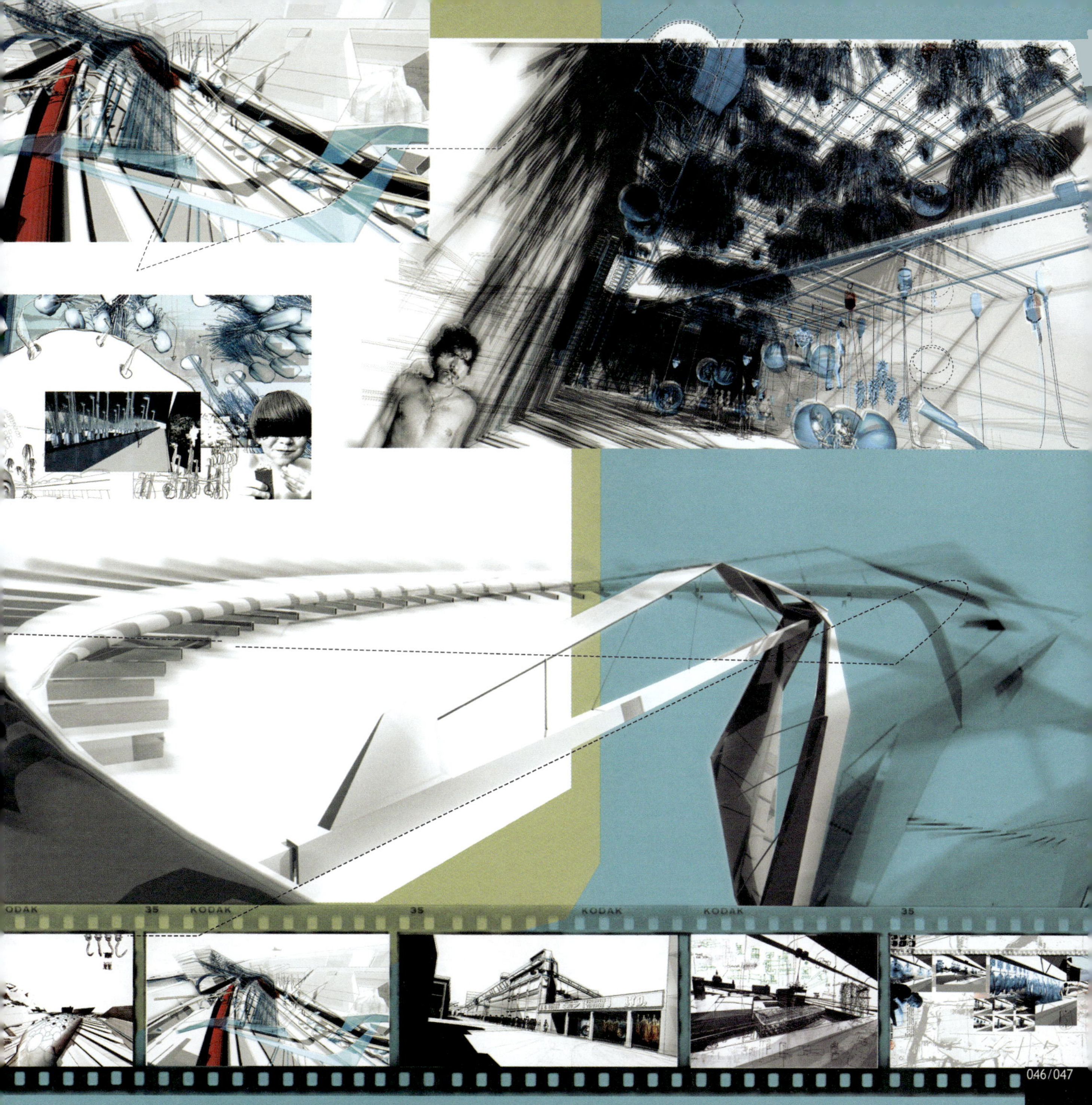
35
KODAK
35
KODAK
KODAK
35

Daniel Dendra was working at various offices in London, Moscow,Düsseldorf and Rotterdam such as A.M.O. (Rem Koolhaas) and Zaha HadidArchitects. Daniel was involved in projects of all scales from regional planning (Hollocore- AMO 2002), master planning, several large scale tower projects in Spain, Dubai and Australia, Architectural Design (shopping centres, offices and residential in Spain, Russia, Germany and UK), Exhibition Design (Spymaker - London and Indianapolis 2006) and small scale installations (Urban Nebula, London, 2007).

In 2007 he established the multidisciplinary design bureau AnotherArchitect.com in Berlin. The office is most recently involved in projects in Germany, Norway, Poland, Australia and Russia.

After leading a studio together with Dr. Jenny Lowe at Bright University(UK) dealing with the problematics of global warming Dan Dendra became Guest Professor at the DIA in 2007. The DIA (Dessau ternational Architecture) is an Englisch speaking postgraduate cou (M.ARCH) situated at the Bauhaus in Dessau with international stude from all over Asia, South and North America, East-Europe. Daniel is lea ing there studio4 which is experimenting with tectonics generated by gorithmic functions. With the help of emergent design tools the stude are asked to taggle problems on local and bi-lateral scale such as aba donedcity airports and the dis-apperance of borders between countr and on global scale such as global warming and world food crisis.

尼尔•丹卓曾在伦敦、莫斯科、杜塞多夫和鹿特丹的各种类型的办公室工过，如雷姆•库哈斯的A.M.O.和扎•哈迪德建筑师事务所。丹尼尔做过的目涉及各种规模，从区域规划（Hol-core，AMO，2002年）、总体规划到西牙、迪拜、澳大利亚的一些大型塔楼目，建筑设计（在西班牙、俄罗斯、国和英国的商场、写字楼和住宅），

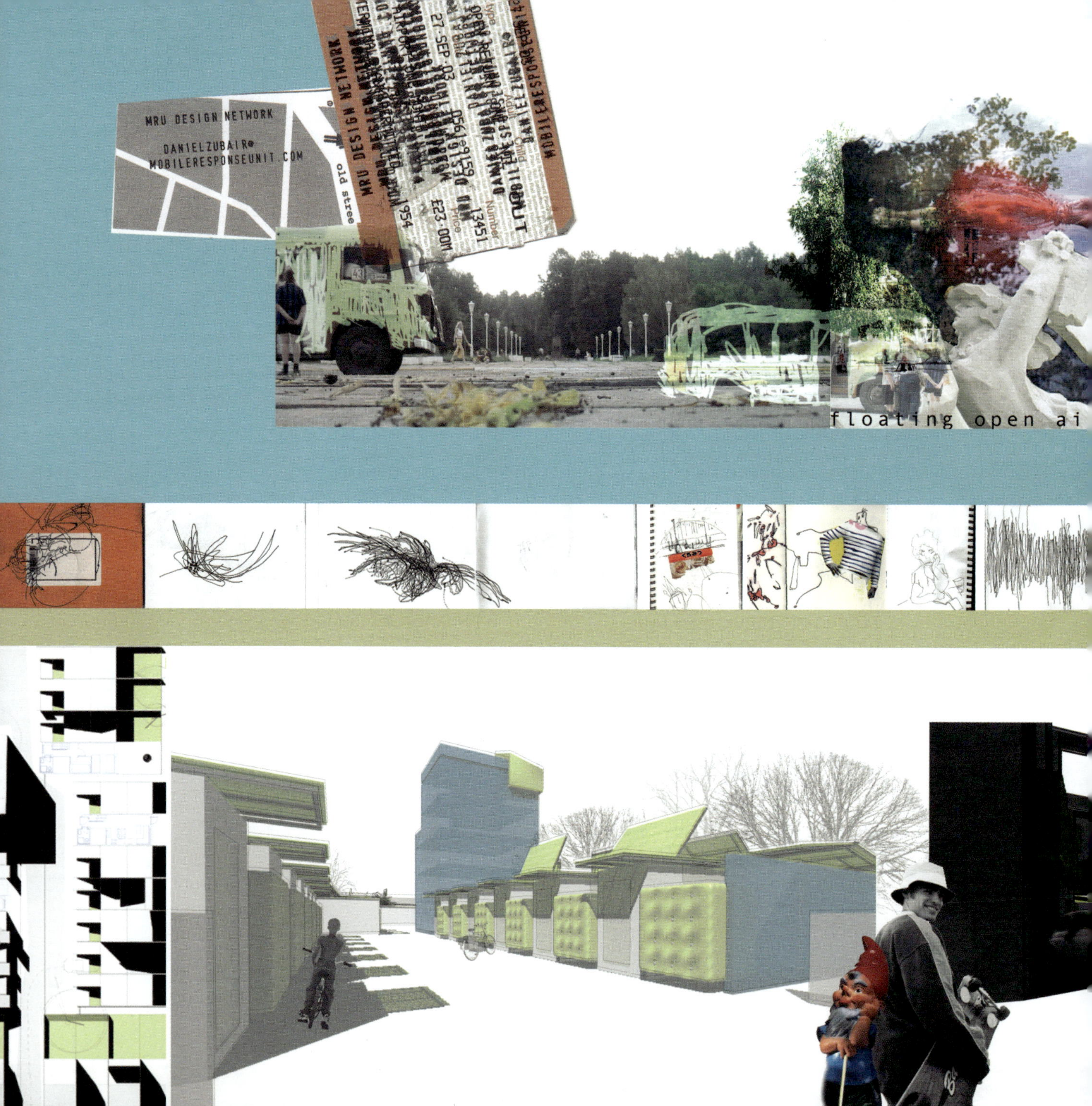
MRU DESIGN NETWORK
DANIELZUBAIR@
MOBILERESPONSEUNIT.COM
MRU DESIGN NETWORK
floating open ai

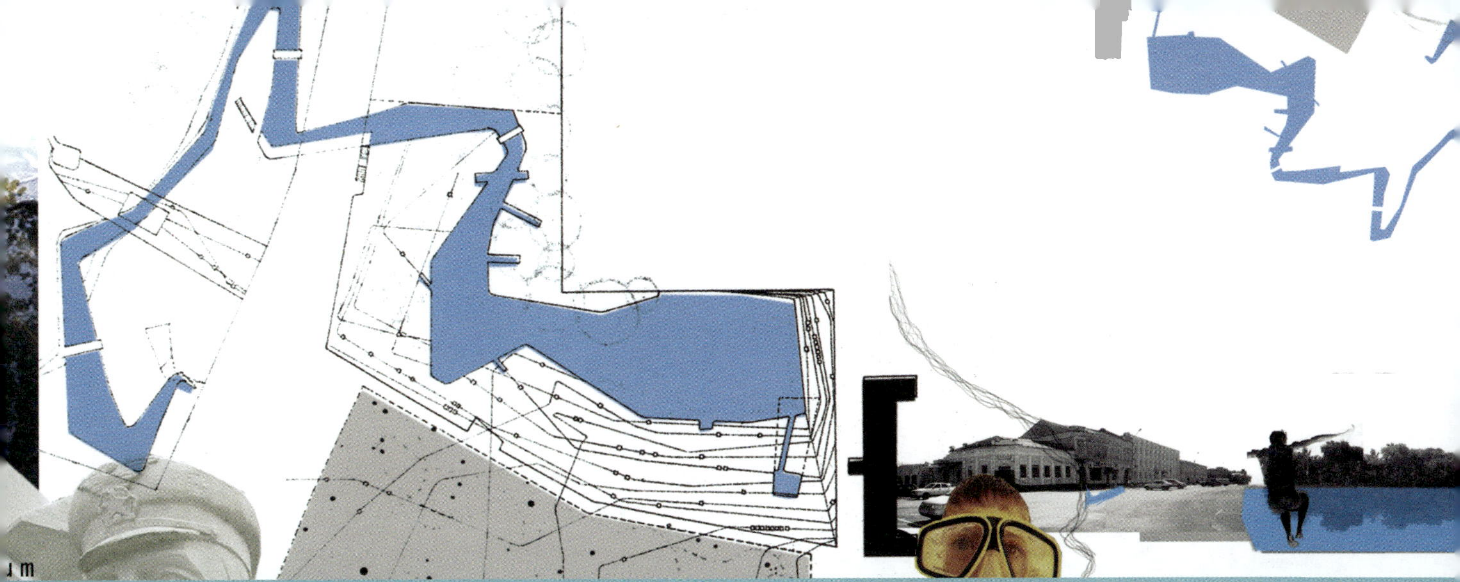

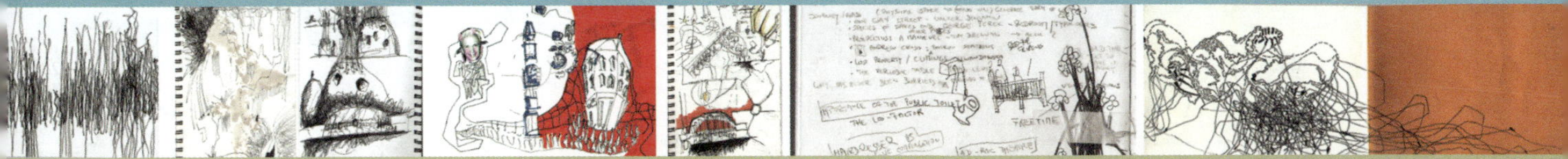

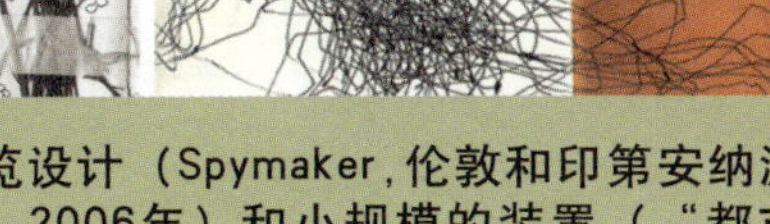

展览设计（Spymaker，伦敦和印第安纳波利斯，2006年）和小规模的装置（“都市星云”，伦敦，2007年）。

2007年，他在柏林成立了多学科的设计机构AnotherArchitect.com，其最近涉及的项目广泛分布在德国、挪威、波兰、澳大利亚和俄罗斯。

在与珍妮•洛威博士在英国布莱顿大学一起领导处理全球变暖问题的工作室后，丹尼尔•丹卓于2007年成为DIA的客座教授。DIA（德绍国际建筑）是一个以英文授课的研究生课程（建筑设计硕士），该课程位于德绍的包豪斯，国际学生来自亚洲、南美、北美、东欧。

丹尼尔带领第4工作室，探索数学生成功能所产生的构造。借助新兴的设计工具，学生被要求从地方的和双边的角度思考问题，比如被遗弃城市机场和国家之间边界的消失，以及在全球范围内的问题，如全球气候变暖和世界粮食危机。

A120(W) B. Stortford, Services
8A
¼m
A120(E)
Stansted
Colchester
Cambridge
Newmarket
M11(N)

阿曼多
赫南德兹
Armando
Hernandez

Armando Hernandez 阿曼多 赫南德兹

DATE OF BIRTH: 01 Jan. 1974	出生日期：1974年1月1日
NATIONALITY: US Citizen	国籍：美国公民
E-Mail address: dogspot@hotmail.com	E–mail地址： dogspot@hotmail.com
Languages: English, Spanish	语言：英语、西班牙语
Education:	教育背景：
Master of Architectural Design	建筑设计硕士
UCL Bartlett School of Architecture	UCL巴特雷特建筑学院
Dates attended: Sept 2002 - Sept 2003	日期： 2002年9月–2003年9月
Bachelor of Architecture	建筑学学士
Woodbury University	伍德伯瑞大学
Dates attended: Sept 1997- May 2002	日期： 1997年9月–2002年5月
Pasadena City College	帕萨迪纳城市学院
Course work focused on fine arts (painting, life drawing, sculpture)	学习重点是美术（如绘画、写生、雕塑）
Dates attended: Fall 1993 - Spring 1997	日期：1993年秋–1997年春
No degree	无学位
Work Experience:	工作经验：
Lean Arch Inc. Los Angeles CA	里恩建筑公司，洛杉矶，加利福尼亚
Aug 2005 - Current	2005年8月至今
Art Center College of Design, Pasadena CA	设计学院艺术中心，帕萨迪纳，加利福尼亚
Office of Program and Planning	规划办公室
Apr 2001 - Aug 2005	2001年4月–2005年8月
Mark Dillon Architect Los Angeles, CA	马克•狄龙建筑师事务所，洛杉矶，加利福尼亚
May1999- Sept 2002	1999年5月–2002年9月
Office of Mobile Design Los Angeles, CA	移动设计办公室，洛杉矶，加利福尼亚
Jan-March 2002	2002年1月到3月

e nomadic lifestyle of the new generation urban dweller will set a v standard of design. This project has its basis on the notion of in- idual living style as pattern. Life patterns are different for every son. Individual space is defined by physical objects, which exude a ecific aesthetic that is comfortable to each person.

e conception of physical form becoming the display of personal pat- ns of living is one that stems from an interest in the notion of auto nerative and auto destructive architecture. The evolution of this ject begins with an expanding and contracting urban fabric through celerated time. To focus on a site is a determining factor in the de- opment of the project, along with programmatic studies, a solution s determined to be random.

set form to the development of a nonexistent living environment. ch individual case would carry its own language and the resultants uld be different each time. Materiality and material performance ve always been an issue in the developing stages of this project. In final installation, the exploration of interaction between the indi- ual and the material was key in development in the final form.

self as the individual being and the materials used were an explora- n in form and reactivity. The questions raised stem from what the formance of the space could be. How does the space created react the existing and how the materials of existing react to the inserted ect?

新一代都市居民的游牧化生活方式将会确立一个设计的新标准。这个项目建立在以个性化的生活方式为模式的内容概念上。每个人的生活模式都各不相同。个性化空间由有形的物质化的实体所定义，这种空间流淌出一种令每个人都感到舒适的特定生活美学。

关于用物质形态来展示个人生活方式的概念来源于对自主生成和自主解构的建筑这一概念的兴趣。项目的演进是随着时间的加速，开始于一种城市肌理的扩展和收缩。集中着眼于一个特定的基地是这个项目发展中的一个决定性要素，伴随程序化的研究，一个解决方案被随机决定下来。

没有针对一个不存在的生活环境而设定的形态。每个个案都有其自己的语汇，并且每次的结果也不尽相同。在这个项目的发展阶段，物质性和物质化的表现已经成为一个关键问题。在最后的装置中，对介于个体和物质之间的互动性的探讨是最后形态发展的关键。

把我自己作为个体，把所用的材料作为在形式和回应上的探索。提出的问题是“空间可以有怎样的表现，空间是如何创造出对现存物的回应的，被创造出的空间是如何回应于现存物的，现存的材料又是如何回应介入的物体的？”

The current state of architecture is one of the static... but this can be no more. Day after day people succumb to the machine that is economy and labour. A society enthralled by the need to own, a need to obtain the material possessions that have no value. There had once existed the idea of design for the masses but this reality has never been realized and the condition will never exist. One has to look upon the world that was created for the individual not the collective that we seek to accept, a rebellion to be one, never a number, never a label.

The guarantee of liberty of each and all is in the value of the game of life freely constructed. The exercise of this ludic recreation is the framework of the only guaranteed equality with non-exploitation of man by man. The liberation of the game, its creative autonomy, supersedes the ancient division between imposed work and passive leisure. (Situationist Manifesto 1960)

I seek to explore the individual as a single entity not the collective of many as I refuse the element of the society, a mere hindrance in the evolution of mankind. A barrier that fosters the machine that makes up the world as it is.
I am against the existing. All around us is a static environment that defines a set pattern of living. The city creates machines that feed the economy.

Against the past and the manipulation of history. The glamorization of events that have ruined ancient societies and dictate the events of the future.

Against the visions of the future as they are predefined by the manipulation of the present and the human machine that lives today, we are here to feed those that have fallen into the idealistic notion of choice.

Against the current situations of everyday life and architecture, we live in this environment that dictates the movement we seek to avoid but never the ability to do so.

1. We must come to a realization that the individual is unique in every way and the notion that generic space can accommodate all is no longer acceptable.

2. The permanence of modern architecture can no longer carry out to the future as it stands.

3. Materials are meant to erode decay and become part of the environment once more. There is no longer any permanence in this world.

4. There are new rules to design and even fewer rules for living. Individuality is the key to development of the new in terms of space.

5. The notion that many are equal and one space for all is part of the past.

6. Design for the masses is merely the simple notion that Ikea is the source for designed items. This no longer works. It is just a fashionable piece that is expendable. Much like the architecture of today.

7. Buildings are designed and built to last only a few years. Let us accelerate this and design the building for the duration of the individual occupant.

宣言

当前的建筑是静态的……但这不能再继续下去。日复一日，人们对机器的屈从是一种经济和劳役。一个被拥有的需求，被攫取物质财富的需求所迷醉的社会是毫无价值的。曾经存在过为大众设计的想法，但这种现实从来就没有存在过，其条件也永远不可能存在。一个人必须将世界看作为个体而非我们试图找寻并接受的集体而创造的。我们要对成为集体中一员进行反叛，人决不是一个数字，决不是一个标签。

每个人和所有人自由的以自由建立的生活游戏的价值为保证。这种戏谑的消遣练习是唯一能保证人与人之间毫无利用的平等关系的构架。游戏的自由，创造性的自治，取代了古代介于强制工作和被动消遣之间的界限（《情景主义者宣言》，1960）。

我想要探索人作为个体而非群体中一员的方式，正如我拒绝这种因素，一种妨碍人类进化的小障碍，一种培育了使世界成为现在如是的机器的障碍。

我反对现存的东西。所环绕我们的一切是一个被界定了生活模式的静态环境。城市创造的是喂养经济的机器。

反对过去和对历史的操纵。活动的全球化英雄化叙事已经摧毁了古代社会，也限定了未来的活动。

反对那些已被现有对当前的操纵和活在今天的人类机器预先设定的未来版本，我们在这儿培养的是那些已经坠入选择的空想主义的概念的人。

反对现有的日常生活和建筑的状况，我们生活在一个规定了我们试图避免的活动却又对此无能为力从来无力避免的环境中。

1. 我们必须接受这样的现实：个体在各方面都是独特的，一般性空间可以适合所有人的概念已不再被接受。

2. 由于现代建筑的立场，当现代建筑矗立起来，其永恒性不再能贯彻到未来。

3. 物质照理要腐烂，并会再次变成环境的一部分。在这个世界上，不再有什么永恒。

4. 存在新的设计法则和几乎没有成规的生活，个性是在空间方面创新发展的关键。

5. 那种大家是均等的，有一个为所有人设计的空间的概念已经是过去式了。

6. 那种宣称宜家是设计产品的来源，为大众而设计的简单概念已不再行得通了。这些只是可消费的时尚品，与当今的建筑很相似。

7. 建筑被设计和建造出来只能存在几年。让我们加速这一进程并设计出为个别居住者而存在的建筑。

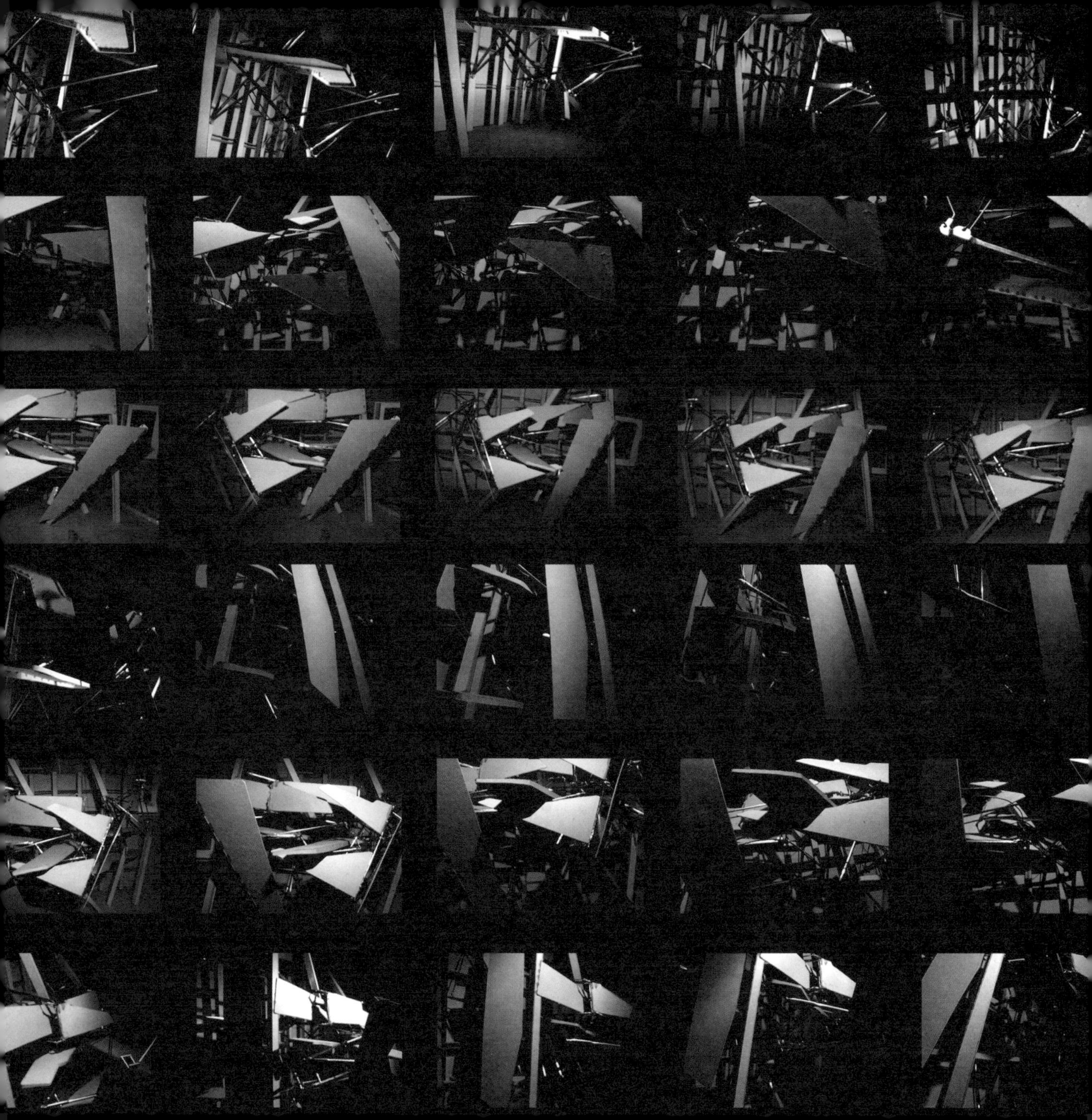

Being asked to tell a story of my year at the Bartlett is rather a difficult task. This was my first year outside of the US for an extended amount of time. But all I can say is that it was one of the memorable times in my life.

My name is Armando Hernandez and I am from Los Angeles, California. I am the youngest of three and I am the first in my family to acquire a master degree. Prior to attending the Bartlett, I received my Bachelors of Architecture at Woodbury University in Burbank California.

There is a drastic difference in the schools of thought between the Bartlett and Woodbury. I wouldn't say there is a Bartlett style but more of a Bartlett mentality. To be at the Bartlett is more to change the way one perceives architecture. You have to be more experimental a bit more extreme in how you think, or even more extreme on how you conceive a building and it surroundings.

My studies at the Bartlett focused on the interaction between architecture and the human. We live according to our surroundings; the way streets are laid out, the way buildings are constructed. Our interaction with the city and its architecture has always been where we are the participants. My research focused on the idea that the city or its buildings could also live and breath according to its population. In the extreme the ideal for my project would be that a city would disappear in the event that the population would be removed from the city. Thus an absence of people would lead to the absence of the city.

The initial idea would have been the extreme, but further into my work at the Bartlett I looked into responsive environments and the effect decay has on the areas we cease to occupy. The initial beginnings of my project where illustrations based on a given program of a hotel for one hundred people. But since the occupation of a building can vary I chose to focus on the idea of fluctuations in the occupation of a building +- 101 project, The variation in the fluctuations in a building can exceed or come under the given population of one hundred inhabitants.

In my illustrations I focused on the growth and collapse of buildings in relation to the amount of space required to accommodate the fluctuating population. Unused spaced collapsed a building leaving hollow spaces and reduction in skylines of the city. The growth aspect would create parasitic protrusions that could in theory connect buildings or overflow into unused portions of the city. As the idea becomes more extreme the illustrations follow. Buildings would tend to defy gravity, floating on cushions of air being attached by a few strands of wire, the structures would have the qualities of malleable plastic appearing to melt and shift in real time thus creating a new language to the city and architecture itself. But this is just the idea.

Following the illustration the idea of a model manifested in the actuality of design. The question arose of how this could be done. This explosion or breakdown of a building to create new space or a new language, the models where used to study the effects of how a machine like structure could remove itself from the skin and become something totally different. The idea of the mechanical movement in combination with the existing structure would have a specific movement but the limitations would be only from the users. If the population increase calls for new space the fragmented explosion would allow for it.

In looking at how space could be used and the belief that liquid movement allows for greater flexibility, I chose to explore the realm of how a building could become an organic liquid element that shaped itself according to the occupation, movement and activity. The idea that space could expand itself from liquid plasma that had no restriction of gravity or space. This was the exploration of the non-real element, a floating non rigid liquid blob structure that shaped itself through use, the idea of a playground that could become a flat or a skate park with the only limitations being set by the user, a literal flexible environment that adjusts itself through use.

Though there are limitations in real life one has to look at what happens when gravity is introduced. What are the reactions of the language when different elements are introduced to the smooth bubble like element of the blob? The construction of the physical model with the same parameters of the digital help define what a space could look like and what kind of external help would be needed to support such kind of structure. The model used was the exploration of creating a real from the digital with the application of a structure.

Further exploration led to a configuration of an actual space at full scale. What are the limits of materials and there reactions to a space? What can be the result of the unknown? This tasked focused on the occupation and the creation full scale working model. The final illustrations the use of all aspects of the project are combined in extent to tell a story of the progression of this project. Through manipulation of photos and the combination of techniques used throughout the process I wanted to see what the definition of space could be. This is the idea of clarification of the question "what does it all look like?"

The perception of the final project is one of subjective ideals. My exploration has led to a different understanding of architecture and an obsession with the "what if?" aspect of human interaction with architecture. My ideas are one of science fiction as the technology of nanobots or liquid solids do not exist as of yet. But in the future there may be the chance to see a progression of architecture that could react to human movement or emotion. Imagine a time where you become the creator of your own space in real time...

Woodbury school of architecture

Project: insertion of a redlight district in the heart of the sunset strip. (A social commentary on the hedonistic lifestyle of the Hollywood socialite.)

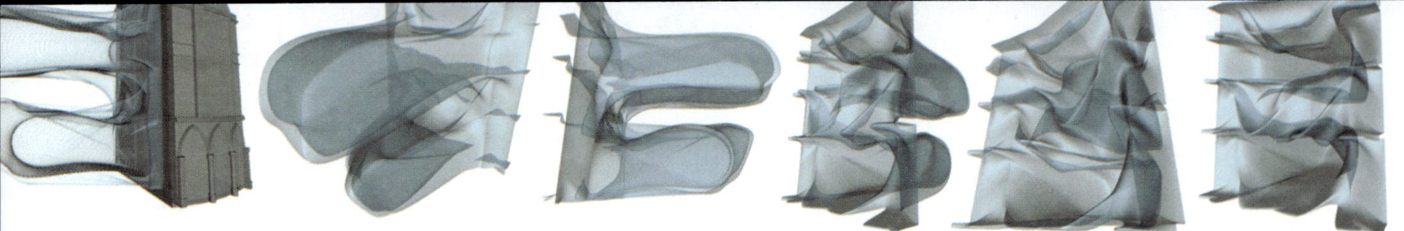

要我讲一个我在巴特雷特学院学习那年的故事，这可真是个相当艰巨的任务。这是我第一次在美国以外的地方呆那么长时间。但我可以说，这是在我生活中一段令人难忘的时光。

我的名字叫阿曼多·赫南德兹。我来自加利福尼亚州的洛杉矶。我是家里最小的孩子，也是这个家里第一个获得硕士学位的人。在来到巴特雷特之前，我取得了位于美国加州伯班克的伍德伯瑞大学的建筑学学士学位。

在巴特雷特和伍德伯瑞的思想流派之间存在着巨大的差异。我不会说存在一种“巴特雷特风格”，更多的是巴特雷特的思想方法。在巴特雷特会改变你认知建筑学的方式。你一定要更加实验一点儿，极端地去思考，甚至更极端地去考虑如何设计一栋建筑物和它周围的环境。

我在巴特雷特的研究着重于建筑和人之间的互动。我们根据我们周围的环境所生活；街道排布的方式，建筑建造的方式与我们的生活息息相关。我们与城市之间相互作用，城市中的建筑一直是我们参与生活的所在地。我的研究侧重于城市或其建筑物与生活在其中的人口同呼吸共命运。在为我的项目所设置的极端假设中，一个城市将会因为所有人口搬出该市而消失。因此，人的缺席会导致城市的缺席。

最初的想法比较极端，但进一步进入到我在巴特雷特的课程设计，我照顾到可回应的环境和我们停止占有的区域的效应衰减。在项目最开始时，我在插图的基础上做出了一间可容纳100人酒店的计划。但建筑物的占有率可以有很大的不同，我选择把重点放在构思波动中占领的+/−101建筑项目的占有率的波动中，建筑物的波动变异使其能容纳超过或低于给定的100的人数。

在我的插图中，我关注建筑物的生长和坍塌，涉及的金额，以容纳起伏不定的人口所需的空间。闲置倒塌的建筑物留下空地及减少了城市天际线。增长的部分会形成寄生凸出物，在理论上可以连接建筑物，或溢流到未使用的城市部分。在接下来的插图中这种概念变得更加极端。建筑物将无视重力，漂浮在绑在几股钢丝的气垫之上，结构将有呈现可延展的可塑外表，从而创造一种关于城市与建筑本身的新语言。不过，这仅仅是个构想。

在插图之后，一个模型的概念体现在实际设计中。问题是如何去做。对建筑物的爆炸或损坏，能创造一个新的空间或一种新的语言，这个模型可以用来研究如何影响一台机器式的结构，可以消除自己的皮肤，并成为完全不同的东西。结合现有的结构构思机械运动，将有一个具体的运动，唯一的限制将是使用者。如果人口增长要求新的空间，片断式的爆炸将允许它发生。

在研究过如何利用空间，并坚信液体运动可以具有更大的灵活性后，我选择了探索建筑如何可能成为一个有机液体元素，根据对空间的占有和其自身的运动来塑造自身形态。这个想法是通过液态参量来扩展自己，没有重力或空间的任何限制。这是对非真实要素的探索，浮动非刚性液体的滴块结构通过使用来塑造自己，一个游戏场地的想法可能是一个公寓或滑板公园，唯一的限制被用户设定的，灵活的环境通过使用来调整自己。

虽然这在现实生活中有局限性，你必须要看看有重力时会发生什么。当不同的要素被引进到像泡泡一样的流体建筑中有什么反应？实体模型的结构具有与数字模型相同的参数，这能够帮助确定空间是什么样子的，还需要什么样的外部支持来支撑这样的结构。所用的模型，是从应用该结构的数字化模型到真实建造的一种探索。

进一步的探索，得出了一个实际的空间配置。有什么限制的材料，并有反应，以一个空间？有什么结果未知？这个任务是着眼于占领和建立全面的工作模式。最后插图使用项目的各个方面结合起来，　在程度上，以说故事的进展，这一项目的开发。通过操纵的照片，并结合使用的技术在整个过程中，我想看看有什么定义空间可以做到。这是理念的阐释的问题——“它看上去都像是什么？”

最后一个关于知觉的项目是我主观想像中的一个。我的探索导致了我对建筑不同的理解和对人与建筑互动方面“假如会怎样？”问题的痴迷。我的想法是科幻的，就像纳米机器人技术或液态固体曾经也不存在。但在将来我们可能会看到建筑学的进步，建筑可以对人的运动或情绪做出反应。想像一下，当有一天你真的能够成为你自己的空间的创造者。

伍德伯瑞建筑学院
项目名称：一个插入日落大道心脏地带的红灯区的设计（对好莱坞社交圈享乐主义生活方式的一种社会评论）。

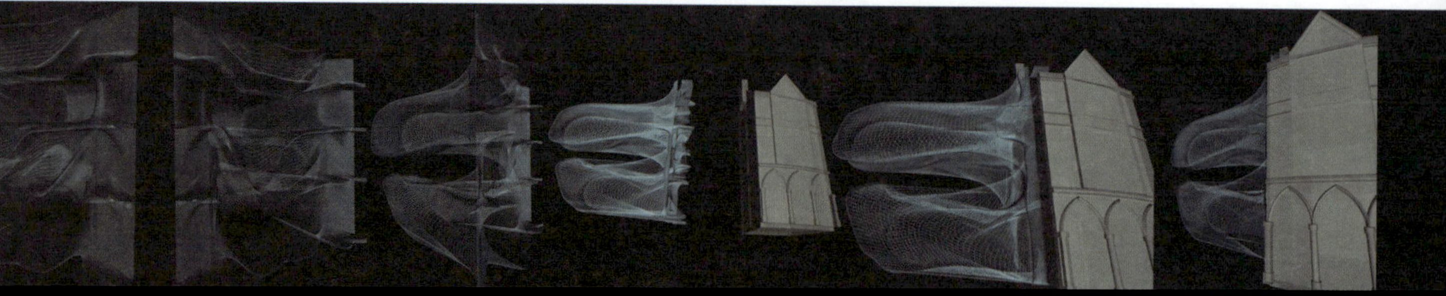

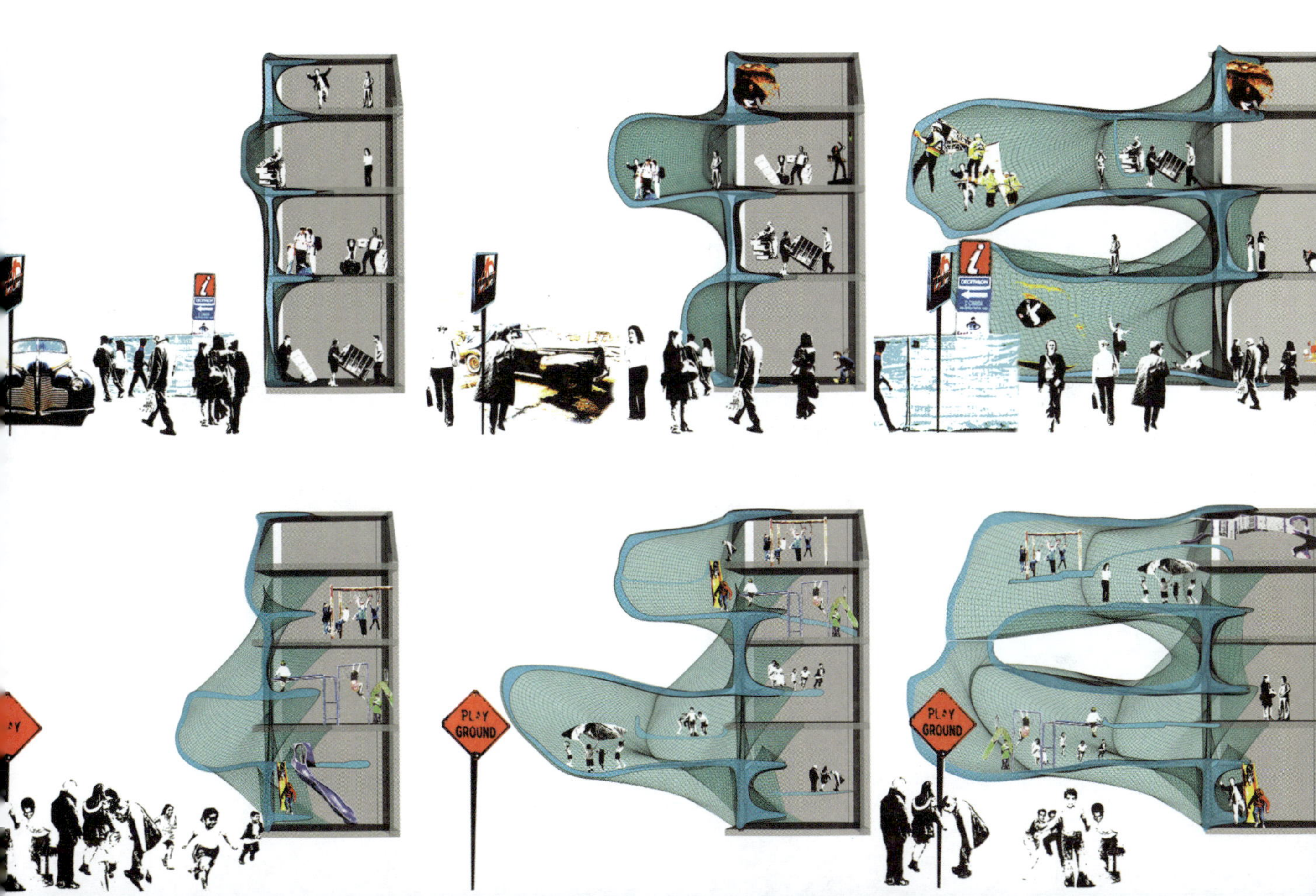
PLAY
GROUND
PLAY
GROUND

Central London (A 11)
Bow, Mile End
Stratford (A 118)
Docklands
Blackwall Tunnel A 12

娜奈德
亚科斯基
Nannete
Jackowski

MArch Nannette Jackowski
建筑学硕士 娜奈德·亚科斯基
n.jackowski@naja-deostos.com

europe
欧洲

germany
德国

halle [saale]
saxony - leipzig
berlin

哈莱［萨莱］
萨克森 莱比锡
柏林

NA.JA.

born in Halle [Germany]
生于哈莱［德国］

29/03/1977

finding roo
寻根

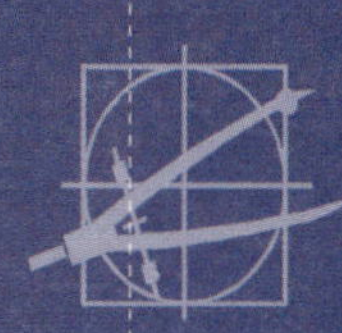

Experiments
实 验

Series
系 列

following traces
追 踪

Approaching
到 达

leaving traces
匿 迹

following traces
追 踪

Explaining
释 意

Defining
定 义

researching
研 究

researching
研 究

following traces
追 踪

study of architecture · University of Applied Science Leipzig [G]
学习建筑—莱比锡应用科学大学

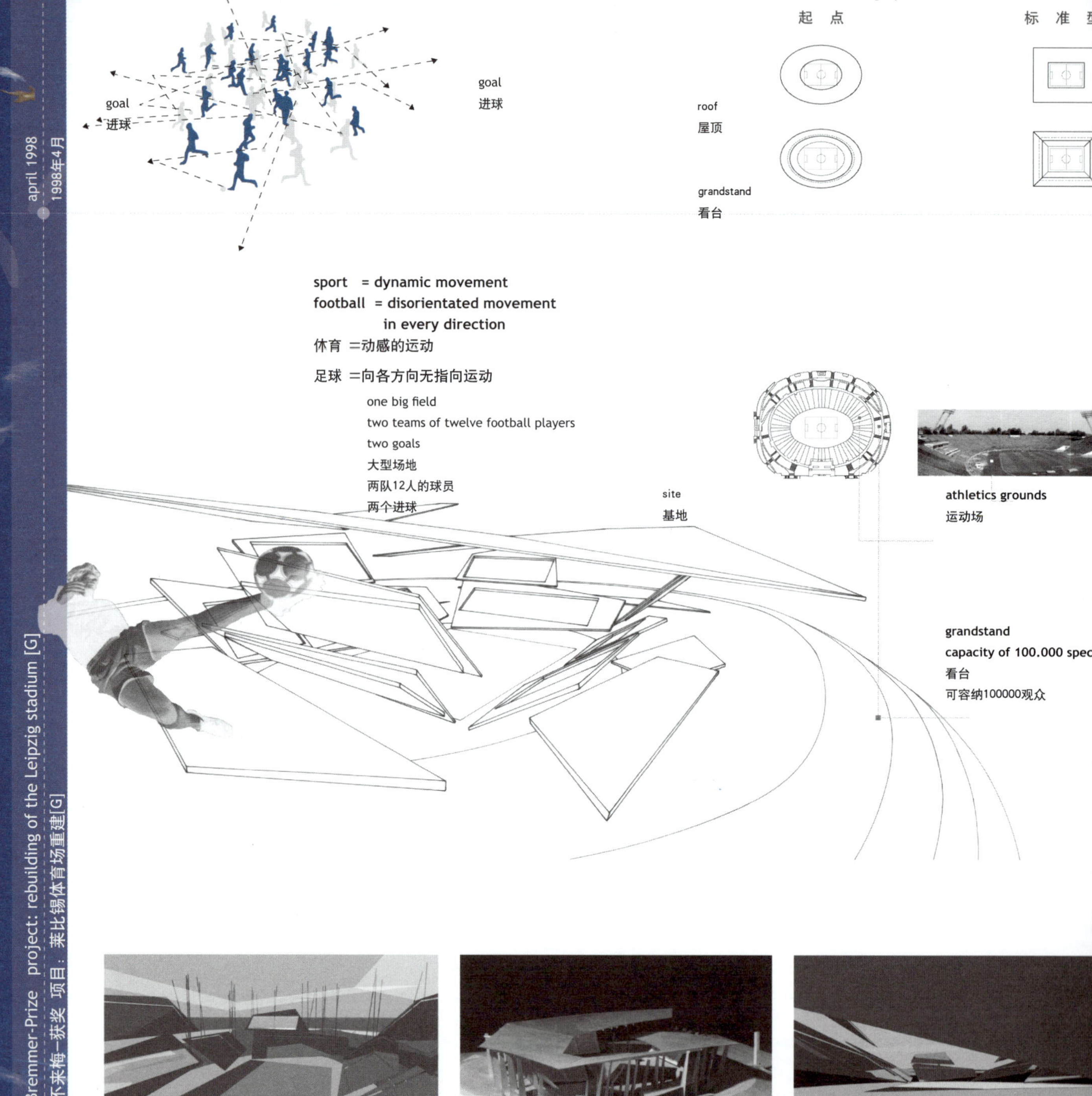

Bremmer-Prize project: rebuilding of the Leipzig stadium [G]
不来梅一获奖 项目：莱比锡体育场重建[G]
april 1998
1998年4月
goal
进球
goal
进球
starting point
起 点
standar
标 准 型
roof
屋顶
grandstand
看台
sport = dynamic movement
football = disorientated movement in every direction
休育 =动感的运动
足球 =向各方向无指向运动
one big field
two teams of twelve football players
two goals
大型场地
两队12人的球员
两个进球
site
基地
athletics grounds
运动场
grandstand
capacity of 100.000 spect
看台
可容纳100000观众

transformation

变 形

result

结 果

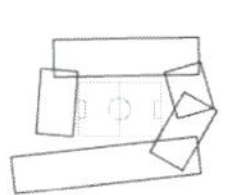
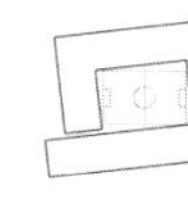
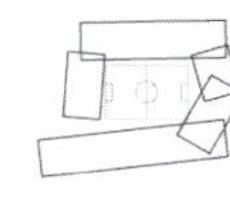
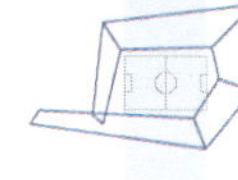
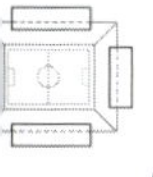
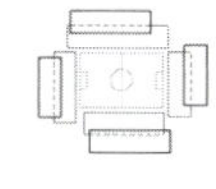
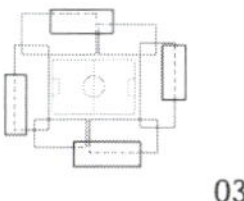
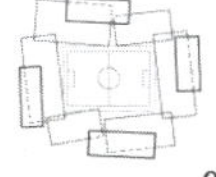
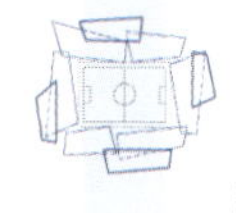

01 02 03 04 05

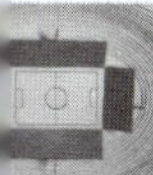

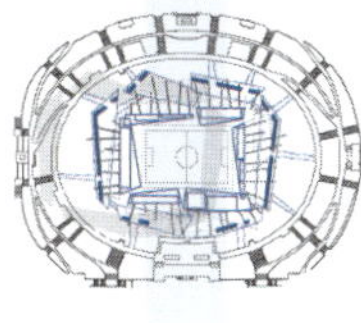

plan level 0

零层平面图

-zig wants to built a football stadium for the World Champion-
- 2006 in Germany. The challenge was it to design a new arena
-de the existing and dilapidated track and dilapidated stadium.

-比锡想要建造2006年德国世界杯体育场。挑战是新体育场
-建造在现存的破损的田径场内。

football stadium
capacity of 45.000 spectators

足球场

容纳45.000观众

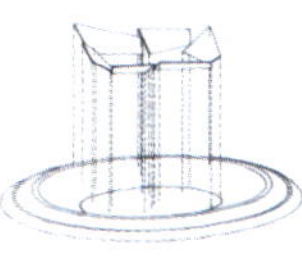
lower stand

低层看台

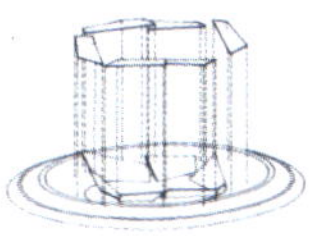
pedestrian walkway

步行道

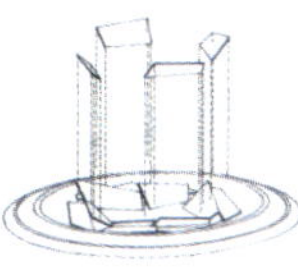
upper stand

上层看台

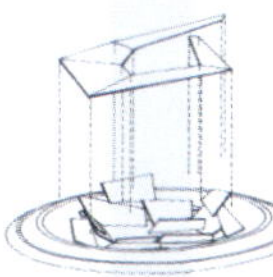
roof

屋顶

october 1998

1998年10月

student assistant of construction and design at the University of Applied Science, Leipzig [G]

学生助理，参加应用科技大学建造与设计，莱比锡[G]

1st Prize
facade design of an office building Leipzig [G]　　november 1998
1st Prize　with further company of execution
Form—Colour—Light, design of the Torgauer Bridge Leipzig [G], completed in october 2000
一等奖
莱比锡某办公建筑立面设计　1998年11月
一等奖　参与公司随后行动
形式—色彩—光，设计莱比锡道格桥，2000年10月建成

1st Prize
Velux-Attic-Award 99
一等奖
威卢克斯—典雅—大奖 1999年

1st Prize
Living space for young and old - life in East German Plattenbau
一等奖
为青年与老年的居住空间—东部德国勃兰登堡的生活

passed with distinction
Diploma: Grandstand structure for a track and field athletics stadium, Leipzig [G]
执业文凭项目：莱比锡田径场看台结构　1999年7月

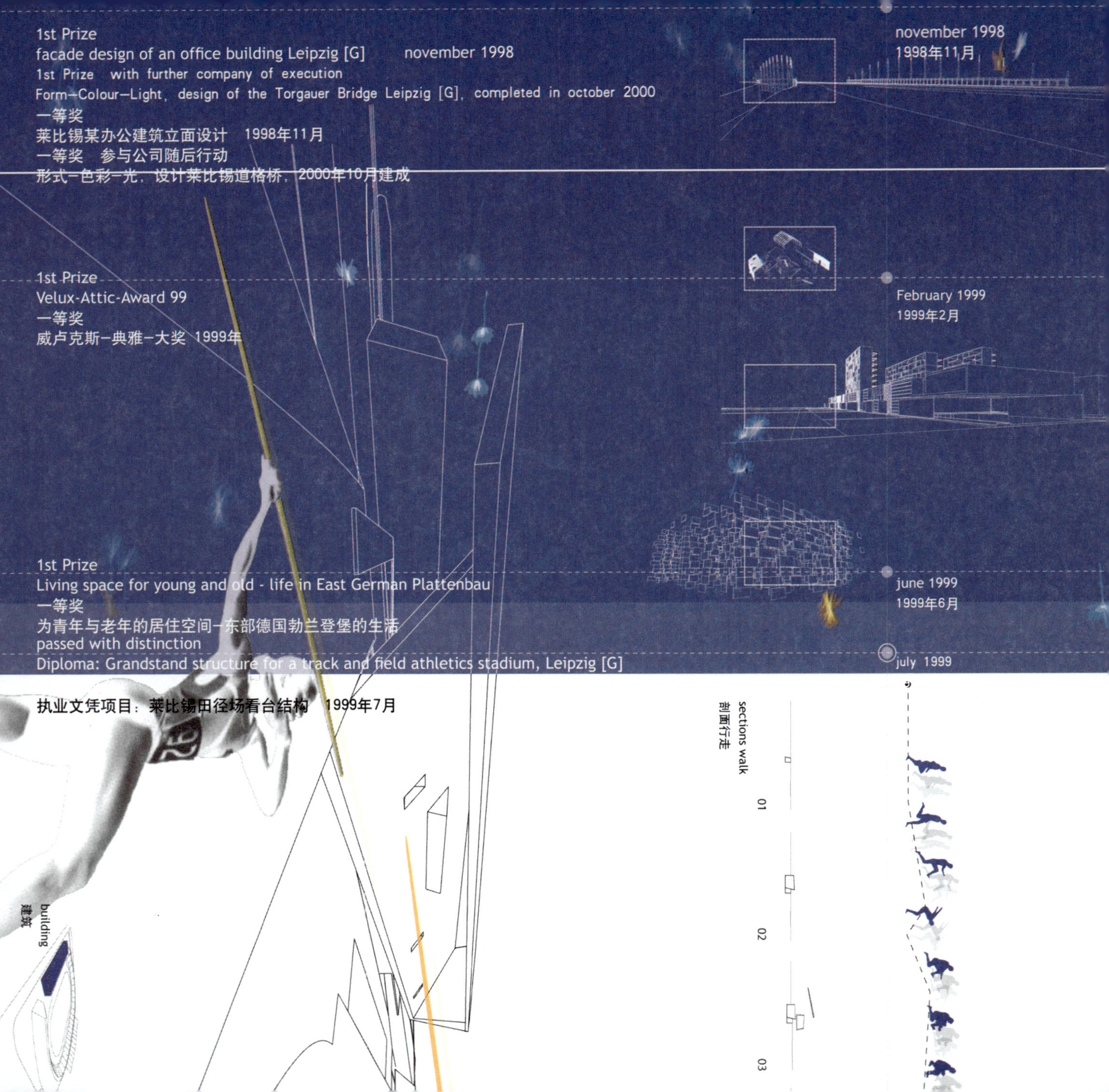

sport = dynamic movement
athletics = linear movement in one direction
体育=动感运动
田径=单方向的线性运动

4 05 06 07 08 09 10 11 12

see detailed section next page
下页见详细剖面图

Because of the new football stadium inside the existing stadium mound the athletics grounds were dropped out. That's why Leipzig athletes need new training and competition places. This design was a chance to think about possible olympic games 2012 in Leipzig already...

因为新足球场要建在原有体育场土堤内，田径场就需要建在外围。这就是为什么莱比锡需要新的训练和竞赛场。这个设计是思考可能的2012年莱比锡奥运会的一个机会（编者：成文时2012年奥运会主办国尚未确定）

Models
模型

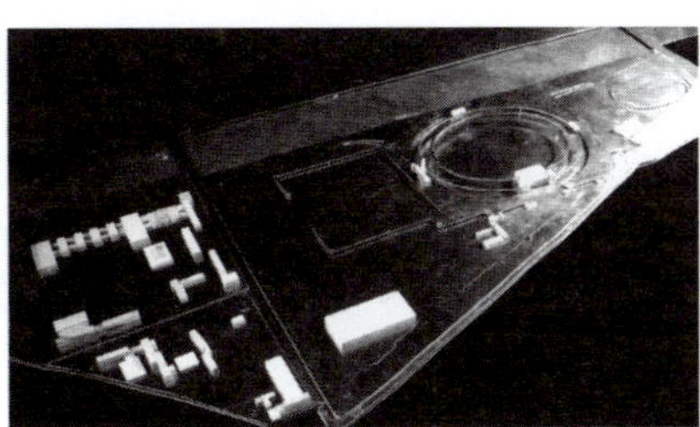
urban integration — sport grounds
城市融合—体育场

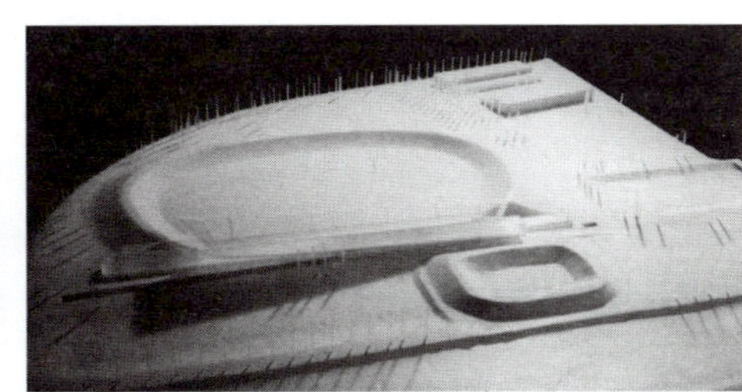
closer environment — north grounds
周边环境—北场地

stand

building
建筑

grandstand roof
看台屋顶

athletes facilities
运动员设施

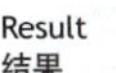
Result
结果

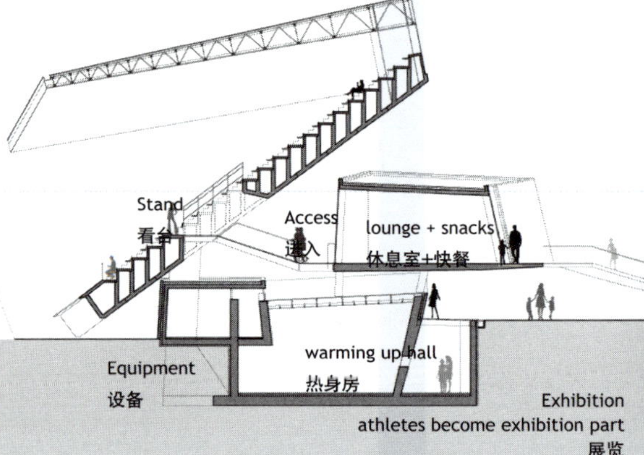

main construction parts
主结构部分

detail section
细节剖面

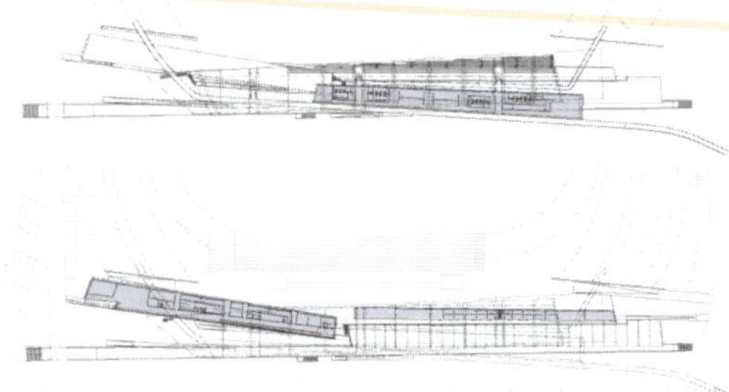

first level PUBLIC 首层 公众
spectators facilities 观众设施
stand access 看台进入

ground level *NON-PUBLIC* 地面层 非公共
athletes facilities 运动员设施

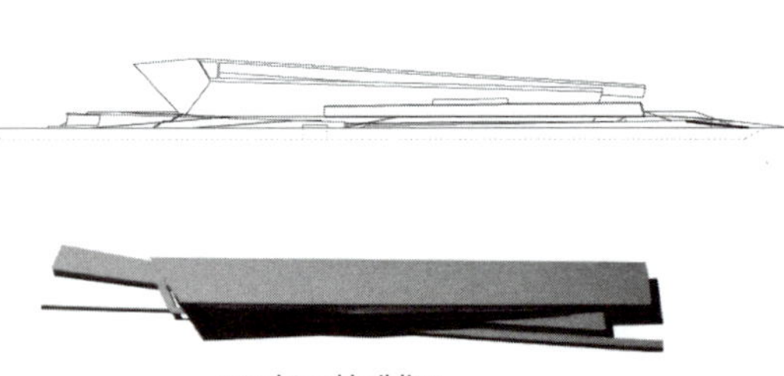

grandstand building
主看台建筑

detail
细部

from october 1999
1999年10月
freelance architect

exhibition Visions for Leipzig – Gallery Beck+Eggeling, Leipzig [G]
展览 莱比锡视觉——贝克+艾格灵画廊，莱比锡[G]

1999年12月

展览　迈尔体育论坛——城市中的体育建筑办公室，莱比锡[G]

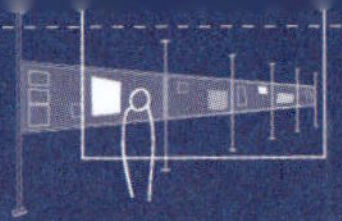

january 2000

2000年1月

preferential and choosen work [invited competition Velux]
Only Roof House
受青睐的作品[受邀请参加威卢克斯竞赛]
只有屋顶的住宅

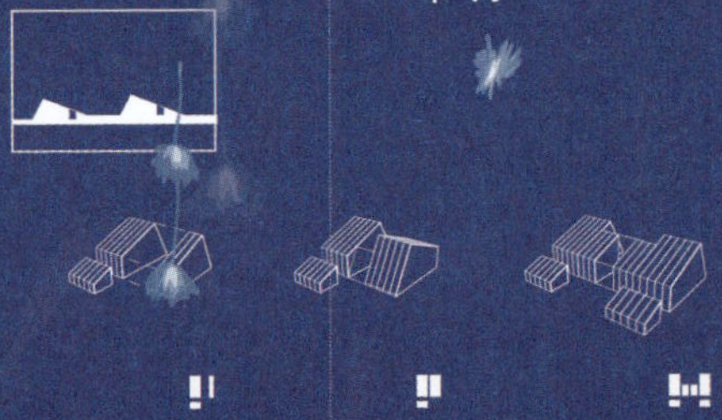

Zentralstadion – the Leipzig stadium – assistance on phases 3–5, 2000–02
in Wirth+Wirth architects, Leipzig [G]
中央车站–莱比锡体育馆——在3～5阶段提供辅助工作，2000–2002
沃斯沃斯建筑师事务所，莱比锡

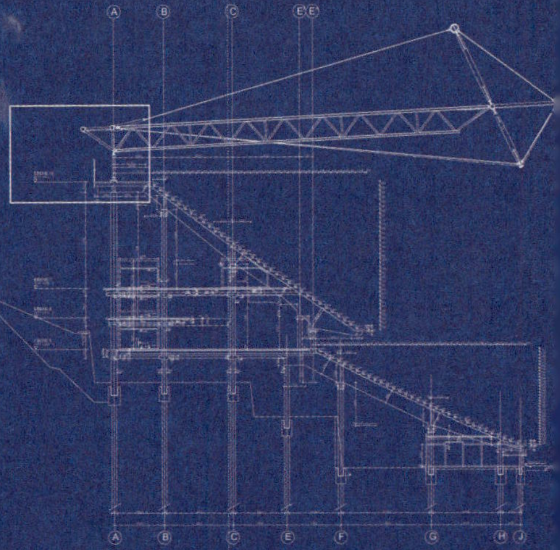

december 2000

2000年12月

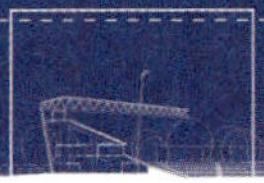

2nd Prize
rebuilding and extension of the Cottbus stadium – on behalf of Wirth+Wirth architects, Leipzig [G]
二等奖
考特巴斯体育馆重建扩建——沃斯沃斯建筑师事务所，莱比锡

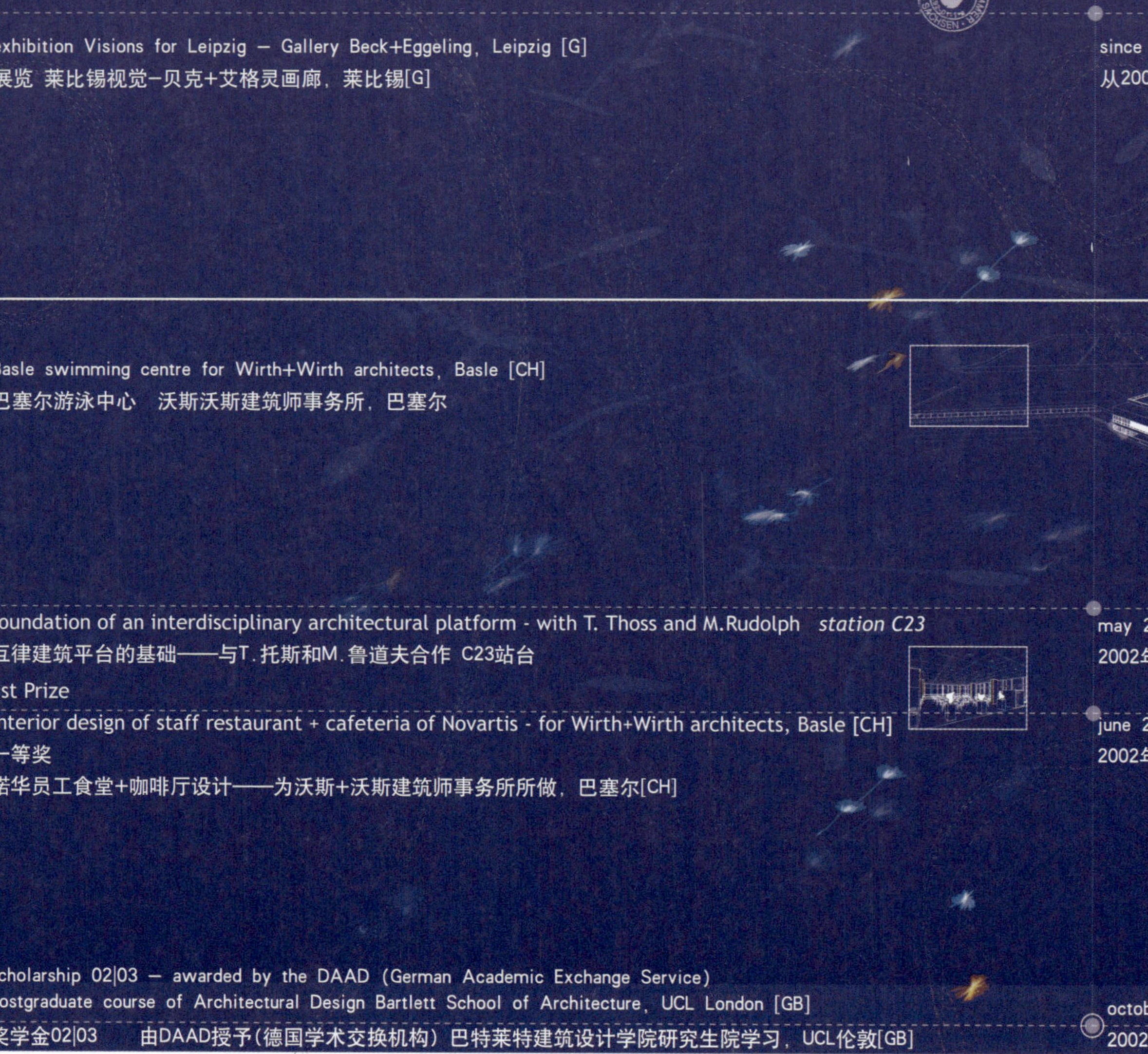

exhibition Visions for Leipzig – Gallery Beck+Eggeling, Leipzig [G]
展览 莱比锡视觉–贝克+艾格灵画廊，莱比锡[G]

since october 2001
从2001年10月起

Basle swimming centre for Wirth+Wirth architects, Basle [CH]
巴塞尔游泳中心 沃斯沃斯建筑师事务所，巴塞尔

foundation of an interdisciplinary architectural platform - with T. Thoss and M.Rudolph *station C23*
互律建筑平台的基础——与T. 托斯和M. 鲁道夫合作 C23站台

may 2002
2002年5月

1st Prize
interior design of staff restaurant + cafeteria of Novartis - for Wirth+Wirth architects, Basle [CH]
一等奖
诺华员工食堂+咖啡厅设计——为沃斯+沃斯建筑师事务所所做，巴塞尔[CH]

june 2002
2002年6月

scholarship 02|03 – awarded by the DAAD (German Academic Exchange Service)
postgraduate course of Architectural Design Bartlett School of Architecture, UCL London [GB]
奖学金02|03 由DAAD授予(德国学术交换机构)巴特莱特建筑设计学院研究生院学习，UCL伦敦[GB]

october 2002
2002年10月

a vessel of water + ink
一船水+墨

one slide filled with acrylic paint oil p
一张幻灯片被装满 丙烯颜料 油画颜料

definition
定 义

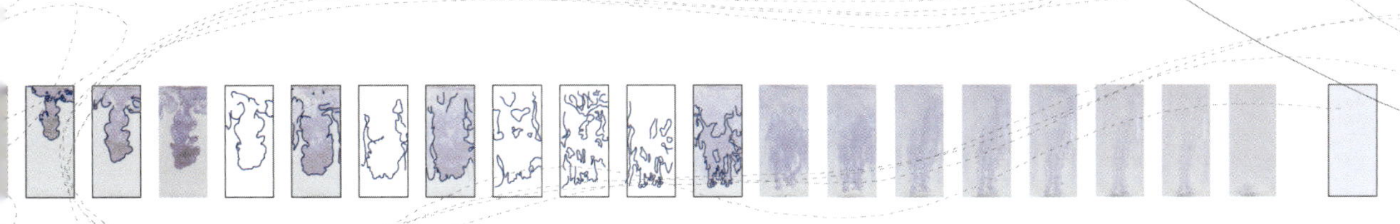

ries of the spreading and the disintegration of ink inside water

-系列墨在水中散开和解体

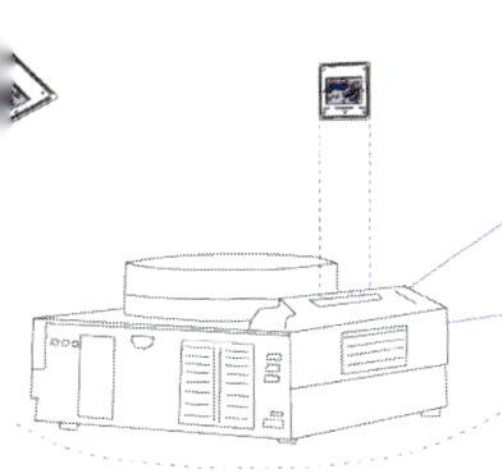

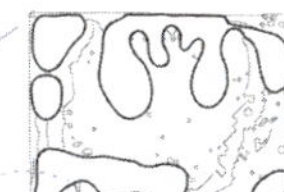

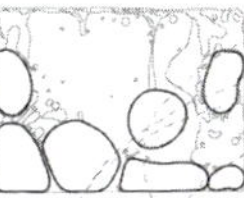

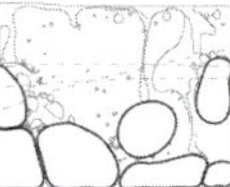

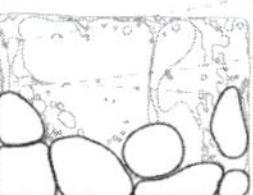

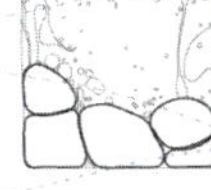

= "movie" - the less liquid the paint the slower the movement the more fascinating the more dreamy the more intense the relaxation

="电影" — 非液态的颜料运动越慢越引人瞩目越梦幻越强烈的舒缓

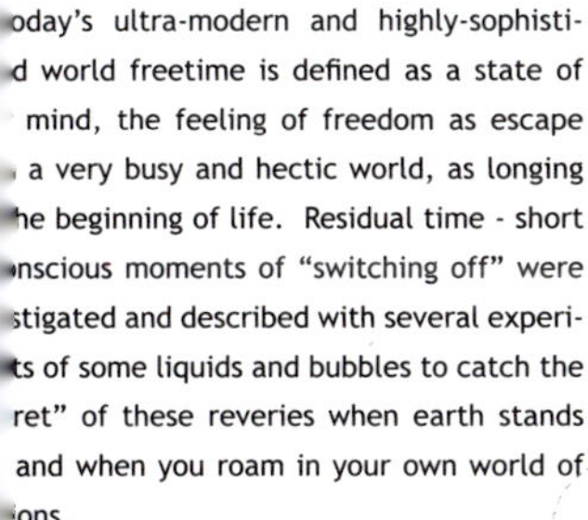

oday's ultra-modern and highly-sophistid world freetime is defined as a state of mind, the feeling of freedom as escape a very busy and hectic world, as longing he beginning of life. Residual time - short nscious moments of "switching off" were stigated and described with several experits of some liquids and bubbles to catch the ret" of these reveries when earth stands and when you roam in your own world of ons.

超级现代和高度复杂的世界闲暇时光被定义的一种精神状态，一种从紧张忙碌中逃脱的，觉得回到生命的初始。剩余时间——短暂关闭"状态的无意识瞬间被研究和用一些液气泡的实验描述，以此来捕捉这些遐想的奥这时，地球静止不动而你在自己的幻想世界哮。

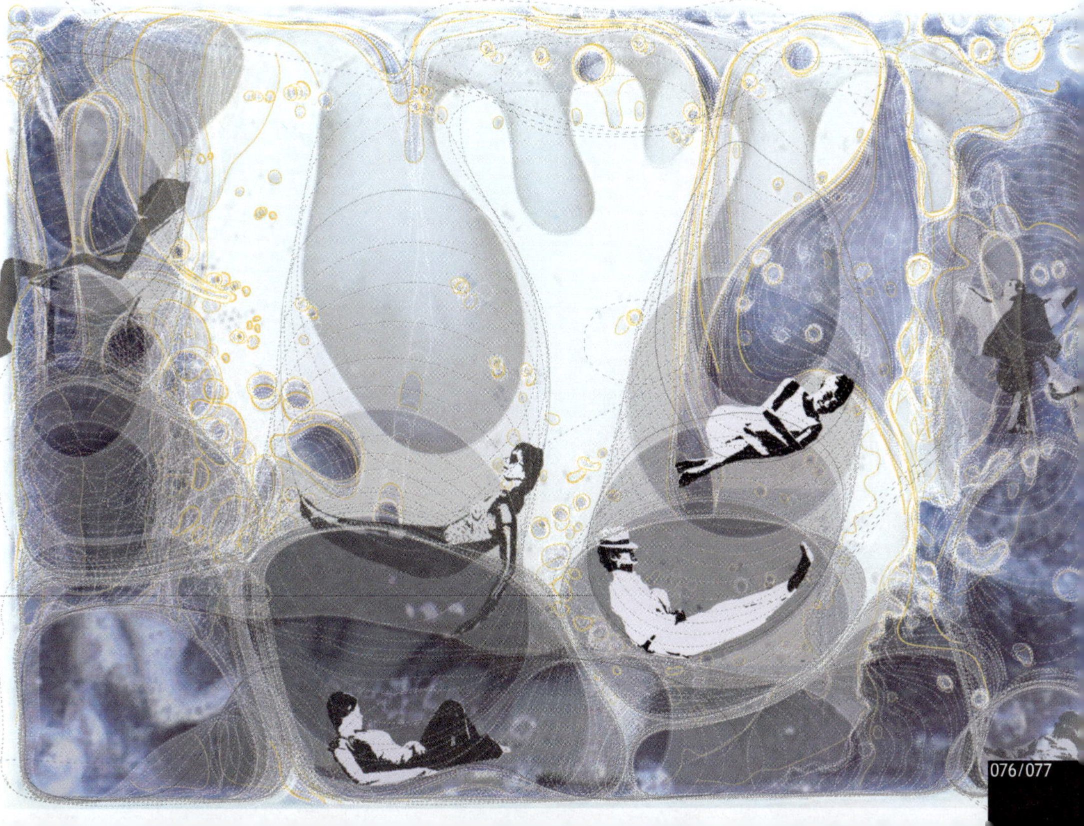

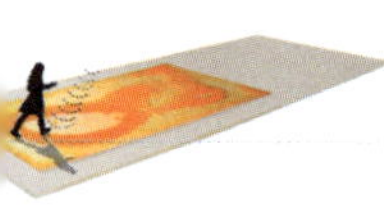

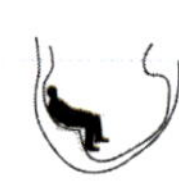
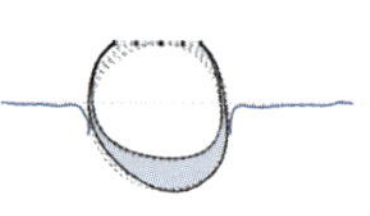
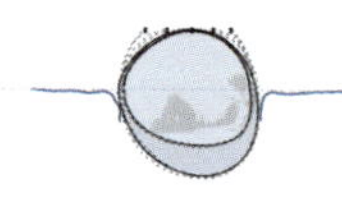
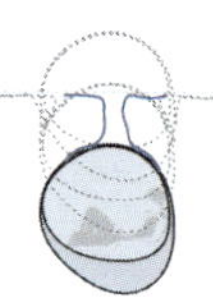

How long does it take?
需要花多长时间？

only some seconds
仅仅几秒钟？

10 - 30 minutes
10～30min

10 - 30 minutes
10 - 30min

some hours
几小时

How does it move?
它如何移动

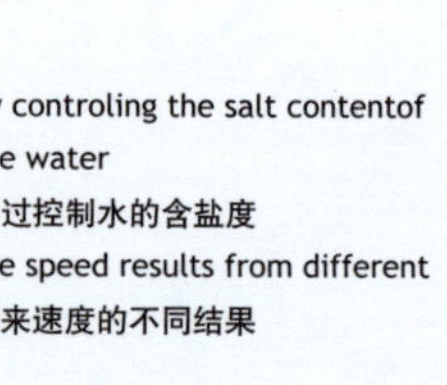

by controling the salt contentof the water
通过控制水的含盐度
the speed results from different
带来速度的不同结果

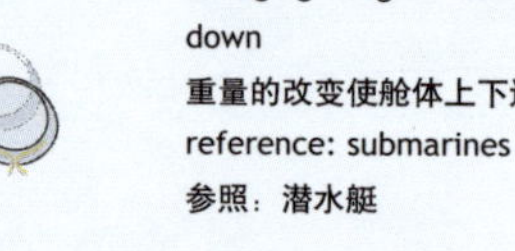

the capsule is wraped by a tank which gets filled by regulating the pressure of the water
舱体被水箱包裹，水箱充满水来提供规律的水压
changing weights let the capsule move up and down
重量的改变使舱体上下运动
reference: submarines
参照：潜水艇

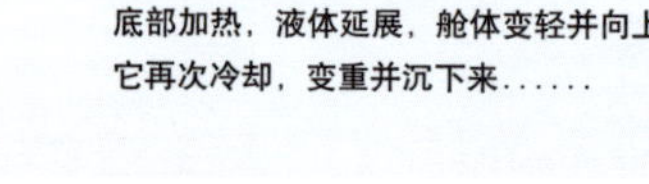

the outer skin of the capsule is filled with a dense liquid than water by warming up the from the bottom the liquid expands, the sule gets lighter and moves up, it cools c again, gets heavier and sinks...
舱体的外层充满比水密度小的液体，通过在底部加热，液体延展，舱体变轻并向上移动，它再次冷却，变重并沉下来……

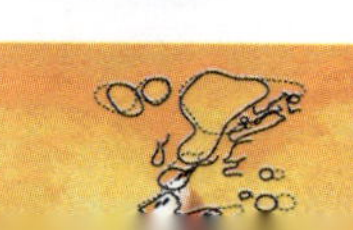
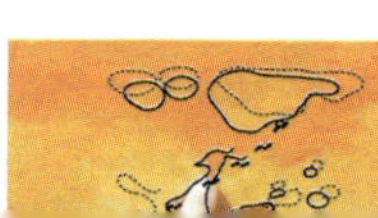
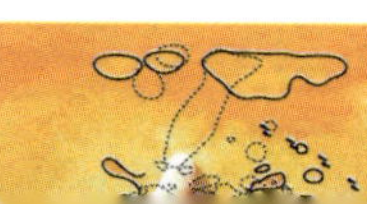

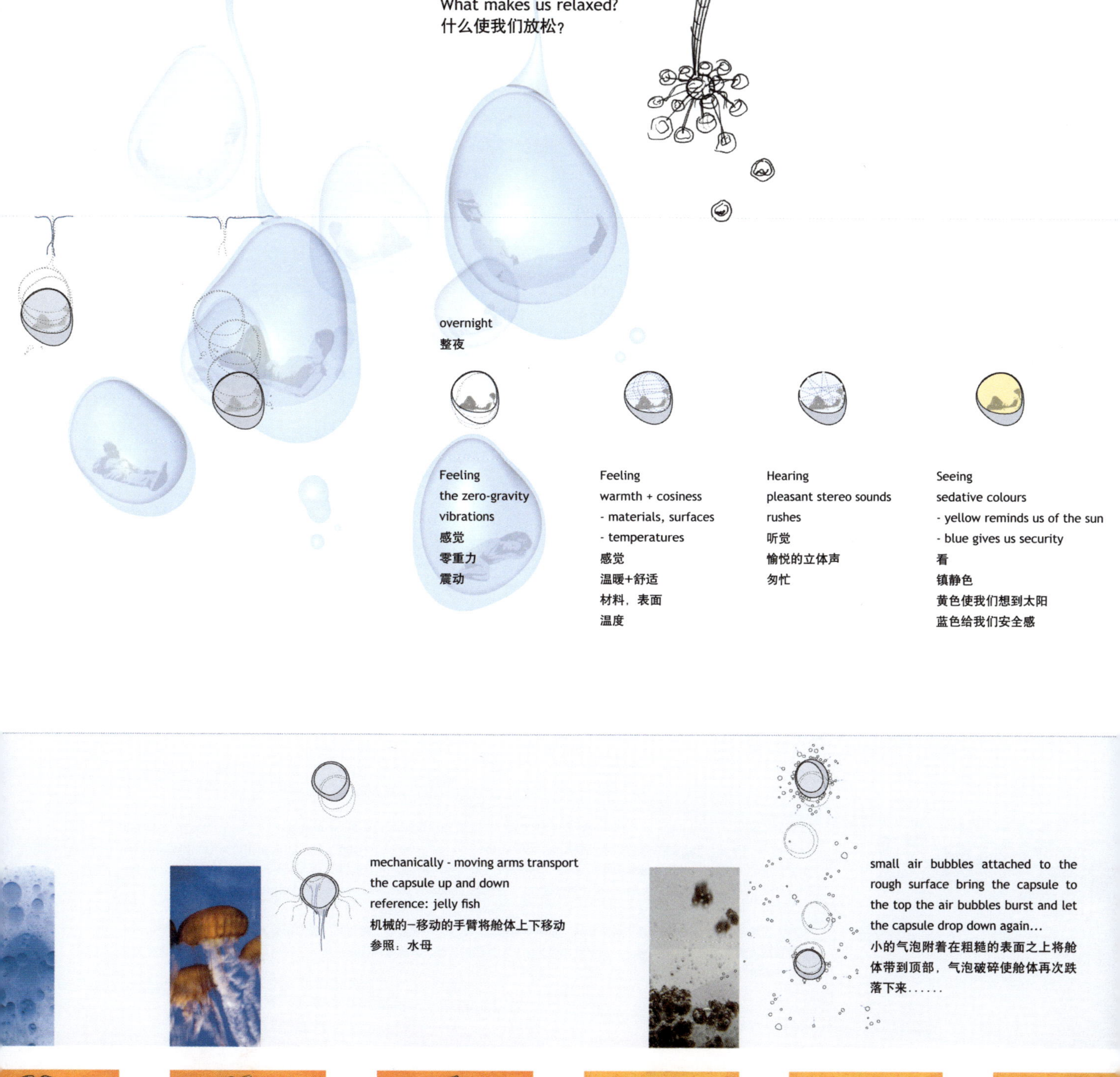

What makes us relaxed?
什么使我们放松?

overnight
整夜

Feeling
the zero-gravity
vibrations
感觉
零重力
震动

Feeling
warmth + cosiness
- materials, surfaces
- temperatures
感觉
温暖+舒适
材料，表面
温度

Hearing
pleasant stereo sounds
rushes
听觉
愉悦的立体声
匆忙

Seeing
sedative colours
- yellow reminds us of the sun
- blue gives us security
看
镇静色
黄色使我们想到太阳
蓝色给我们安全感

mechanically - moving arms transport the capsule up and down
reference: jelly fish
机械的—移动的手臂将舱体上下移动
参照：水母

small air bubbles attached to the rough surface bring the capsule to the top the air bubbles burst and let the capsule drop down again...
小的气泡附着在粗糙的表面之上将舱体带到顶部，气泡破碎使舱体再次跌落下来......

Bank Area - the origin of London
河岸地区–伦敦的源始

1

2

3

4

5

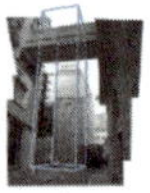

Liverpool Street Station
利物浦街车站
1
2
3
4
5
6
St. Paul Cathedral
圣保罗大教堂
Thames
泰晤士河
Thames
泰晤士河
London Bridge
伦敦桥
main site
主基地

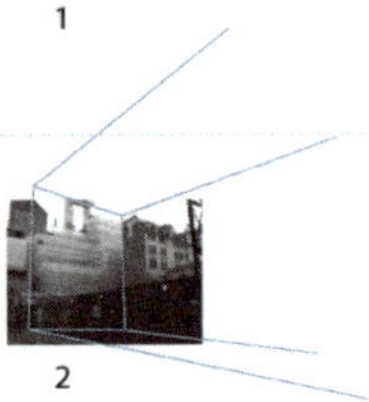

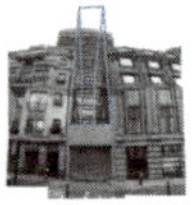

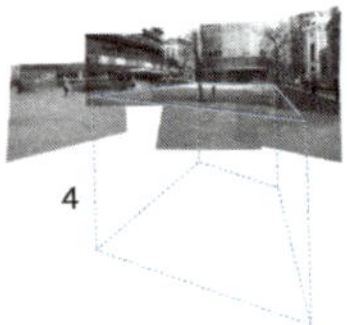

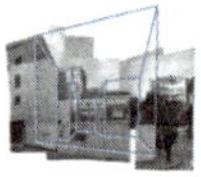

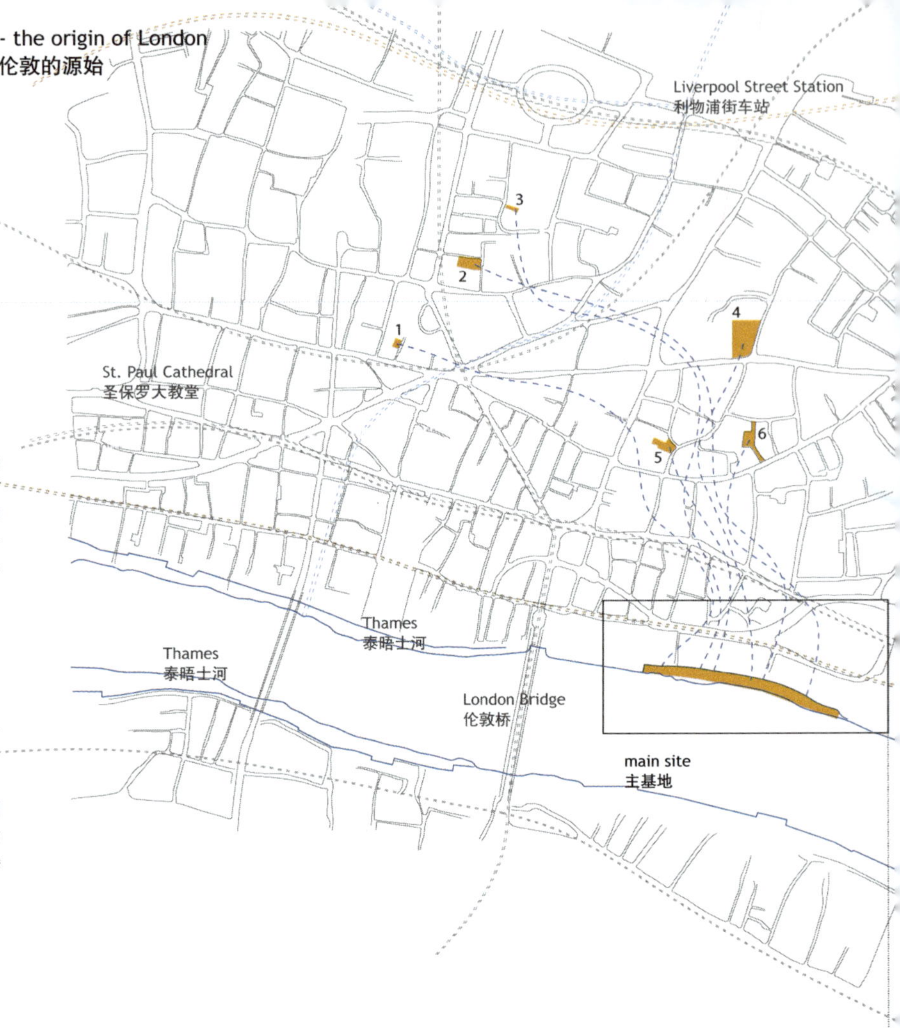

To remind oneself of the original idea of freetime some spaces in the midst of busy areas in London were occupied and transformed into inactivity spaces to create a direct confrontation between stress and relaxation.

sorts of stress:

- shopping stress, tourism places
- traffic stress, speed + noise
- business
- waiting time = waste of time (eg. airports)

提醒自己休闲时间的源初想法： 在伦敦闹市区中的一些空间被占用和转换成非活动性空间来创造紧张压力和放松之间的直接对立。

紧张的种类：

购物紧张，旅行紧张

交通紧张，速度+噪声

6

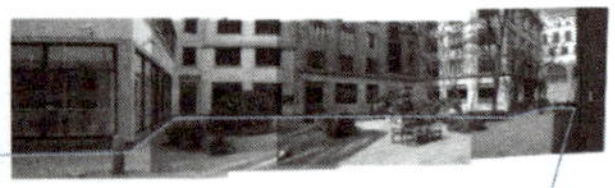

Business people carry loads of documents run from meeting to meeting across the streets day by day, calling or eating in between... Black suits, white shirts and ties determine the “powerhouse of London”. The area of the original London has developed into the worlds biggest financial business centre and into one of the most hectic and busy places as well. Only few small old buildings surrounded by modern high-rises and huge company buildings still remind of the origin.

商务人士携带大量文件天天奔波于各个会场，间隙中用餐或打电话……黑西装、白衬衫、领带定义了“伦敦动力”。伦敦源始区发展成为世界上最大的商务中心，同时变成最为活跃和忙碌的区域。只有被现代高层和大公司建筑围绕的几个小型老建筑是原来就有的。

existing bank way
现存河岸路

extension
扩建

main functions
主要功能

relaxation
放松

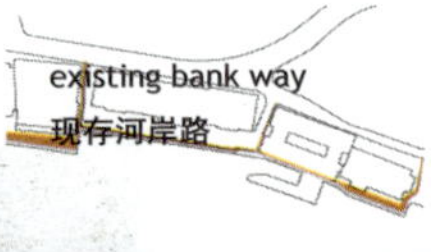

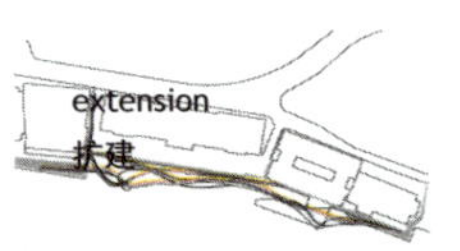

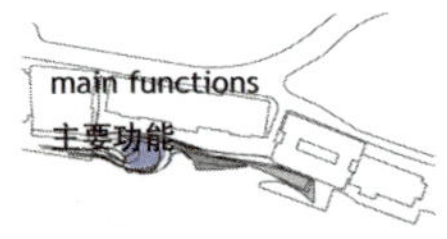

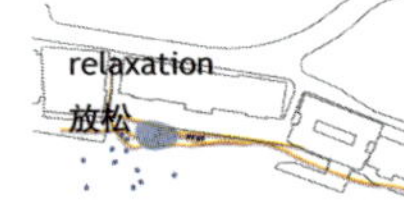

conceptual plan
概念平面图

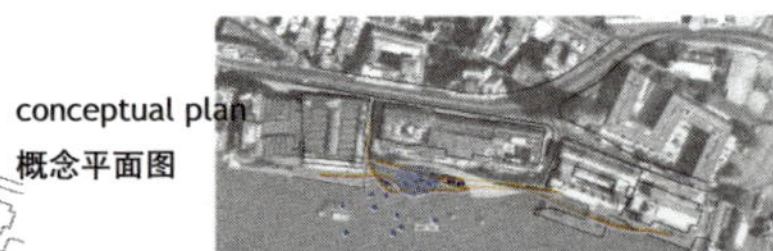

3rd Prize rebuilding and extension of the spa Bad Gollingen - on behalf of Wirth+Wirth architects, Kundl [A]
三等奖 格林根浴场的重建与扩建—代表沃斯沃斯建筑师事务所，坤多[A]

March 2003
2003年3月

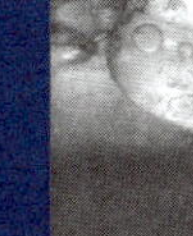

experiments
试 验

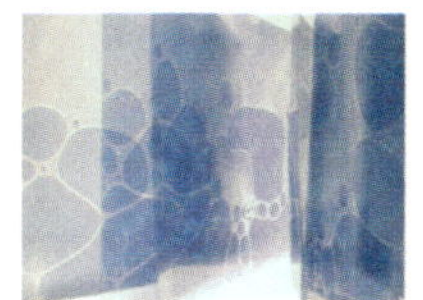
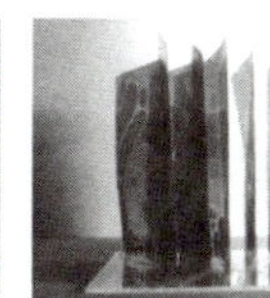

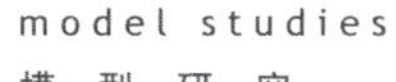

model studies
模 型 研 究

00

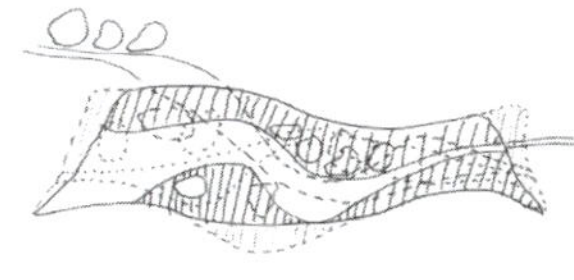
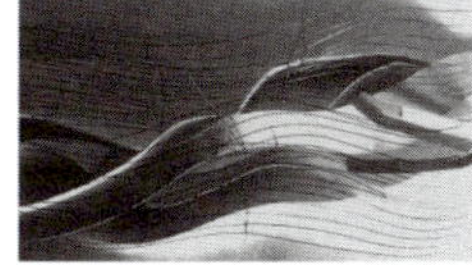

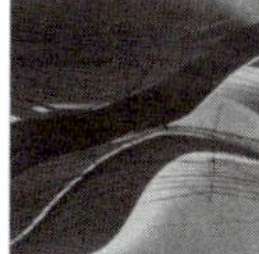

01

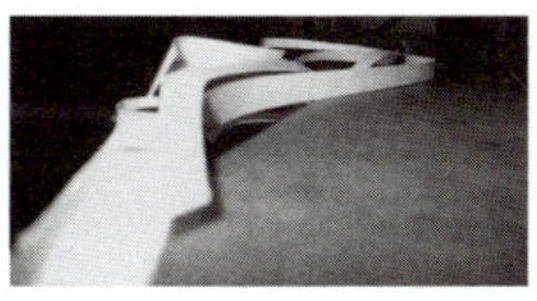

02

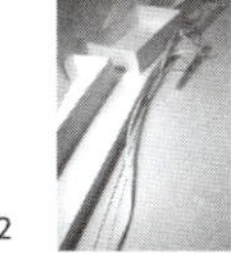

03

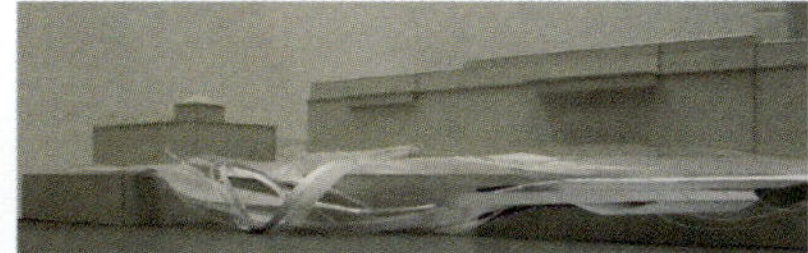
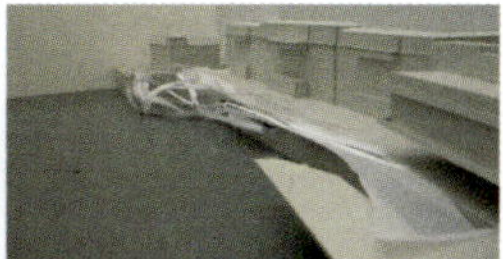

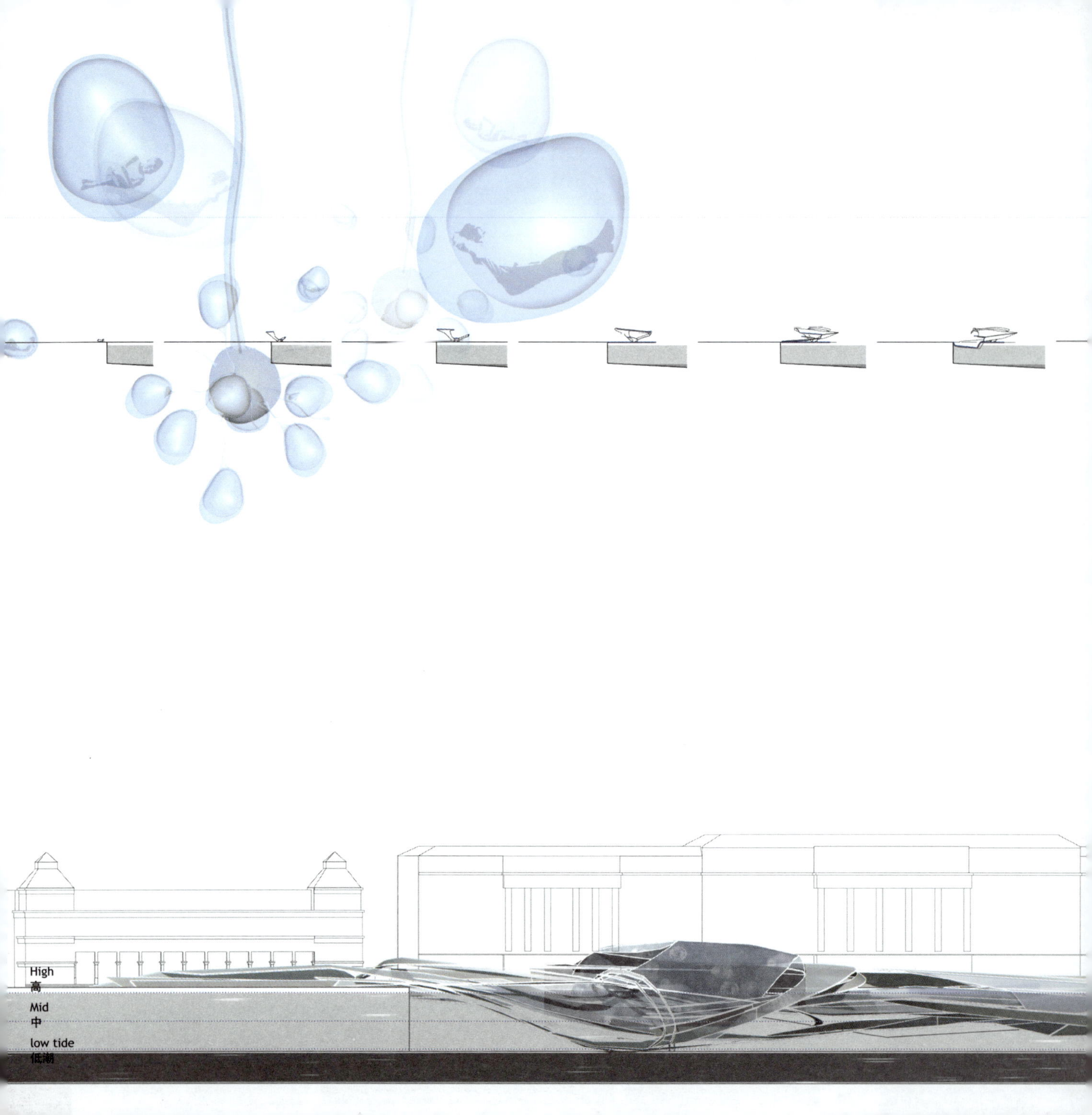
High
高
Mid
中
low tide
低潮

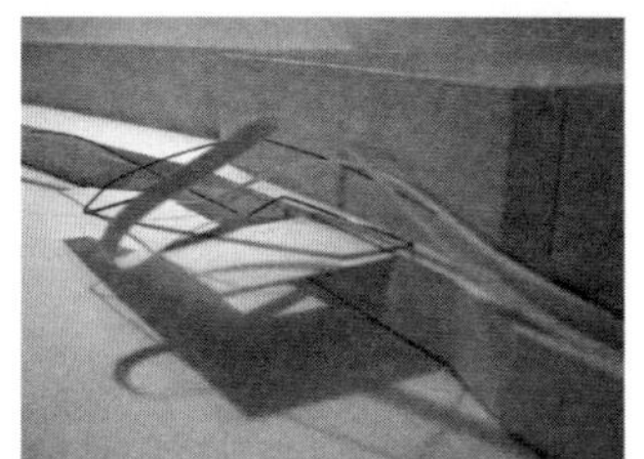

floatation hotel
漂浮酒店

main floatation tank
主浮箱

single floatation capsules
单个漂浮舱体

relaxation spa
休闲spa

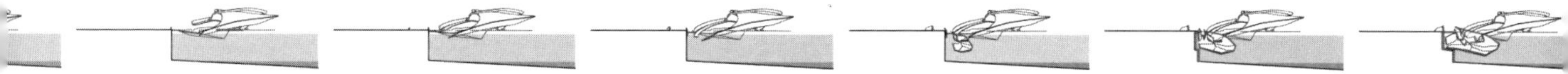

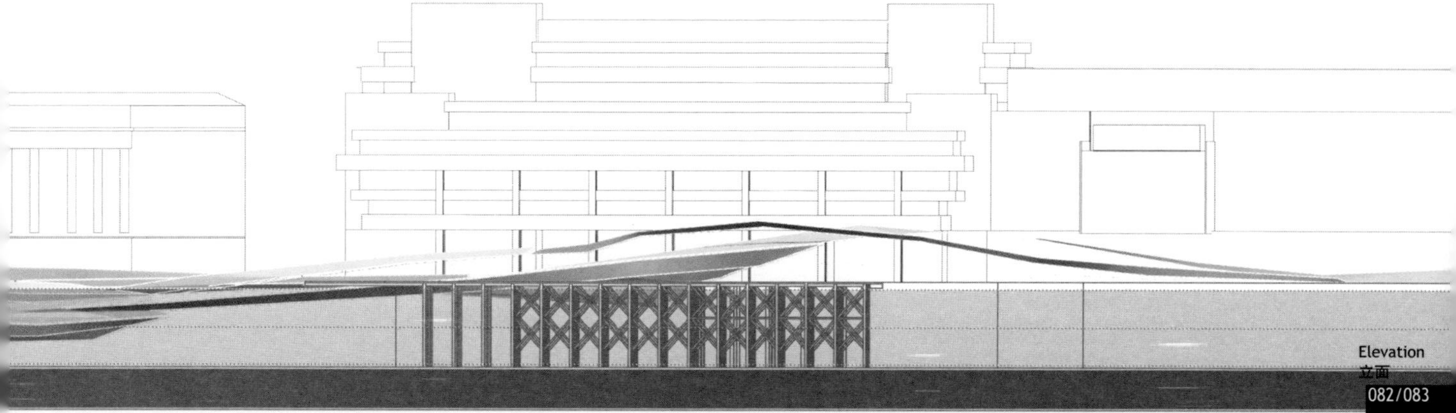

Elevation
立面

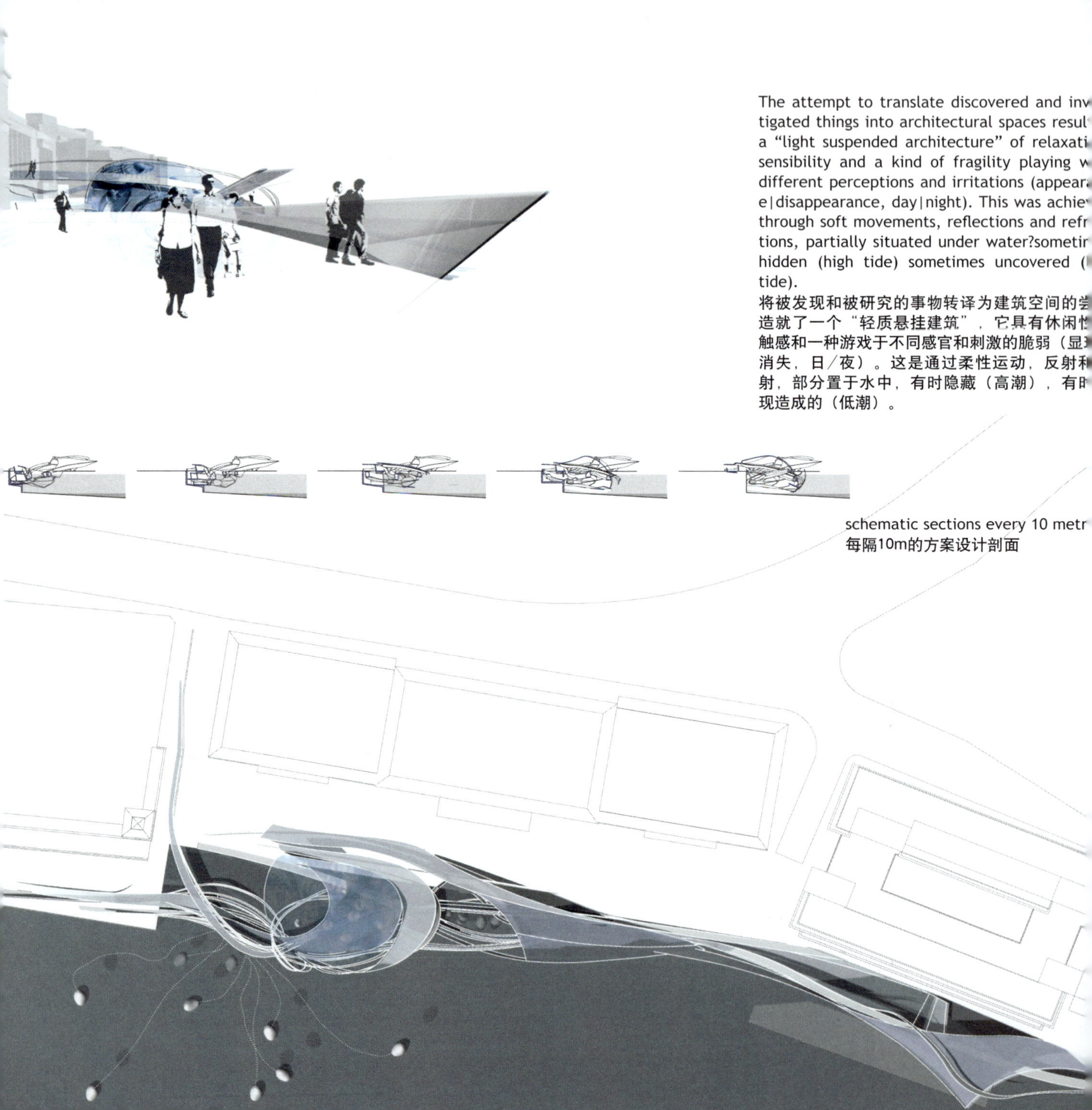

The attempt to translate discovered and inv
tigated things into architectural spaces resul
a “light suspended architecture” of relaxati
sensibility and a kind of fragility playing w
different perceptions and irritations (appear
e|disappearance, day|night). This was achie
through soft movements, reflections and refr
tions, partially situated under water?sometir
hidden (high tide) sometimes uncovered (
tide).

将被发现和被研究的事物转译为建筑空间的尝
造就了一个“轻质悬挂建筑”，它具有休闲性
触感和一种游戏于不同感官和刺激的脆弱（显
消失，日／夜）。这是通过柔性运动，反射和
射，部分置于水中，有时隐藏（高潮），有时
现造成的（低潮）。

schematic sections every 10 metr
每隔10m的方案设计剖面

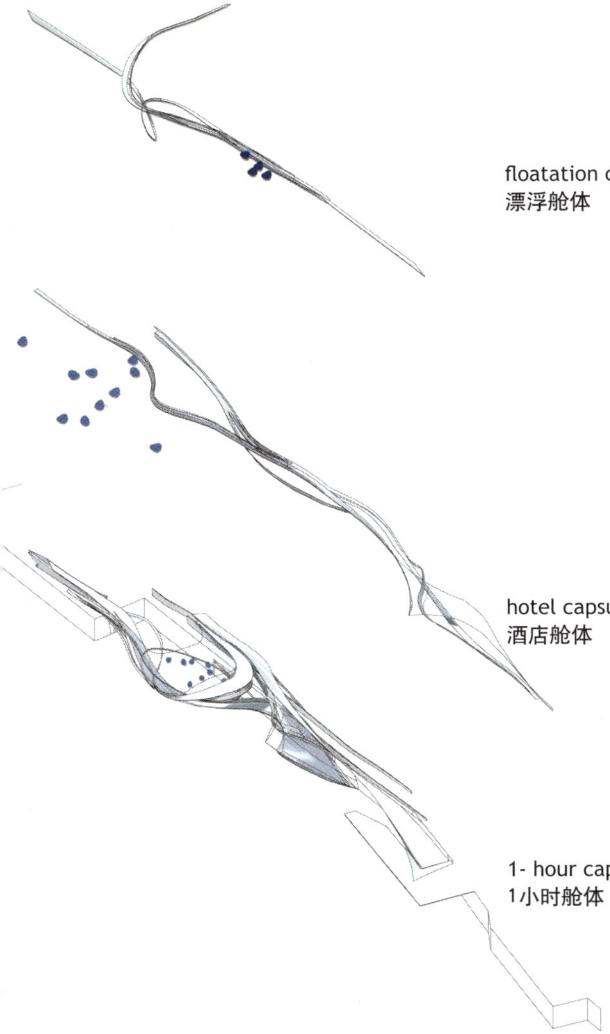

floatation capsules
漂浮舱体

hotel capsules
酒店舱体

1- hour capsules
1小时舱体

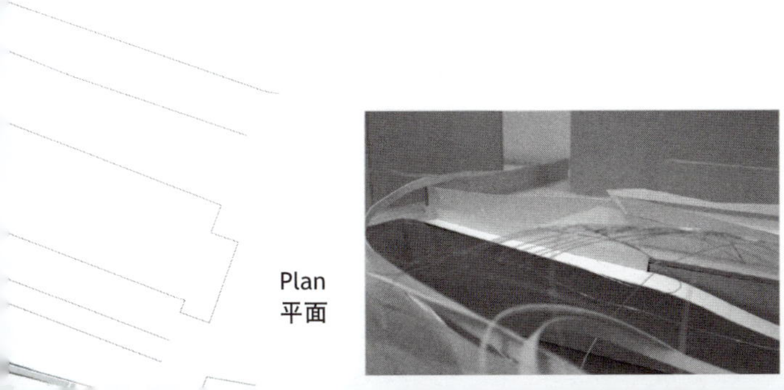

Plan
平面

september 2003
2003年9月

passed with commendation
M.Arch [Architectural Design]
优良毕业
建筑学硕士 [建筑设计]

december 2003
2003年12月

evening lecture of station C23 at *TU Dresden*, series spann-weiten - young architects in Germany
TU 德累斯顿C23车站夜间演讲系列，跨度系列—德国青年建筑师2003

march 2004
2004年3月

co-author of Sportforum *Leipzig - history and future*, editor: pro Leipzig
莱比锡体育论坛共同署名——历史与未来，编辑：莱比锡专业协会

Cambridge
(M 11 (N))
B. Stortford
A 120
Non - motorway traffic
London
Harlow
M 11 (M 25)
Emergency vehicles
Y584 FEV
POLICE
50

卡利玛尼
Maro
Kalimani

I invested seven years of my 26-year life in studying in architectural schools. One more year in learning during professional employment... Not quite a long experience so as to express consciously an honest written opinion on a borderless subject like architecture. However, I felt sometimes in my life architecturally strong. (Didn't I?). I recall the end of my 6-year studies in the Faculty of Architecture in the National Technical University of Athens having created a visionary young architect ready to change the world with no compromises.

My beloved school taught me through the ritual process of design how to persistently investigate the fundamental creative forces of the concept and give solutions to any problems that my daring fiction had. Meanwhile, the professional practice helped me observe in real life terms the tangible characteristics of the matter. And that power feels so good that you want to share the feeling with people outside the narrow borders of your school, your city, your country (even if you hadn't realised that narrowness, yet). And you "win" a trip in London...just for you. No friends, no family...only you and your ideas.

That should be enough. (Shouldn't it?) And then comes the moment that your power evaporates by the time you meet the multiethnic cocktail of architectural ideas. And you absorb all that knowledge from a school like Bartlett and from a city like London that you can't help seeing yourself, your country, your future, your architecture from an unexpectedly fresh point of view. And yes! You regain the power...but now double the size and the maturity. So, now you have learned that there is another hidden reality behind your reality. That evolution continues. And this time, maybe the journey starts from China...who knows...

在我26年的人生中，我投入了7年的时间在建筑学院中学习。一年多一点的时间在职业工作中学习……并非很长的经历，并不足以对像建筑这样一个无边的话题作出有意识的忠诚的书面表达。然而，有时我感到自己人生中的建筑性是那么强烈。（不是吗？）我回忆起在雅典国立技术大学建筑学院6年学习的最后时光，一个满怀梦想的年轻建筑师，预备毫不妥协地改变这个世界。

我心爱的学校教会我如何通过那些仪式性的设计步骤来坚持研究概念中基本的创造性力量，并对那些我大胆的想像所带来的问题给出解决方案。同时，专业的训练帮助我在现实生活中如何观察事物切实的特征。这种力量感觉真好，以至于你想要超越狭窄的边界，和你的学校以外，城市以外，国家以外的人共同分享这种感觉（即使你还没有察觉到这种狭窄）。而且你“赢”得了去伦敦的旅行……只是你一人，没有朋友、没有家人……只有你和你的观念。

这应该够了。（难道不是吗？）然而接下来，当你遭遇多族裔的鸡尾酒式的建筑思想时，你的这种力量蒸发了。从伦敦这样的城市，从巴特雷特这样的学院吸取所有的知识无可阻挡地使你从一个毫无先验的全新角度看到你自己，你的国家、你的未来、你的建筑。太棒了！你又重新获得了这种力量……但是现在是效力倍增的，并且更趋成熟的。所以，现在你意识到了在你的真实之后还有另一个隐藏着的真实。那是进化在继续着。这一次，也许旅程从中国开始……谁又知道呢……

2006-present FOSTER+PARTNERS, London, U.K
2007-present ARB Registered architect, U.K
2004-2006 YAEL REISNER-COOK, architect, London, U.K
2006 SANA Pan-Hellenic Architectural Design Exhibition
2004-2005 ATELIER ONE LTD, structural engineers, London, UK
2005 ROYAL ACADEMY OF ARTS summer exhibition of the Master diploma thesis
2002-2003 UNIVERSITY COLLEGE LONDON, THE BARTLETT SCHOOL OF ARCHITECTURE
Master in Architectural Design [MArch] with Distinction
2002-2003 ANNUAL SCHOLARSHIP from Evgenides Foundation (Greece) for postgraduate studies
2003 Published the diploma thesis project in the annual review of "Architecture in Greece", issue no.3
2001 VOZANI ARIADNI architects, Athens-Greece
2001-present Individually contracted projects as a freelancer in Greece
2001-present Greek Chamber of Engineers Registered architect, Greece
2001-present Member of the S.A.D.A.S-P.E.A [Pan-hellenic architectural Society]
2001 Award for the NTUA competition "Conversion of a boiler laboratory into a student restaurant"
1995-2001 NATIONAL TECHNICAL UNIVERCITY OF ATHENS [NTUA], FACULTY of ARCHITECTURE
Diploma of Architecture-Engineering with Distinction
1999-2000 "GREEK CHAMBER OF ENGINEERS" award for excellent course performance
1995-1996 "CHRISTOS PAPAKIRIAKOPOULOS" award for excellent course performance

2006年至今 福斯特及其合伙人建筑师事务所，建筑师，英国伦敦
2007年至今 ARB注册建筑师，英国
2004–2006年 耶尔•瑞瑟–库克建筑师事务所，建筑师，英国伦敦
2006年 参加萨纳泛希腊建筑设计展
2004–2005年 第一画室有限公司，结构工程师，英国伦敦
2005年 参加英国皇家美术学院硕士文凭论文夏季展
2002–2003年 英国伦敦大学学院，巴特雷特建筑学院
建筑设计硕士，优秀毕业
2002–2003年 研究生课程学习获得Evgenides基金会奖学金（希腊）
2003年 在《希腊建筑学》第3期“年度回顾”中发表本科论文
2001年 VOZANI ARIADNI建筑师事务所，建筑师，希腊雅典
2001年至今 作为一个自由职业者在希腊单独承担工程
2001年至今 希腊工程师协会，注册建筑师，希腊
2001年至今 萨纳泛希腊建筑社团成员
2001年 “从锅炉实验室到学生餐厅的改造”项目
在雅典国立科技大学竞赛中获奖
1995–2001年 雅典国立科技大学，建筑学院
建筑工程文凭，优秀毕业
1999–2000年 因优秀科目表现获“希腊工程师协会”奖
1995–1996年 因优秀科目表现获“CHRISTOS PAPAKIRIAKOPOULOS”奖

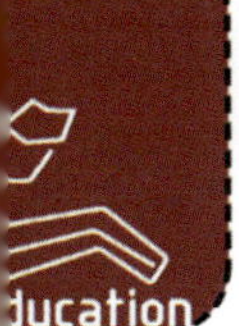

FOR PRE_BARTLETT: diploma thesis (2001). Lavrion conferencenter. In collaboration with Eleni Koubli and Olga Xarhoulakou

The initial intention of this project is to provide an architectural solution of the issues posed by a conference centre as a space of knowledge exchange and to study the influence on the design in an area having a significant historical past, on a site adjoining with the town's microscale, the port's trans-local nature, and the natural landscape. In interpreting the concept, the conference center is seen as actively participating in the process of transmitting and disseminating information, contributing to the research on all scientific fields. It is the temple of knowledge, a perpetual dynamic course that never stops developing, since its objective (absolute knowledge) is never fully attained.

The structural rule of the arrangement is a smooth hill descending or ascending path, which connects the town with its natural boundary, the sea. This crooked line is not only the primary motion of the person wishing to go up or down an elevation but also the path on which the building is being gradually discovered, similar to the gradual acquisition of knowledge. The beginning of that path and the peak of the structure are stated by a tower-observatory.

This is a point where the user, having eye contact with all the surroundings, receives a large number of blurred messages. Through the path of the building, the users filter the stimuli they have received. The path ends upon the contact with the water. It is the moment or purification, where man assimilates all he has acquired, which are now the things making up the base for the beginning of the personal journey of discovering knowledge.

前巴特雷特时期：职业文凭论文（2001）拉夫龙会议中心 与埃里尼·寇博利和奥尔加·撒霍拉考合作

这个项目的最初意图是为作为一个知识交换的场所会议中心所带来的问题提供一个建筑上的解答，并且对具有显著的历史背景的地区对于设计的影响进行研究：一个连接市镇的微观尺度，港口的地域交换的特质和自然景观的基地。在解读概念时，会议中心被看作积极的参加进传送和散布信息的过程，为各领域的科学研究作出贡献。这是知识的殿堂，一项永远不会停止发展的恒久的动态运程，因为它的客观性（绝对知识）永远不会完全地被获得。

这样布置的结构性要点是一个光滑山坡上的向下或向上的路径，将城镇与它的自然边界——大海相连。这条弯弯曲曲的线不仅是人沿着高地上上下下的主要运动线路，也是建筑物被逐渐发现的路径，就像知识的逐渐获取一样。路径的开端和结构的顶端开始于一个天文台。

在这一点上，使用者与所有的周围环境发生目光接触，接受大量的模糊信息。通过建筑中的路径，使用者筛选他们接收到的刺激。路径在与水相接的地方终结。这是一个特别的时刻或一种净化，人吸收了他所获得的一切。而这一切奠定了开始探索知识的个人之旅的基础。

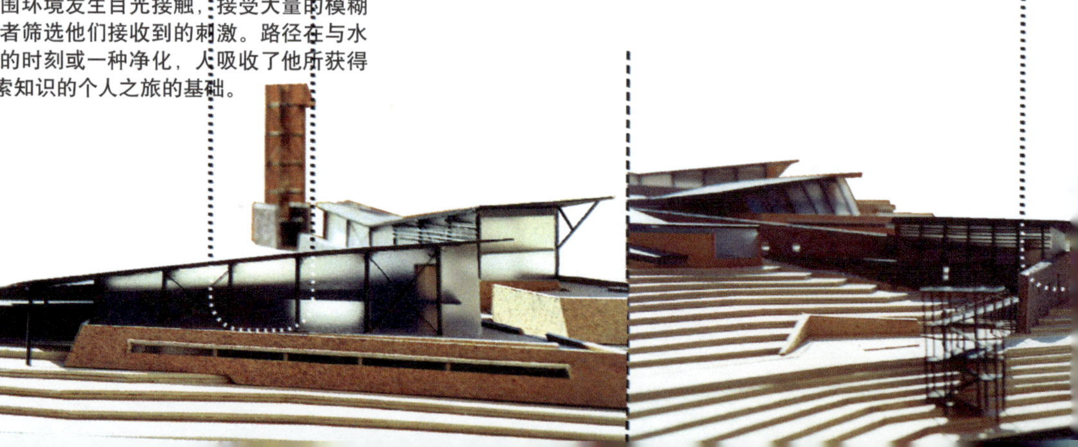

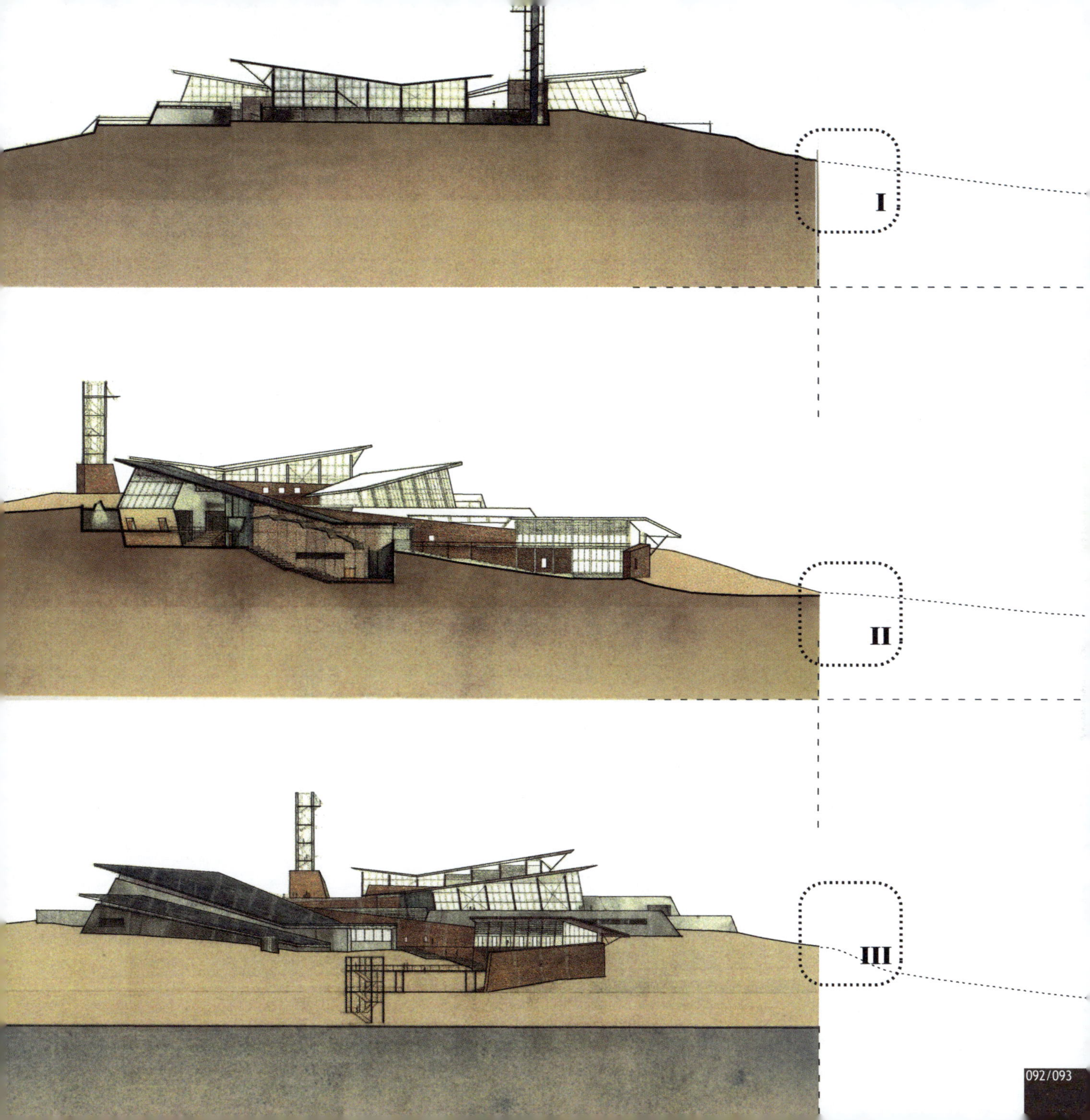

I
II
III

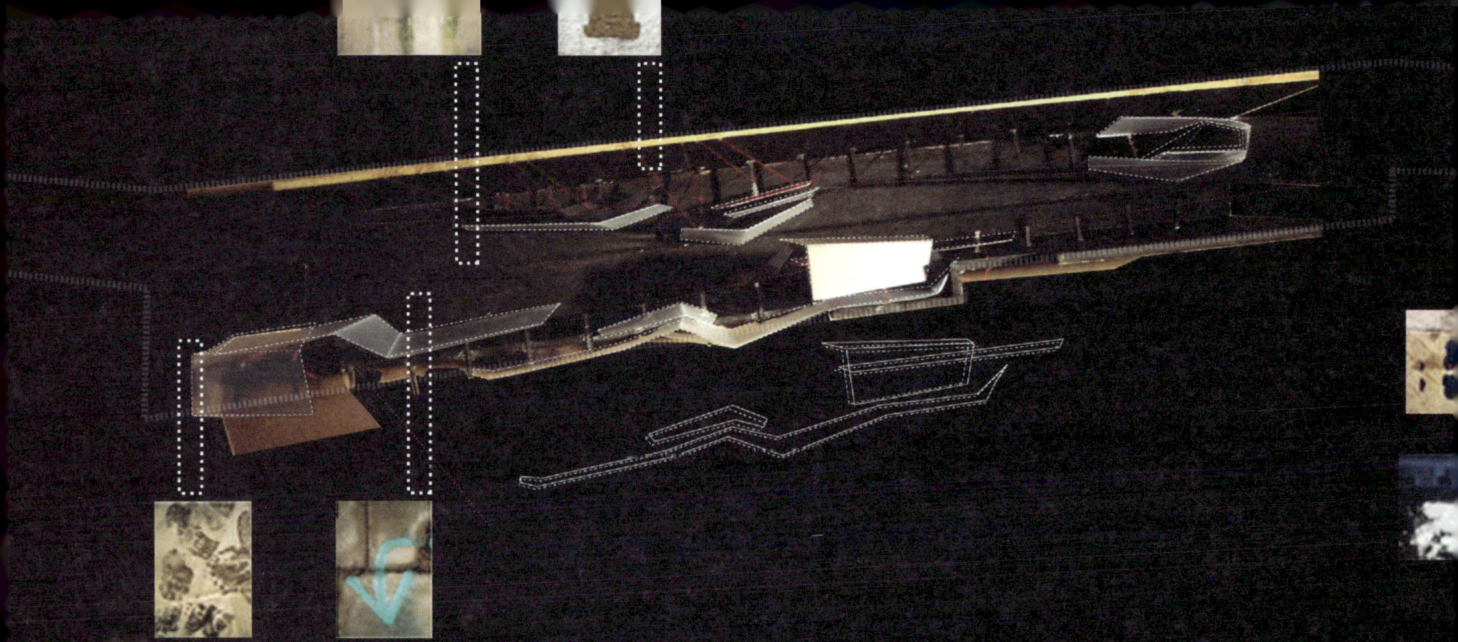

FOR MASTER: Microssing

<< DO I PLAY FOOTBALL ON THE STREETS?
DO I SEAT ON THE STEPS OF THE HOUSES' STAIRCASES?
DO I SUNBATH ON THE HOUSES' TERRACES?
NO: I'M JOGGING IN ORDER TO BE ON TIME FOR MY JOB.
I AM SUNBATHING WHILE WAITING FOR TAXI.
I AM DOING SAUNA IN THE BUS...

硕士课题：微观交互

我能在大街上踢足球吗?
我能坐在楼梯的台阶上吗?
我能在屋顶露台上做日光浴吗?
不：我慢跑是为了准时上班。
我在等出租车的时候做日光浴。
我在公共汽车上做桑拿……

1.Unknown writer. Graffiti on a house's wall in Athens, 1

1.无名作者，雅典一所房子墙壁上的涂鸦，1

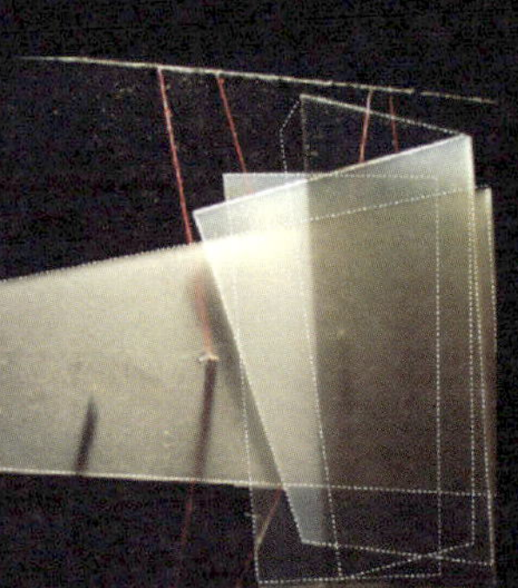

The need to define a personal rhythm of living in the contemporary metropolis induced the search of those concealed moments, which could give nuance to the everyday urban experience. A closer look on the micro scale of the urban environment unveiled those microelements, which could vacillate between a chink on the wall that reveals a hidden garden, or an odour in an alley that carries a memory of a previous age. These fragments of space and time act as moments of discontinuity in the urban homogeneity and disconnect the user from his pace.

The design proposal is initially developed as the spatial organizational system consisted of architectural fragments spread in the city, forming the conditions for this alternative way of engaging to the veiled urban moments. The construction in Covent Garden, being a fragment of the system, is materialised as a ramp hooked on the perimetric walls of the adjacent buildings, transforming the crossing of Bull-Inn court into a ritual. It forms a kind of labyrinth where the user has to calibrate his path, without losing his desire for discovering every aspect of his "journey".

在当代都市生活中，定义个人节奏的需求引发了对那些体现日常都市体验的细微差别的隐秘瞬间的探究。在微观尺度上对都市环境的近距离观照，使那些细微的要素呈现在大众面前。它们摇摆在墙壁的缝隙里，暴露出一个隐藏着的花园，或是游荡在小巷的气味中，传递着久远年代的记忆。这些时空的片断在都市的同一性中充当了间断性的瞬间，并把使用者从他原有的步调中分离出来。

这个设计意图最初被发展成为一个空间的组织系统，系统由散布在城市中的建筑片断所构成，形成了以另一种的方式迷醉于被遮掩的都市瞬间的条件。作为这个系统的一个片断，考文园的结构被物化为一个挂在邻近建筑物的围墙上的斜坡，并把公牛旅店庭院的交叉口变成一个仪式性场所。它形成了一种迷宫，在这里使用者必须校准他的路径，而又不失探索"旅程"中每一部分的渴望。

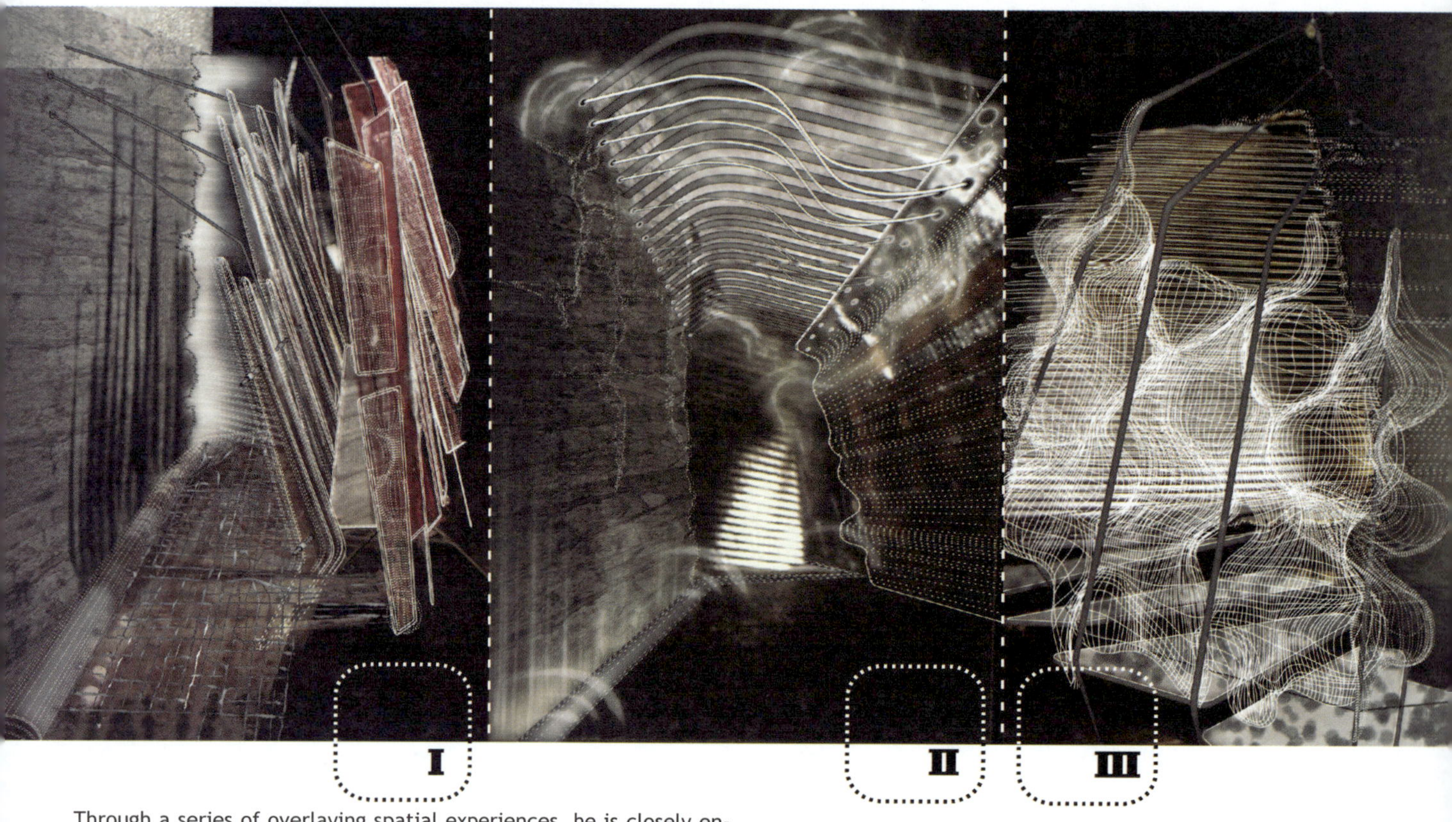

Through a series of overlaying spatial experiences, he is closely engaged to the elements of the micro-site reconsidering his network of references and creating his own interpretation of walking. Remnants of people's interference, marks of nature's power, shadows of throwaways of the past, textures and temperatures of the matter, sounds but also the conditions related to the sunlight or the rainwater, influence the design of the architectural elements. Finally, all together contribute to a micro-choreography generating an event that changes from day to day.

通过一系列的相层叠的空间体验，角色贴近地使用微观基地的元素，重新考虑他的参照系，并创造出他自己对步行的解读。人的介入残迹，自然力的痕迹，过去宣传品的投影，事物的质地与温度，声音与阳光或雨水相关的条件，都在影响着建筑要素的设计。最后，综合所有作用因素的微型舞蹈制造出一个事件，日复一日地变幻着。

IV
V
VI

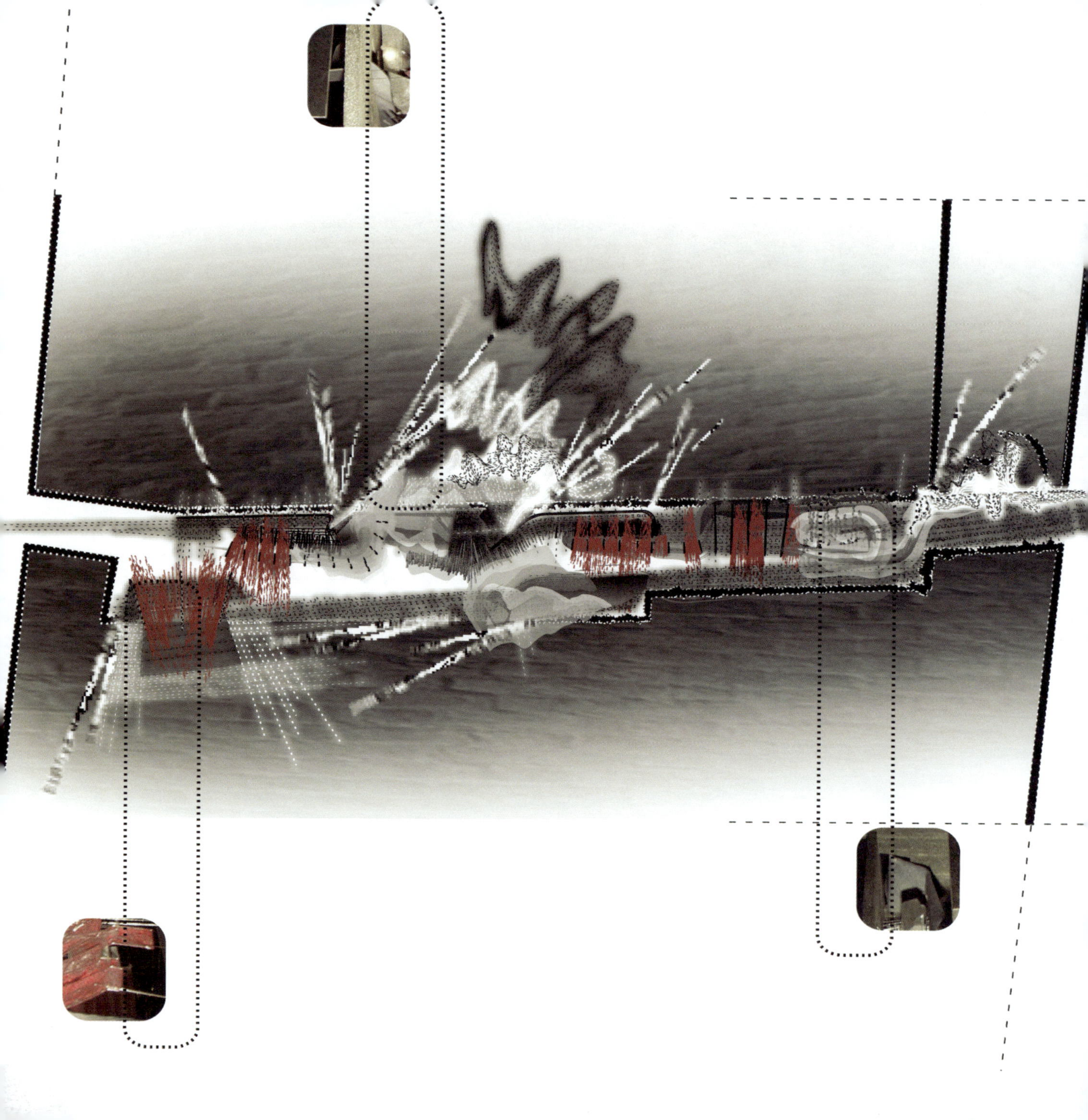

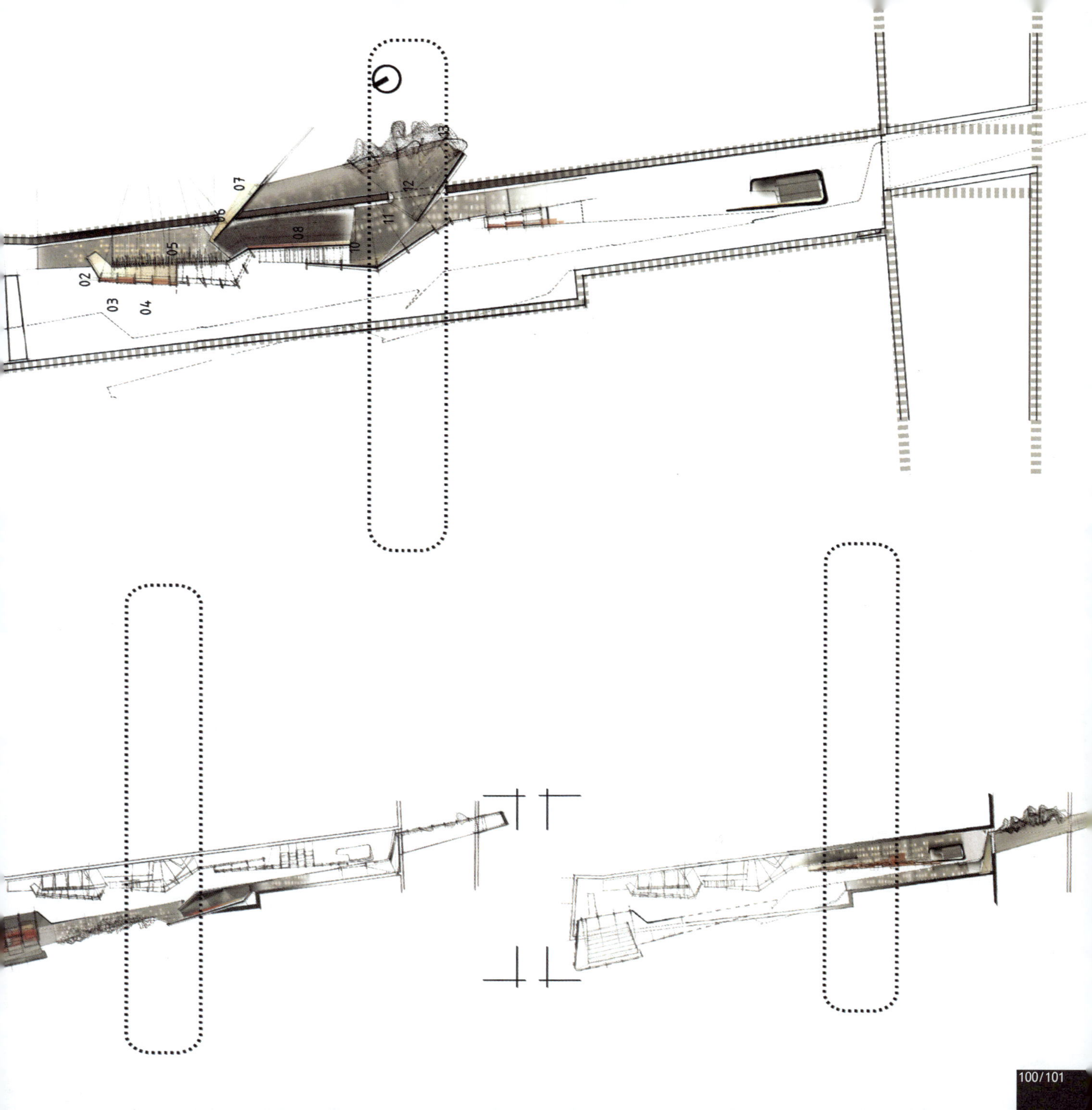
02
03
04
05
06
07
08
10
11
12
13

stion
ing
Fri
es
61758

昆资
Raoul
Kunz

Illustrations

To illustrate the process, a diagrammatical presentation technique is chosen. 'In architecture the diagram is historically understood in two ways: as an explanatory or analytical device and as a generative device... Thus, what would be seen in an architectural object is both the first perceptual stimulus, the object itself, along with its aesthetic and iconic qualities, and another layer, the trace, a written index that would supplement this perception. '(Eisenman) A well-designed architectural diagram is both, abstract and concrete.

Simple drawings doesn't arrogate by pretending a final appearance but instead the will to discuss a tectonic subject. In this simple but effective way, a series of these icons can be read like a storyboard. They are capable to tell an entire narrative. As a result, it is not just a medium for presentation, but also for analysis. With regards to Deleuze and Somol, as well as Dirrida and Freud, Peter Eisenman defines this architectural tool in the following way: 'The diagram acts as an agency which focuses on the relationship between an authorial subject, an architectural object, and a receiving subject; it is the strata that exist between them.' (Eisenman) The diagram bridges between the idea and the final design.

Therefore, the technique is applied in order to reveal a certain process. Like Lukács said, 'we stand between two worlds, one that we don't acknowledge, the other that does not yet exists.' (Andreotti + Costa) But the variety of different diagrammatical presentation-techniques is wide. The preferred method of this work is based on the opportunities that result from contemporary computer applications. The rendering machine is not understood as a medium to enchant the spectator but instead to reveal facts.

Lionel March offers a plan scheme as an example of how to illustrate the idea of coding and defining generating rules of this concept. 'Conventional drawings consist of projections of three-dimensional objects onto various plans. ... In contrast, the computer-based model can be arranged as a three-dimensional model, not a collection of two-dimensional projections. All drawings are then generated by appropriate transformations of the three dimensional model.' (March)

图解

为了阐明过程，选择图解的表述技巧。“在建筑学中，图示被历史性地理解为两种用途：作为说明性与分析性的工具，和作为生成的工具……因而，在建筑对象中应该被关注的既是第一知觉刺激物，对象本身，及其美学的和标识化的品质；并且在另一层面上，又是图解纪录，作为书面的索引补充这种知觉。”（艾森曼）一个设计良好的建筑图示是兼具抽象与具体的。

单纯的绘图并不僭越而充作最终表现，而是旨在探讨建构物体。以这种简单而有效的方式，一系列这样的标识可以像情节串联板一样被予以解读。他们有能力讲述一个完整的叙事。作为结果，这不应该只是一个表述的媒介，还应该用于分析。出于对德勒兹和索莫，还有德里达和弗洛伊德的关注，彼得•艾森曼用以下的方式来定义这种建筑工具：“图解行为像是一个致力于建筑对象和接受物之间关系的中介机构；它是存在于他们之间的层次。”（艾森曼）图解在观念与最终设计之间铺设了桥梁。

因而，这一技巧被用于展示特定的过程。正如卢卡奇所说：“我们位于两个世界之间，一个我们不予承认，另一个至今还不存在。”（安德雷奥蒂+科斯塔）但是各不相同的图解表述技巧是多样化的。这项工作的首选方法是基于当代计算机的应用而造成的机遇。这个表现的机器不应被理解为对观众施法的媒介，而应是展现事实。

作为一个例子，莱昂内尔•马奇提出的一个设计方法解释了如何展现编码的想法和定义生成这一观念的法则。“传统的制图包括三维物体在各个平面的投影……相反，电脑生成的模型可以被改编成一个三维模型，而非一个二维投影的集合。然后所有的图通过恰当的转译由一个三维模型生成而来。”（马奇）

联系方式：		洛乌 昆兹
		raoulkunz@gmx.net
		0044(0)7903820634
格林威治大学	05.2006	设计课教师/讲师
埃利斯与莫瑞森建筑师事务所		项目负责人/建筑师
		海奥码头－英国托特纳姆.海奥
		住宅建筑委托（500个单元）
		皇家阿森纳－英国沃尔维奇
		总体规划委托（3700个住宅单元）
		能源－英国克瑞考伍德
		废品处理设施和热力电力联合系统委托项目
伦敦建筑联盟	02.2007	注册建筑师
职业实践课程		德国建筑师协会
		英国建筑师注册委员会
泰瑞•帕森建筑师事务所－伦敦	04.2006	音乐剧院－奥地利林兹
		剧院竞赛项目（获胜者）
		艺术中心－爱尔兰卡娄
		美术馆与剧院设计竞赛（获胜者）
		小住宅－英国温布尔登
		住宅委托项目（发表在GA第80期上）
迪米特瓦格拉德政府	07.2003	受邀参与者
		庆祝改造－迪米特瓦格拉德（俄罗斯）
		城市规划研讨会
巴特雷特建筑学院－伦敦大学学院	09.2002	建筑设计硕士
论文：异物	09.2003	推荐奖毕业
指导人：彼得•库克/魏拉德斯拉瓦•科尔皮车夫		DAAD奖学金获得者
沃尔克 斯塔布建筑师事务所－柏林	02.2001	自由实践
	09.2001	交通行政－当地政府项目
		办公楼设计竞赛（获胜者）
明斯特建筑学院	09.1996	学士学位/艺术硕士
论文：艺术－公共转接器	08.2002	建筑学
指导人：朱莉包勒斯－威尔森/沃尔克•斯塔布		
巴登菲尔德挤压技术	08.1993	工程绘图员
	09.1995	荣誉

Contact:		Raoul Kunz raoul kunz@gmx.net 0044(0)7903820634
University of Green wich	09.2007-	Design Tutor / Lecturer
Allies and Morrison Architects-London	05.2006-	Project Lead / Architect
		Hale Wharf-Tottenham Hale(UK) commissionforresidentialbuilding(500units)
		Royal Arsenal-Woolwich(UK) commissionformasterplan(3700resi.units)
		Energy-Cricklewood(UK) commissionforwastehandlingfacility and C H P
AA-School London Professional Practice Course	02.2007-	Registered Architect Architektenkammer Germany Architects Registration Board U K
Terry Pawson Architects-London	04.2006-	Musiktheater-Linz(Austria) competitionforoperahouse(winner)
		Art Centre-Carlow(Irland) competitionforartgallery and theatre(winner)
		Small House-Wimbledon(U K) comissionfordwelling(published in GAHouses80)
Authority of Dimitrovgrad	07.2003	Invited Participant Celebrate Transformation-Dimitrovgrad(Russia) Urban Planning Seminar
The Bartlett-University College London Thesis : Strange Objects Supervis or : Peter Cook / Vladislav Kirpichev	09.2002- 09.2003-	Master of Architectural Design with commendation DAAD-ScholarshipHolder
Volker Staab Architekten-Berlin	02.2001- 09.2001-	Freelancing Traffic Administration-Local Authority Passau competitionforofficebuilding(winner)
Münster School of Architecture Thesis : Adapter Art-Public Supervis or : Julia Bolles-Wilson / Volker Staab	09.2096- 08.2002-	Bachelor/Master of Arts inArchitecture
Battenfeld Extrusionstechnik	08.1992- 09.1995-	Engineering Draftsman withhonours

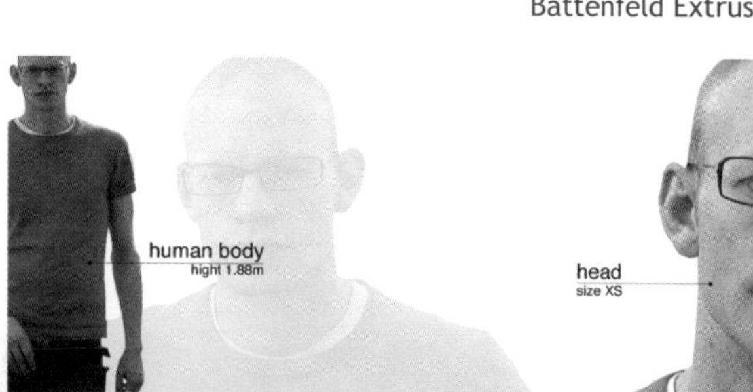

Rational Confusion

Looking back at my relatively short architectural career I am somehow reminded of the times when growing up really wasn't easy. First you deal with everything that is thrown at you, later you rebel against everything and everybody on one hand while trying to be accepted on the other and after all you are being told it's best to be yourself.

Shaped by experience and education, by our immediate social network, cultural heritage and the values of our family, personalities develop, aspiration are formed. That's when things get in comfortable order and just as you'll might think to have put any confusion behind you it starts all the same over again: with the training as an architect.

My architectural approach is intensely shaped by my studies and professional experience in Germany and Switzerland where I adopted a sensibility for detail and developed a taste for austere, restraint and minimal architectural language. With the spirit of the "Swiss school" in mind there was always a clear set of rules amongst the architectural community – a general aesthetic consensus.

Those where the grounds I was used, the rules of the trade that had become second nature for me when I stepped onto the architectural landscape of the Bartlett School of Architecture, ready for a new set of rules to be thrown at me. While it's always hard to part from the learned, the obvious it is also about the willingness to put achieved qualities into question not to put them aside. That's where architectural training not only becomes an intake of information but also a testing ground.

Interested in transformation and process it has become the nucleus of my doing. The outcome of this process is the product of what one's exposed to in the environment - on the surface - and depending on the severity of impact is channeled to the core to be assimilated, manipulated – digested if you will. This is energy to generate new ideas. That's where ideas can be put into form - form that doesn't merely follow function but form follows inspiration.

Being creative is to be able to adapt, is to inspired by means to assimilate ideas, to marry the conventional and the experimental, the known with the unknown. To deal with architectural approach is analysis and intuition and the evaluation of such influences as Zeitgeist and heritage.

And that's where I should revoke the mere acceptance of timelessness as a quality in architecture. . It's not a question of complexity or simplicity not of ornament is crime or less is a bore (or more??). Making a statement is to breathe soul into architecture - going out there and getting whipped. How better can one do this as by being inspired? Acceptance is good. Being oneself is better. The goal is to search for qualities and opportunities in architecture. Or as John Frazer puts it: an important characteristic of … influences is they are never prescriptive or static. They are plans for action, not plans of form… (John Frazer).

理性的迷茫

每当我回顾自己相对短暂的建筑生涯,不知为何，时常让我回想起困难的成长岁月。最初，你得应付向你迎面掷来的一切。接着你一面反抗所有人所有事，一面试图被接受。最终，你被告知，最好还是做你自己。

被经历与教育塑造着，被当下社会网络、文化承传和家庭的价值观塑造着，我们的个性得以发展，志向得以形成。当一切变得井然有序，你正打算把所有疑虑抛置脑后时，随着作为建筑师训练的开始，一切的烦恼又重头再来。

我的建筑学方法，极大程度上成形于我在德国与瑞士期间求学工作的经历。在那里，我培养出了一种对细节的敏感以及对严谨、克制与极简的建筑语汇的偏好。深植于我内心的“瑞士学派”的精神始终有一套清晰的法则，一种建筑界普遍的美学共识。

当我步入巴特雷特建筑学院的建筑图景，等待一套全新法则迎面而来的时候，作为我曾经的基础的行业规则已经成为我的第二天性。使自己从过去所学的东西之中分离出来总是很难。当然，这是基于你更愿意把自己已获的品质纳入提问之中，而非把它们置之一旁。在这里，建筑学的训练不仅仅是相关知识的接入口，更是一个试验场。

对于转译与过程的兴趣成为我行动的核心。如果你愿意的话，这一过程的成果会是一个人暴露在这种环境中的产物——发生于表面——凭借碰撞的激烈性，这样的激烈性成就通向吸收、处理与消化的核心渠道。这是一种催生新观念的能量。这是观念能被转化为形式的关键之所在——形式不只是追随功能，形式追随灵感。

创造力其实是一种适应能力，是吸收各类思想来激发灵感，联姻固有的与试验的，已知的与未知的。研究建筑学的方法是分析与灵感以及对诸如时代思潮与传统遗产的影响力的求解。

在这里，我应该摒弃对于建筑永恒品质的简单认同……这不是复杂与简洁的问题，也无关于“装饰即罪恶”，“少即是乏味”（或“少即是多”）。表达应该是为建筑注入灵魂——走过去获得鞭策。关于这一点，获得灵感的人能做得有多好？认同是不错的，葆有本色更好。目标是在建筑中寻找品质与机遇。或者正如约翰　弗雷泽所表述的：“……（这些）影响的一个重要特质是它们既不说明也不缄默。它们为行为而生，而非为形而生……”

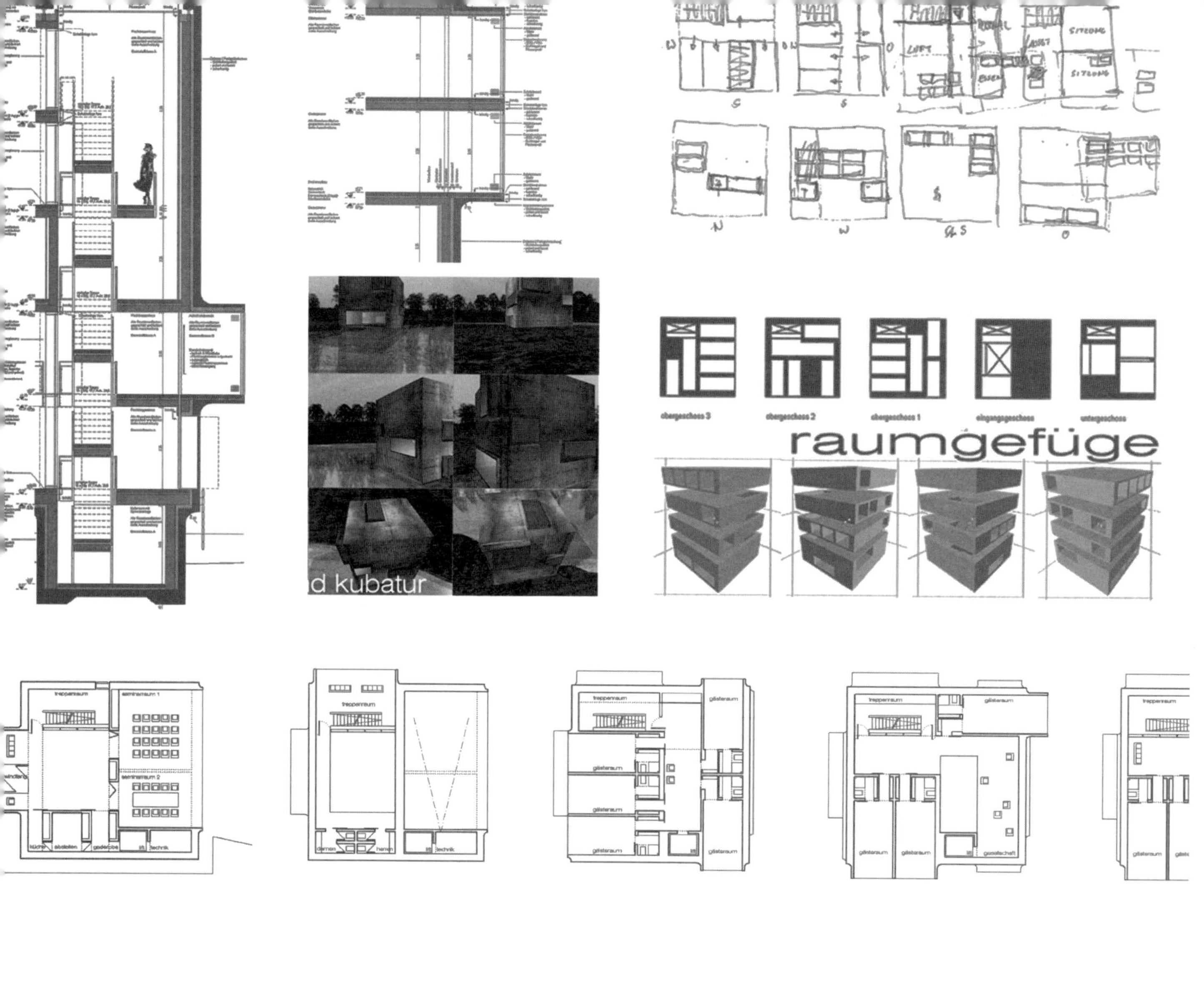

made of concrete

guest - house

KUBUS
DER INNRAUM
DER DIALOG
DER DYNAMISCHE PROZESS
GLASTÜREN
FESTVERGLASUNG
INNENRAUM
KLIMAPUFFER
made of concrete
OBERGESCHOSS 3
UNTERGESCHOSS
ein guss

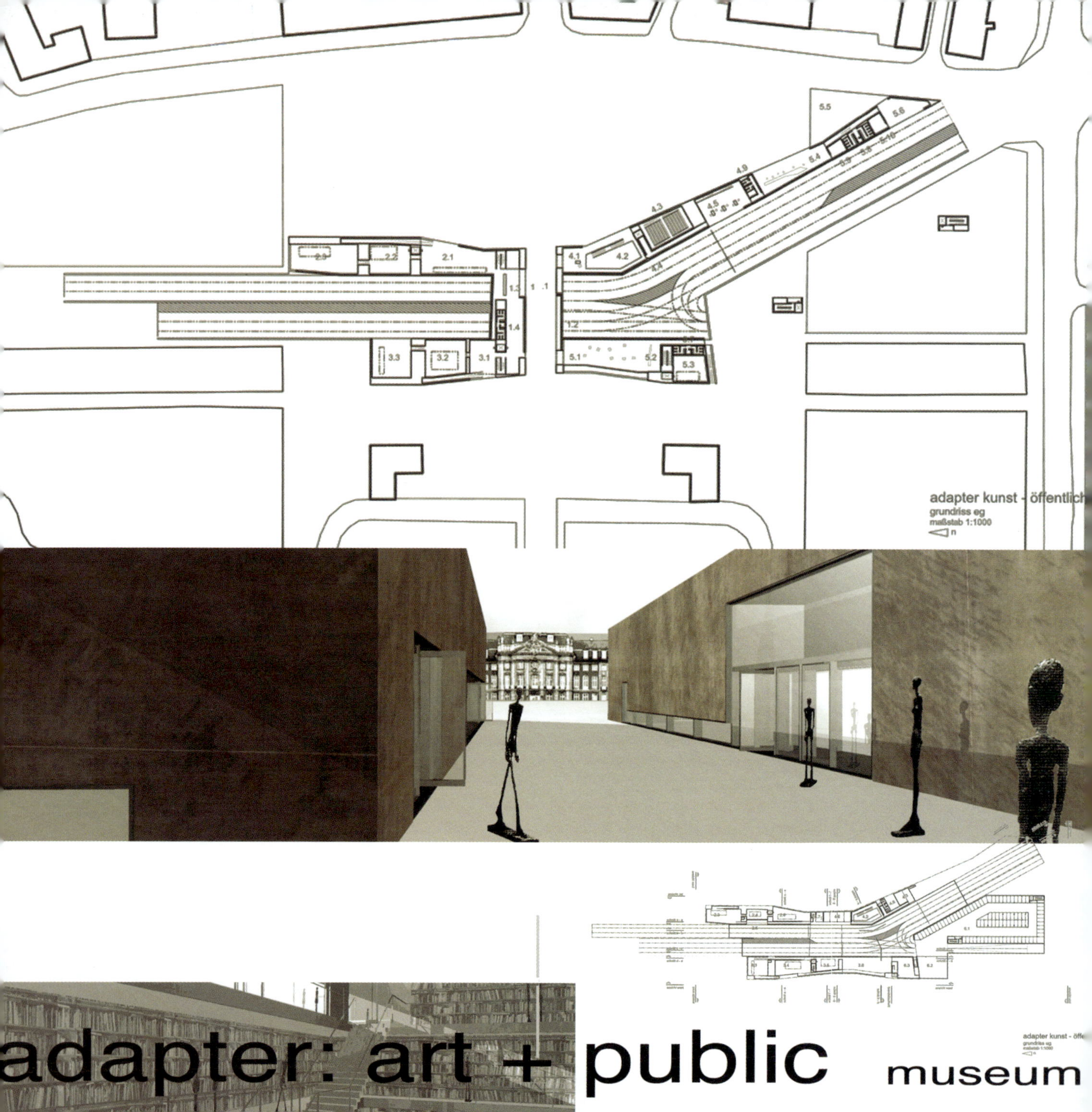
adapter kunst - öffentlich
grundriss eg
maßstab 1:1000
n
adapter: art + public
museum

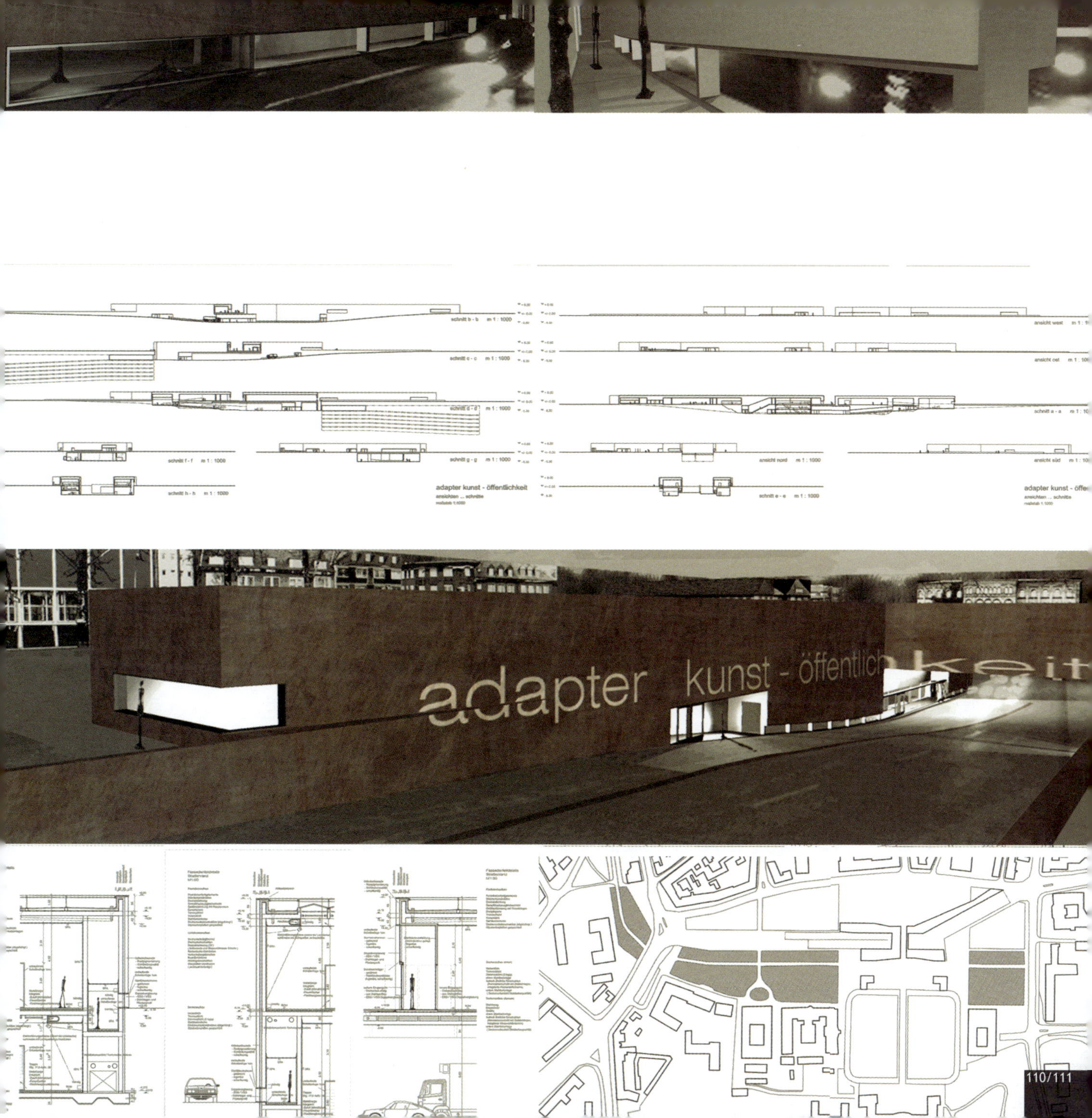
schnitt b - b m 1 : 1000
schnitt c - c m 1 : 1000
schnitt d - d m 1 : 1000
schnitt f - f m 1 : 1000
schnitt g - g m 1 : 1000
schnitt h - h m 1 : 1000
adapter kunst - öffentlichkeit
ansichten ... schnitte
ansicht west
ansicht ost
schnitt a - a
ansicht nord m 1 : 1000
ansicht süd
schnitt e - e m 1 : 1000
adapter kunst - öffentlich keit

Proposal

'Narrative art is para-pictographic in sense that it represents a story by a sequence of distinct events arranged in a certain order.' (Lebesque + Van Vlissingen about Yona Friedmann). Everyday there is a certain amount of factors, which influence our behaviour. Some of them mark a point or a situation we attempt to avoid or to reach. Some influence our consciousness in an unexpected way. Some are unpredictable, others seem to be predictable but in fact they are not. And then again, there are some factors, which when proven to be as predicted, turn out to be disappointing. To deal with these situations means to deal with the unfamiliar, the inconstant and ever changing.

The subject 'strange objects' is chosen with regards to the following definitions. Strange: 'Belonging to some other place or neighbourhood; unknown to the particular locality specified.' (A New English Dictionary on Historical Principles) Today, 'strange' is an often-used term in common society. It can have several different meanings. In this thesis it is understood in close sense to 'stranger', which literature defines as something or someone which/who seems not to be a part of its environment. Object: 'To place (something) before the eyes or other organs of sense, or the mind; to present or offer to the sight, perception, understanding ...' (A New English Dictionary on Historical Principles) In other words, a task of an object is to stimulate the mind. It stands in dialogue with the people. In order to push these definitions to its limit, a series of living organism has to be designed. The 'strange objects' I am looking for are meant to be capable to read and to respond to their surroundings and the human senses.

In order to determine an architectural language, designers might look for the issues and opportunities of every specific project: What are the site's characteristics? To what extent is the program already defined? Beyond this, what is the social meaning of this intervention? And what is the team of planers and clients intrigued by? Then, those influences are combined to form a final result. In order to reveal the factors that influence with this specific work, the design process is decoded and illustrated. Therefore it is split into three parts: In the first part, a series of 'devices' explores different influential subjects. Each of them is translated into an architectural artefact, which is in the second part going to be combined to a catalogue of 'applications'. Two samples are taken for further development. As 'prototypes', they are meant to proof a possible functional translation.

device 01
description
understanding the object
folding the facade
looking for functions
a range of situations

device 02
description
destruct the tube
disorder the shape
rearranging the fragments
observing the light

device 03
description
prefabrication
repetition: an aesthetical effect
adapting the structure
initiating with light

device 04
description
range of density
connecting functions
pipeline supply
observing streams

application catalogue
assembling devices to each other
formal approach between device01 and device02
applying the technique of device02 on device01
applying the technique of device01 on device02
assembling applications to each other

提案

“叙事艺术在感官上是象形的，它是以一定的顺序，再现一个由一序列特定事件构成的故事。”（勒贝格+凡•弗利辛恩谈关于犹纳弗里德曼）。在日常生活中，有相当数量的影响我们行为的因素。他们中的一些标定了我们试图避免或达到的位置或点；一些以一种意想不到的方式影响着我们的意识；一些不可预测，另一些似乎可以预测，但事实上并非如此；然后，还有一些因素，一度被证实正如预测那样，旋即又被发觉是令人失望的。处理这样的情形意味着处理那些我们不熟悉的、变化无常的事物。

基于对以下几种定义的考虑，我选择了“异物”这个课题。奇异：“属于其他地方的或邻近的；不为特定所指的当地所知的。”（一本关于历史法则的新英语辞典）今天，在公共社会中“奇异”是一个在社会中被经常使用的术语。它可以有几种不同的含义。在这篇文章中，它被近似地理解为“异类”，像是文艺作品中定义的那样：某事或某人似乎与其所处的环境格格不入。物体：“置于眼或其他器官之前而被感觉到存在（的事物）； 表现、呈现于视野、感知、理解……”（一本关于历史法则的新英语辞典）换句话说，物体的任务是激起关注。它存在于与人的对话之中。为了将这些定义推向极限，一系列的活的有机体必须被设计出来。我所寻找那些“异物”意味着读取和回应它们所处的环境和人的知觉的能力。

为了形成一种建筑语言，设计者们会寻找每一个项目的问题与机遇：基地的特征是什么？在多大程度上，程序已经是被给定的了？除此之外，这种干预的社会意义是什么？设计者与客户被什么激起兴趣？然后，所有的影响被结合形成最终的结果。为了展示影响这一特定工作的因素，设计过程被予以解码与图解。因此，它被分解成为三部分：在第一部分中，是一系列探讨不同影响物的装置。它们中的每一个都被转译成为一个建筑作品，并将在第二部分，被合成为一个“应用”的目录。其中两个范例被选取用作进一步发展。作为“原型”，它们试图验证一种可能的功能转译。

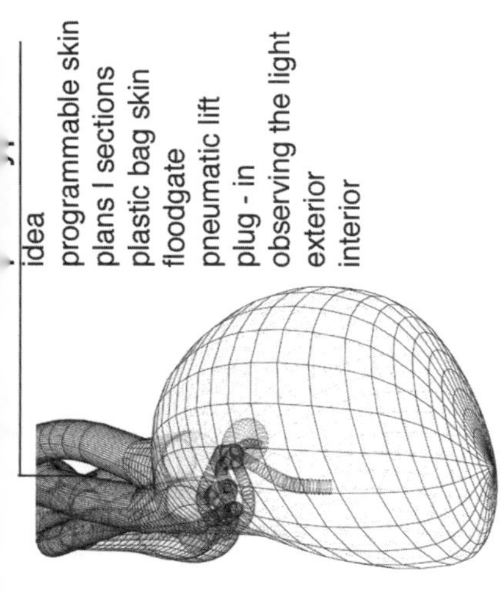

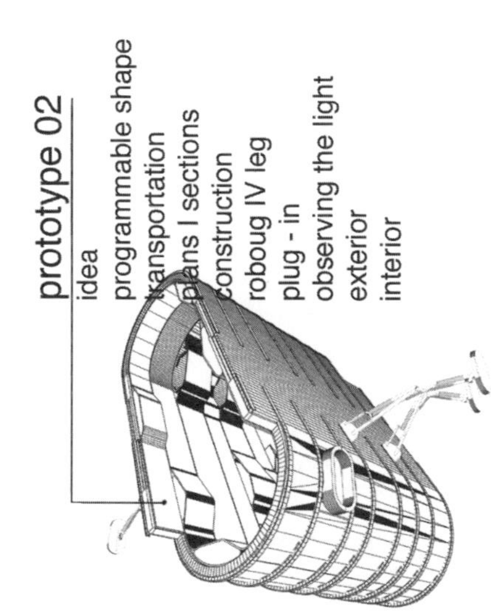

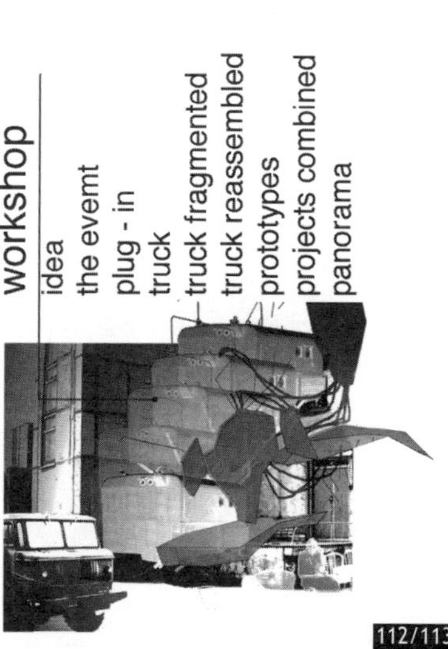

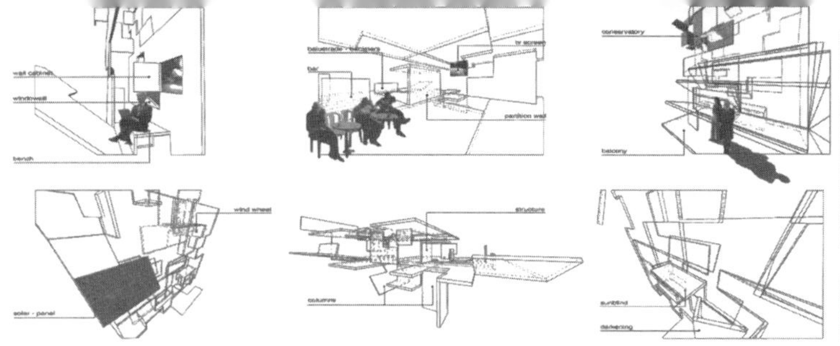

wall cabinet
windowsill
bench
balustrade - banisters
bar
partition wall
conservatory
balcony
wind wheel
solar - panel
structure
columns
sunblind
darkening

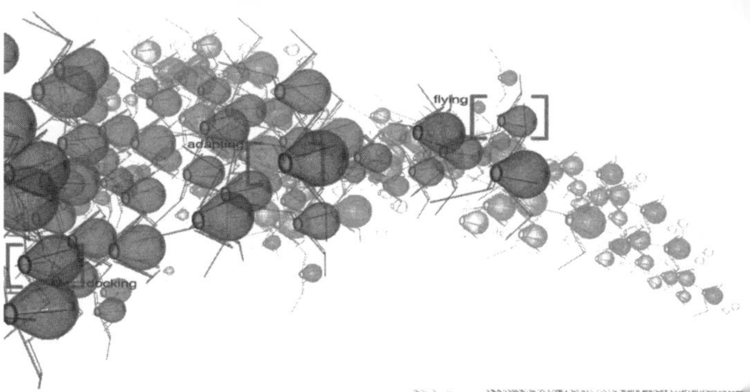

flying
adapting
docking

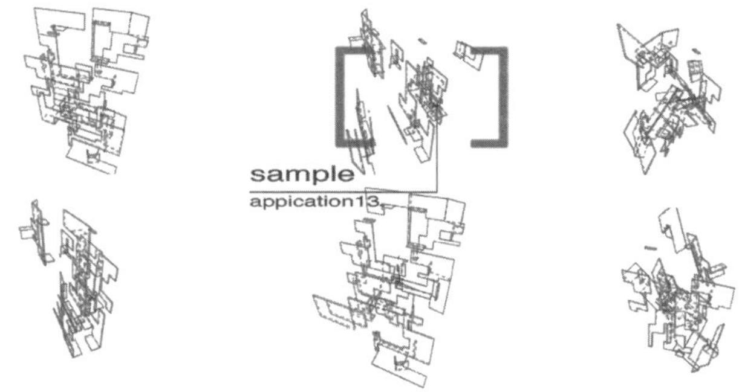

sample
appication13

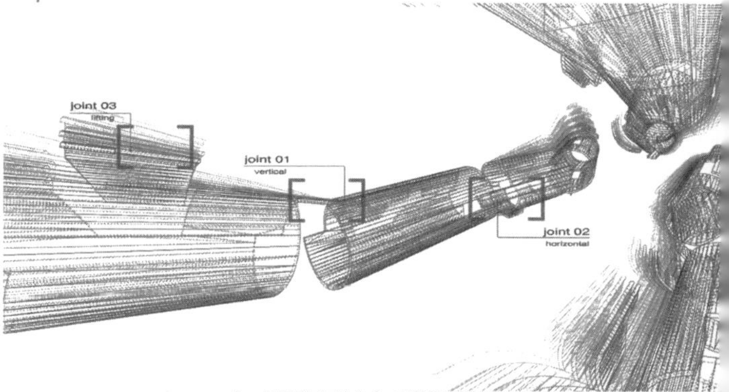

joint 03
lifting
joint 01
vertical
joint 02
horizontal

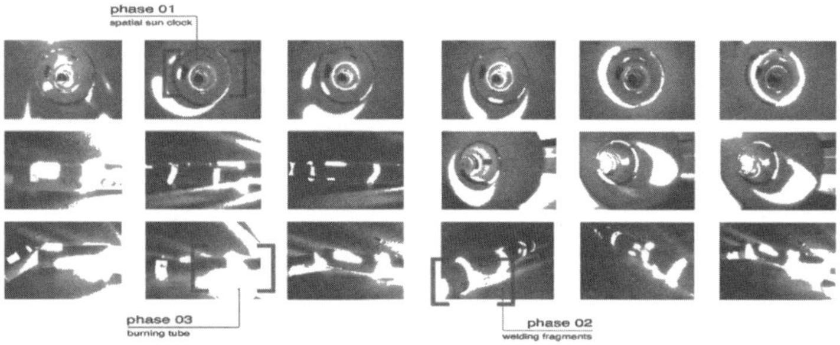

phase 01
spatial sun clock
phase 03
burning tube
phase 02
welding fragments

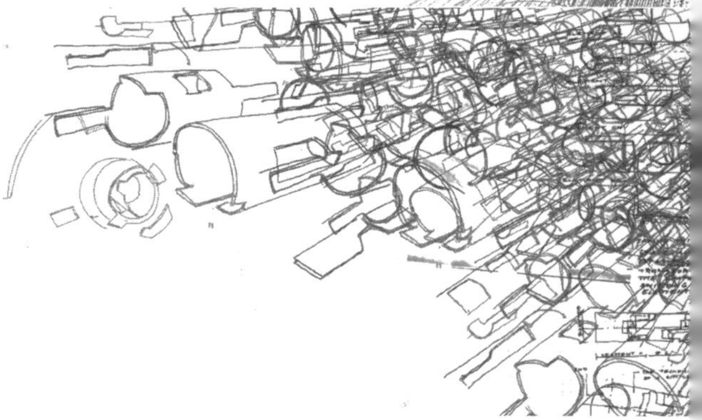

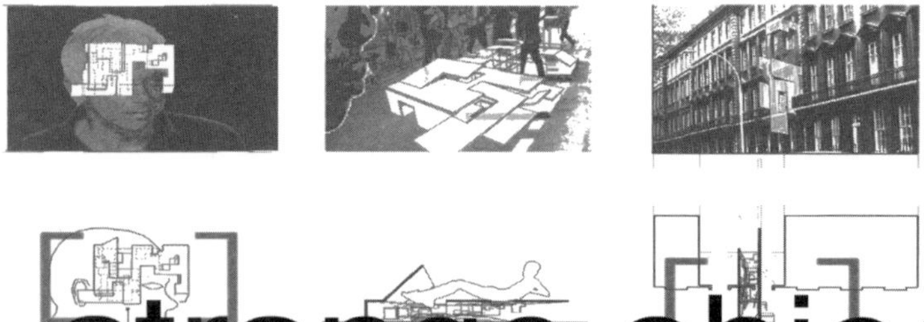

01. device studies 02. 03. 04.

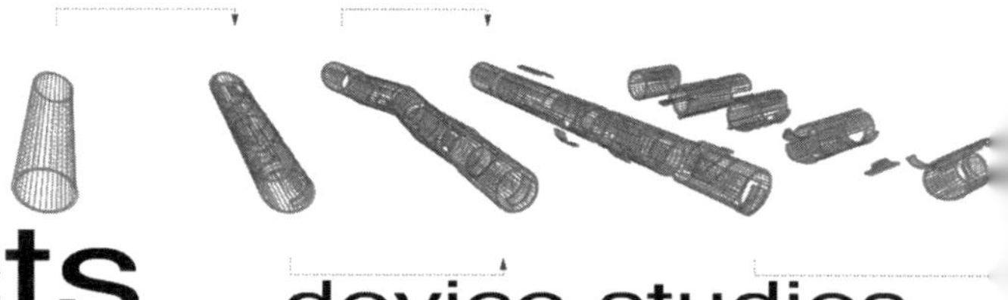

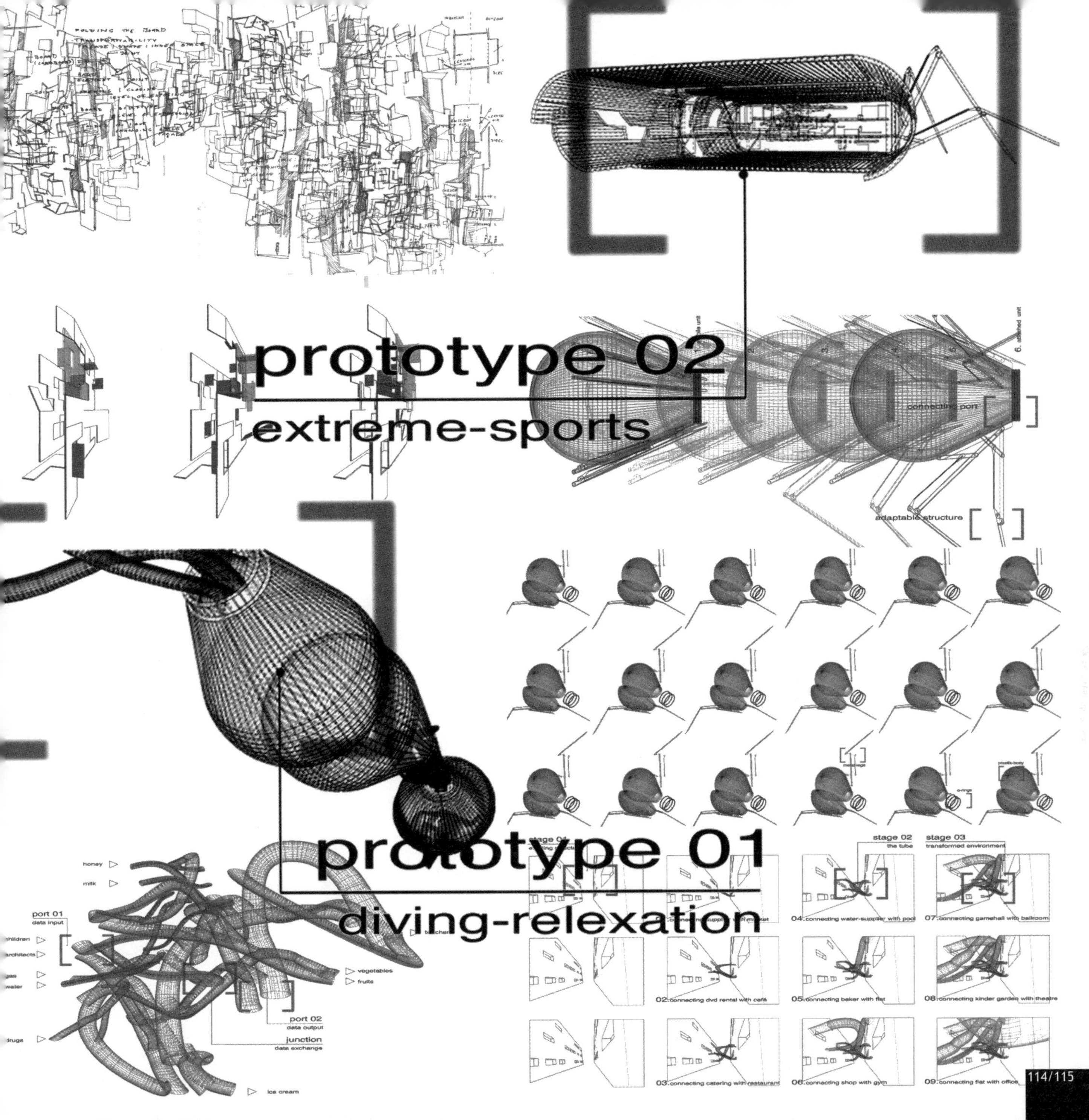

prototype 02
extreme-sports
6. attached unit
connecting port
adaptable structure
prototype 01
diving-relexation
stage 02
the tube
stage 03
transformed environment
04.connecting water-supplier with pool
07.connecting gamehall with ballroom
02.connecting dvd rental with café
05.connecting baker with flat
08.connecting kinder garden with theatre
03.connecting catering with restaurant
06.connecting shop with gym
09.connecting flat with office
honey
milk
port 01
data input
children
architects
gas
water
drugs
teacher
vegetables
fruits
port 02
data output
junction
data exchange
ice cream

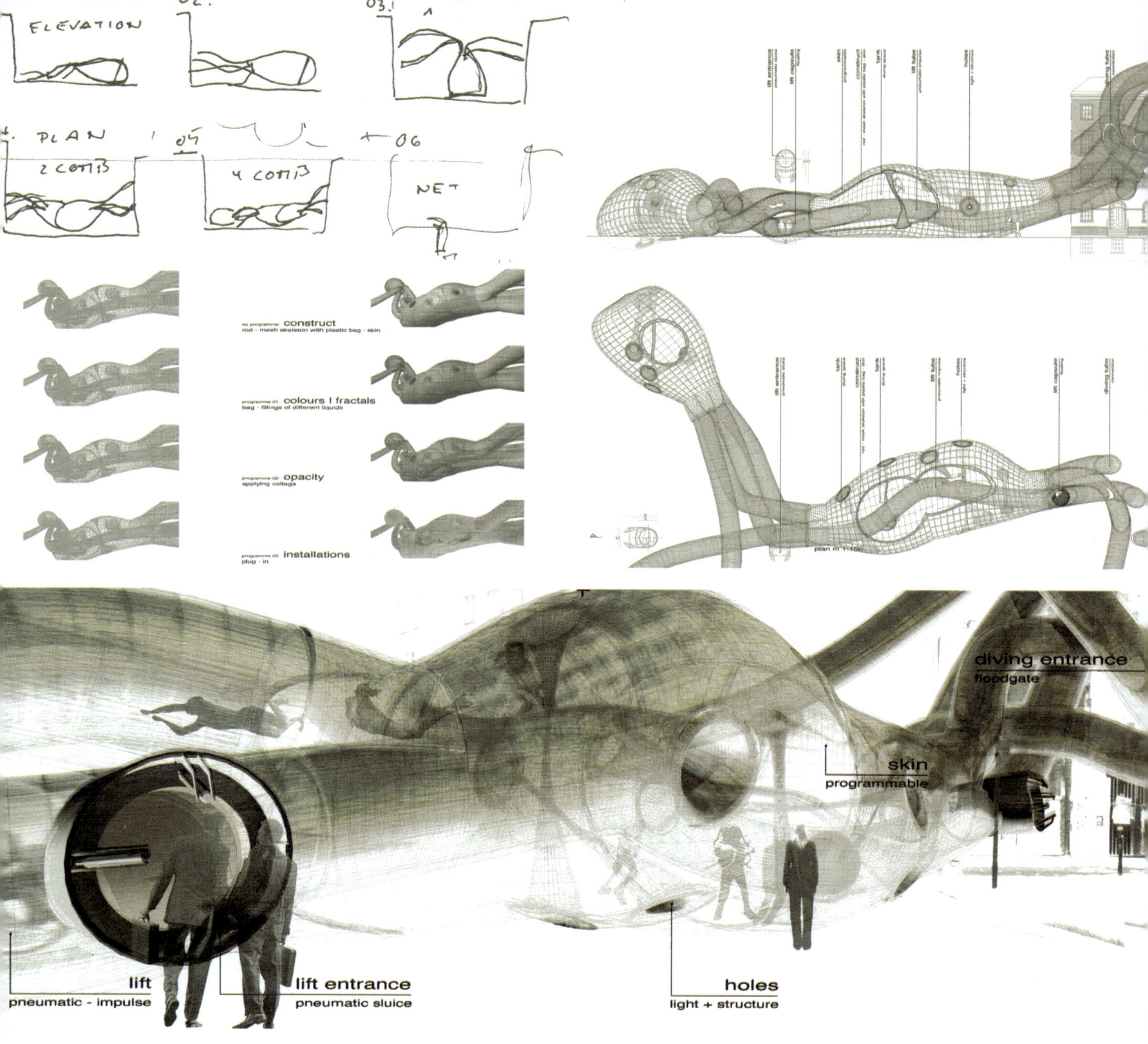

prototype 01

diving - relaxation

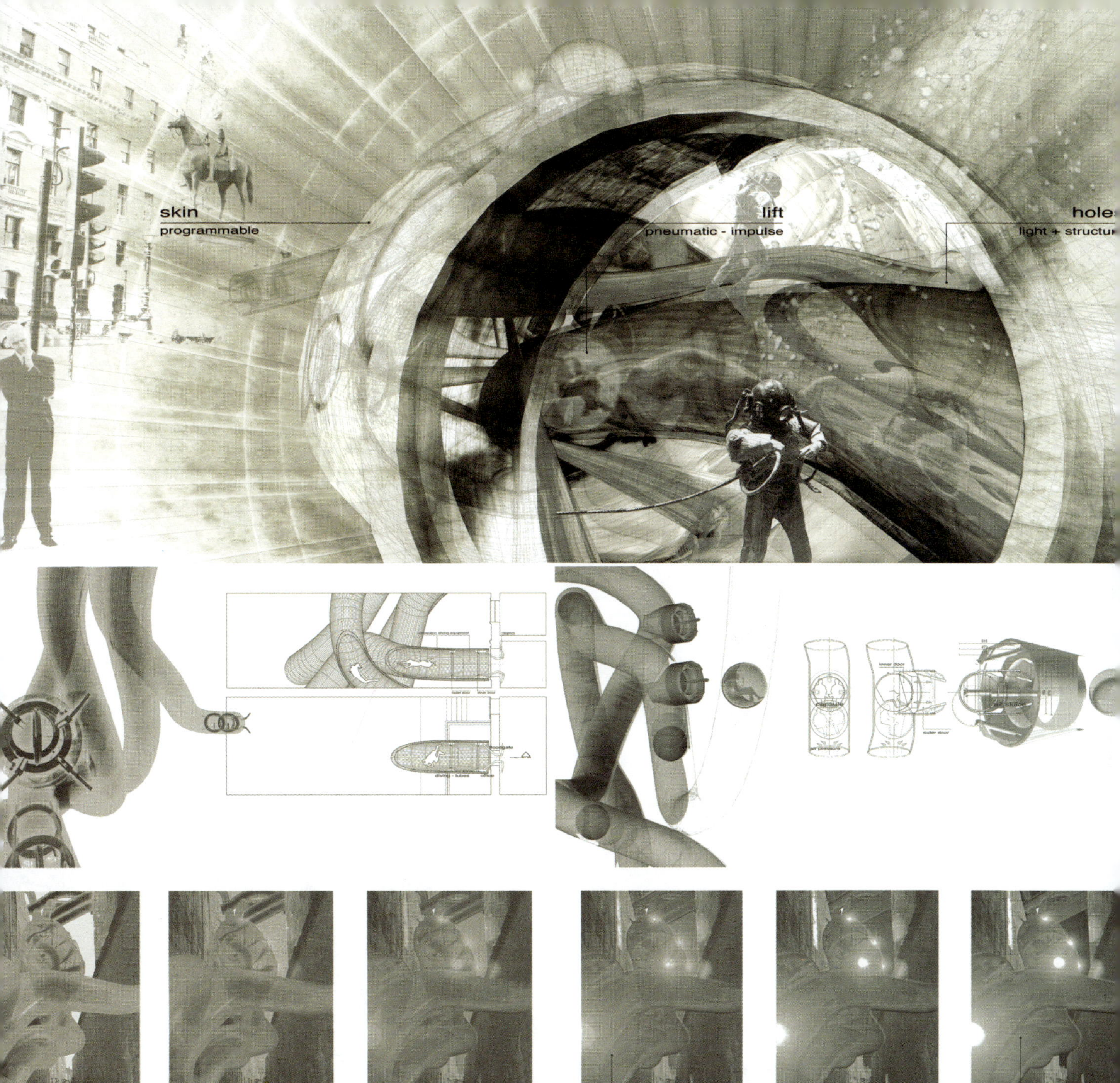
skin
programmable
lift
pneumatic - impulse
light + structu
floodgate
diving - tubes
office
inner door
capsule
air pressure
outer door

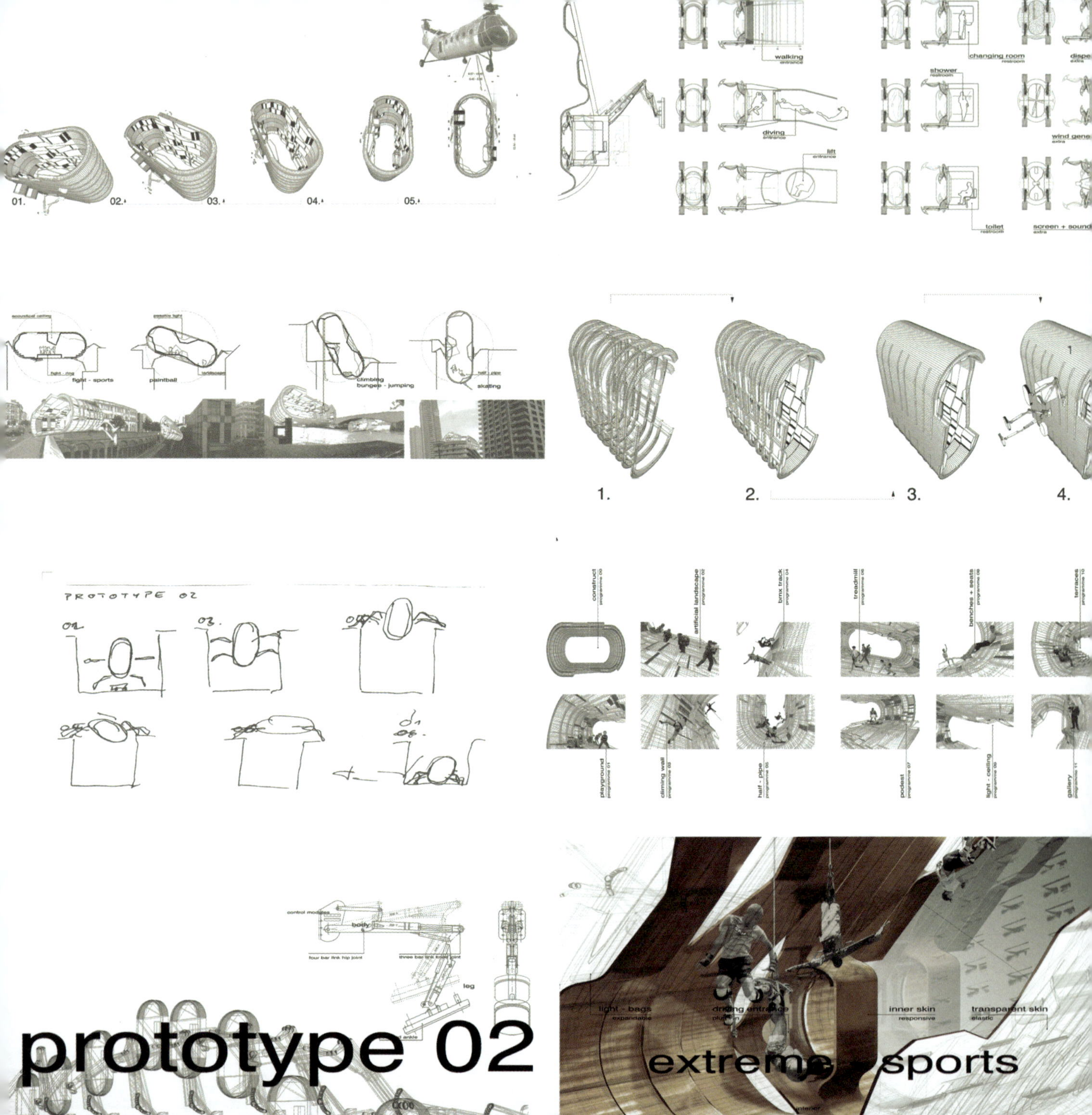

01.
02.
03.
04.
05.
walking
entrance
diving
entrance
lift
entrance
changing room
restroom
shower
restroom
toilet
restroom
screen + sound
extra
acoustical ceiling
fight - ring
fight - sports
paintball
landscape
climbing
bungee - jumping
half - pipe
skating
1.
2.
3.
4.
PROTOTYPE 02
construct
programme 00
artificial landscape
programme 02
bmx track
programme 04
treadmill
programme 06
benches + seats
programme 08
terraces
programme 10
playground
programme 01
climing wall
programme 03
half - pipe
programme 05
podest
programme 07
light - ceiling
programme 09
gallery
programme 11
control modules
body
four bar link hip joint
leg
light - bags
expandable
inner skin
responsive
transparent skin
elastic
prototype 02
extreme sports

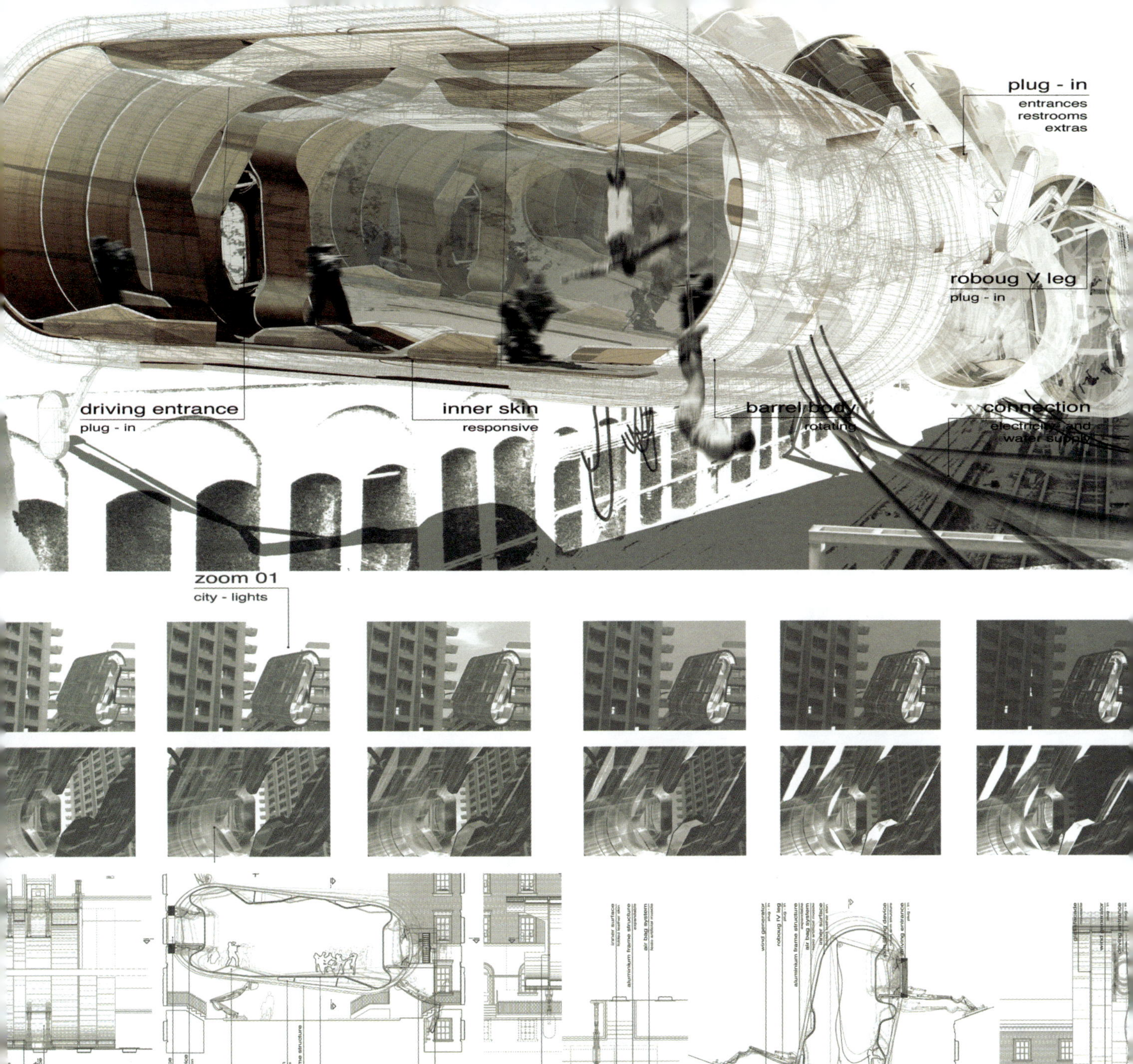
plug - in
entrances
restrooms
extras
roboug V leg
plug - in
driving entrance
plug - in
inner skin
responsive
barrel body
rotating
connection
electricity and
water supply
zoom 01
city - lights

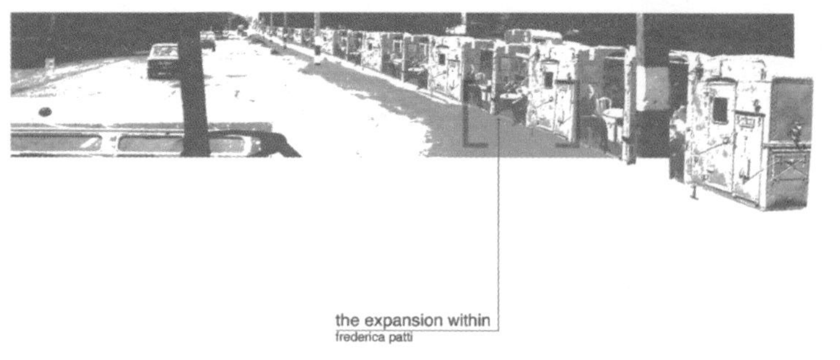

the expansion within
frederica patti

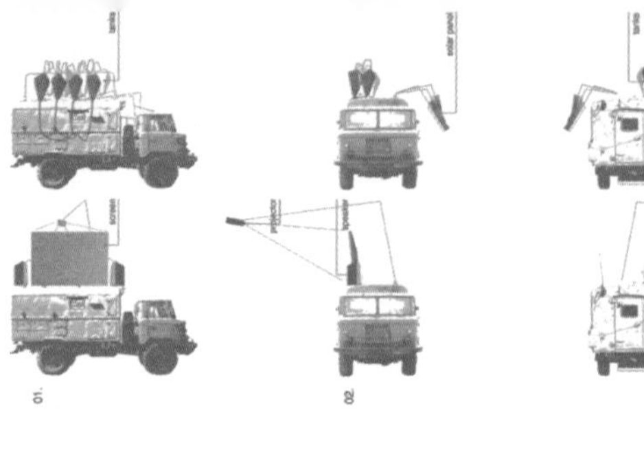

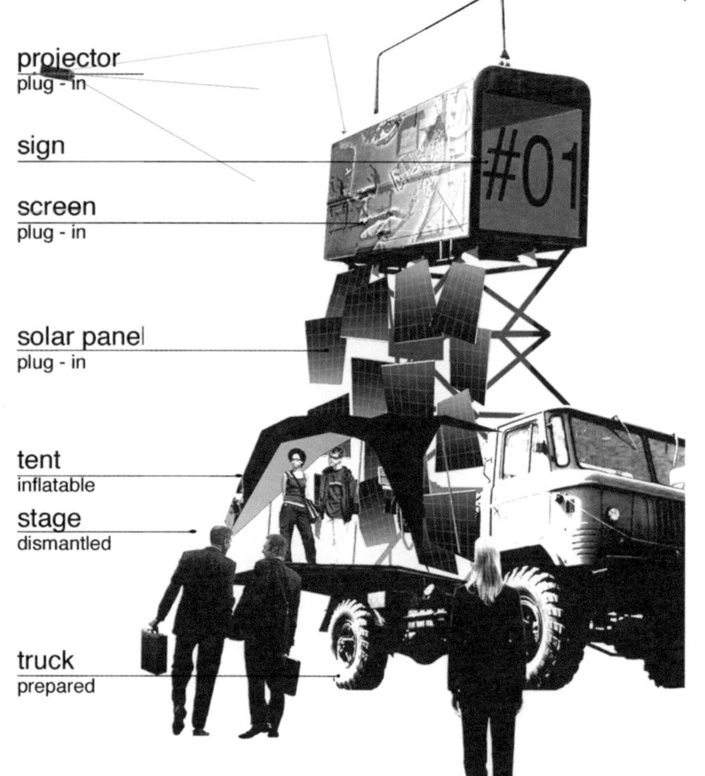

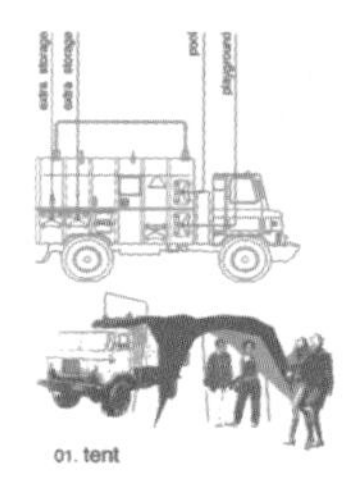

01. stage

02. bar

03. podium

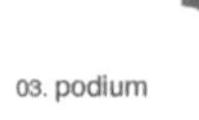

fish resort
fillip pschik

димитровград seminar

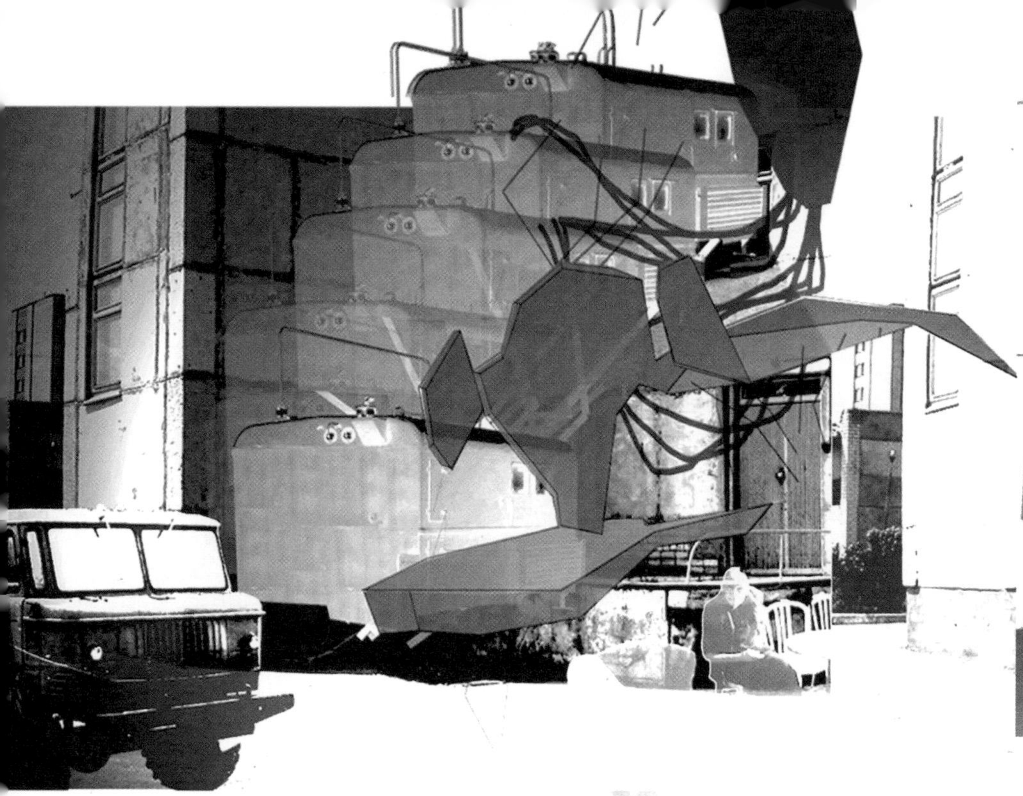

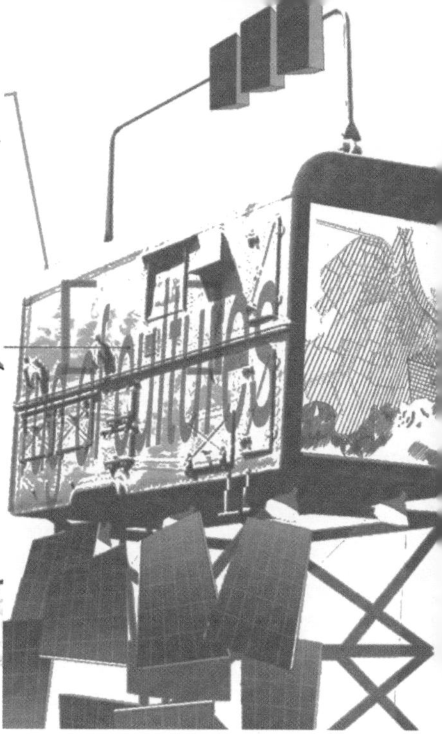

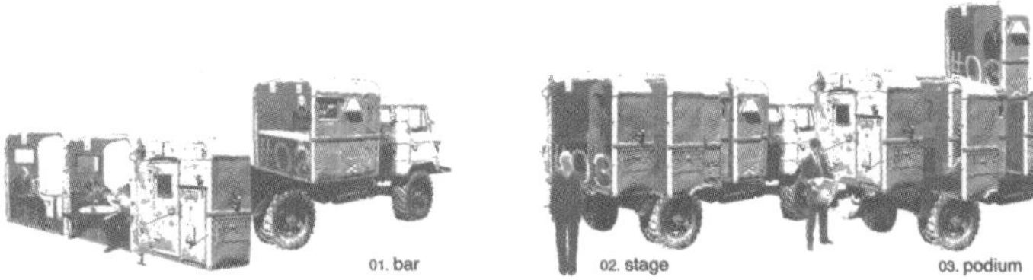

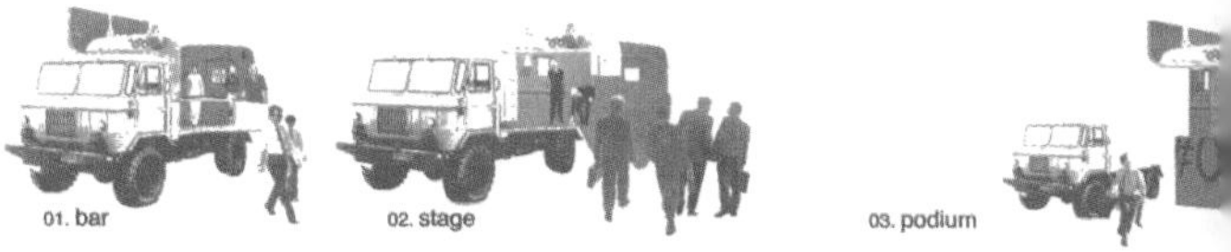

01. bar

02. stage

03. podium

01. bar

02. stage

03. podium

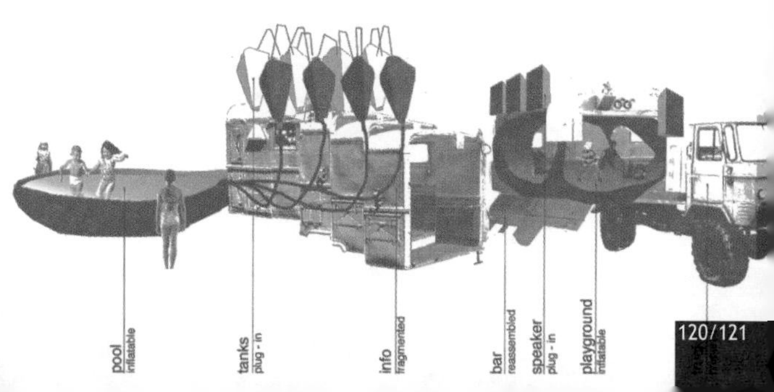

Dalston, Hackney
A 106
International
Freight Depots
New Spitalfields Market
1/3 m
Central London, Docklands
Blackwall Tunnel
A 12

里奥纳多
拉塔瓦
Leonardo
Lattavor

Leonardo Lattavo – CV

1989-1995 Windsurfing and Sunbathing on a daily basis in the beaches of Rio de Janeiro and on the spare time studying Architecture and Urban Design at St. Ursula University, Rio de Janeiro, Brazil.

1995-1997 Working like a dog with my own practice in Rio de Janeiro mainly in residential, retail and competition projects and not making any money at all.

1997-2000 Living and working in London in prestigious retail projects in Italy, Germany, Spain, Portugal and England and making a little bit of money too.

2001-2002 Spending all my savings travelling around the world and doing a Masters course in Architectural Design at the Bartlett School of Architecture.

2003-2004 Working in a prestigious architectural office in London and on competition projects. And also always looking forward for some holidays.

里奥纳多•拉塔瓦

1989—1995 每天在里约热内卢的海滩进行帆板运动和日光浴。业余时间在巴西里约热内卢的圣・厄休拉大学学习建筑与城市设计。

1995—1997 在里约热内卢做住宅兼零售业竞赛项目，累得像条狗，没赚到什么钱。

1997—2000 在伦敦生活工作，设计了位于意大利、德国、西班牙、葡萄牙、英格兰的名店项目，赚了点小钱。

2001—2002 花光了所有积蓄周游世界，并在巴特雷特建筑学院攻读建筑设计的硕士课程。

2003—2004 在伦敦一个享有盛名的建筑师事务所工作，做些竞赛项目，也一直期盼着假期。

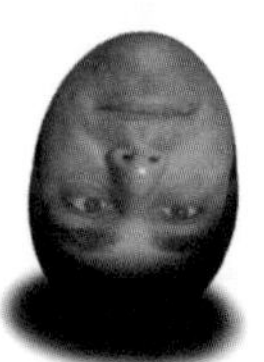

My move from Rio, my home town, to London happened very spontaneously and suddenly - without much plan. Almost as if I got the wrong airplane and landed in an unknown city. I never visited or thought about going to London before. Once I arrived my plan was to stay 'no more than 3 months'. That was almost seven years ago. I am still in London.

What attaches me to London is the cosmopolitan environment. In London I feel in the center of the world. Once I did a 'little big' project that expressed this feeling: I decided to design a hole for the world. And I did it. It is a virtual hole that runs through the centre of the earth linking opposite locations of the globe. So, for example, people looking through the hole in Peking would be able to see and exchange greetings with people in its opposing city - Buenos Aires. The 'Hole' can likewise link other cities with their respective antipodal cities: Tokyo could then be linked to Sao Paulo in Brazil; Hong Kong, China to La Paz, Bolivia; Chunking, China to Lima, Peru; Dunedin, New Zealand to Barcelona, Spain; Hawaii, USA to Johannesburg, South Africa; Jakarta, Indonesia to Caracas, Venezuela; etc.

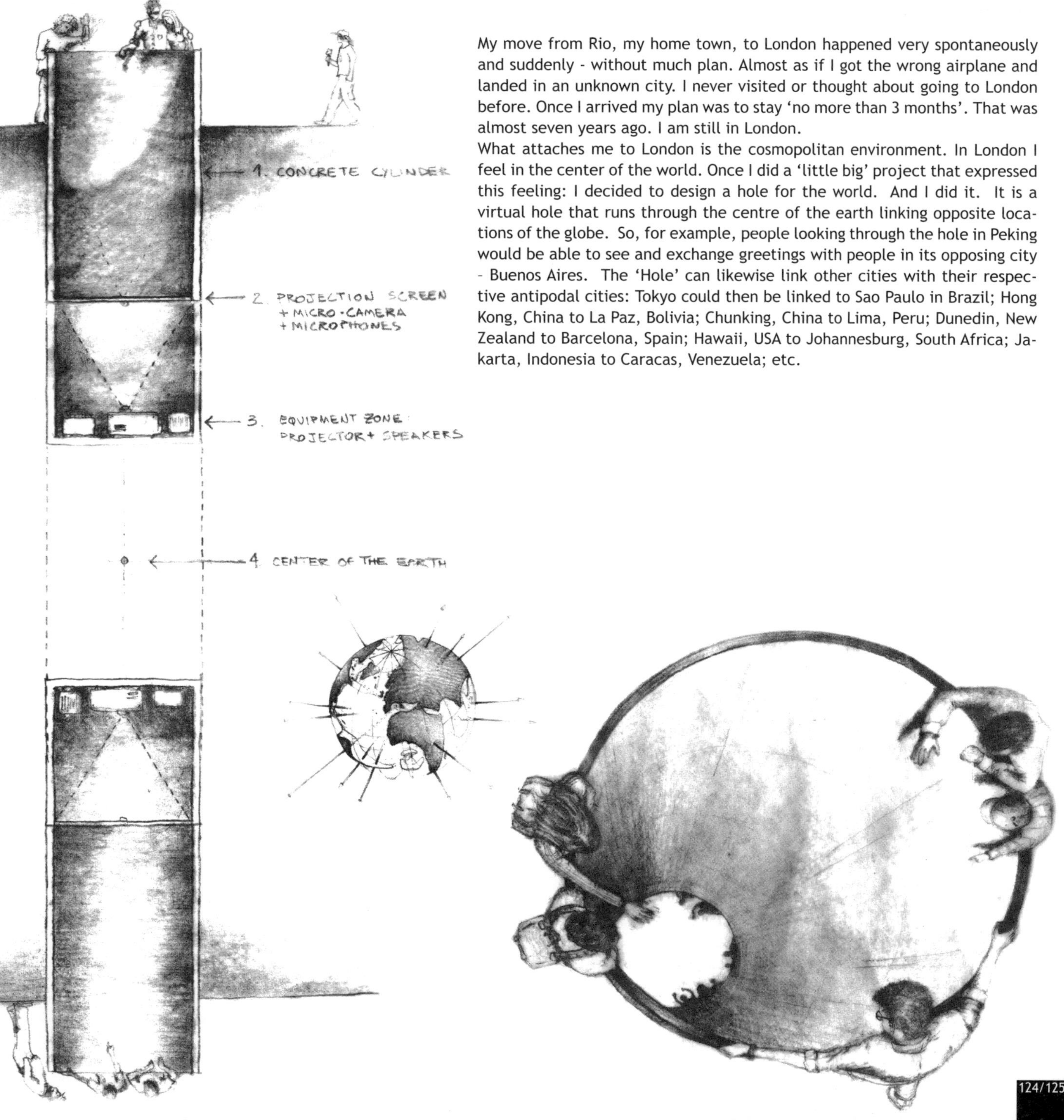

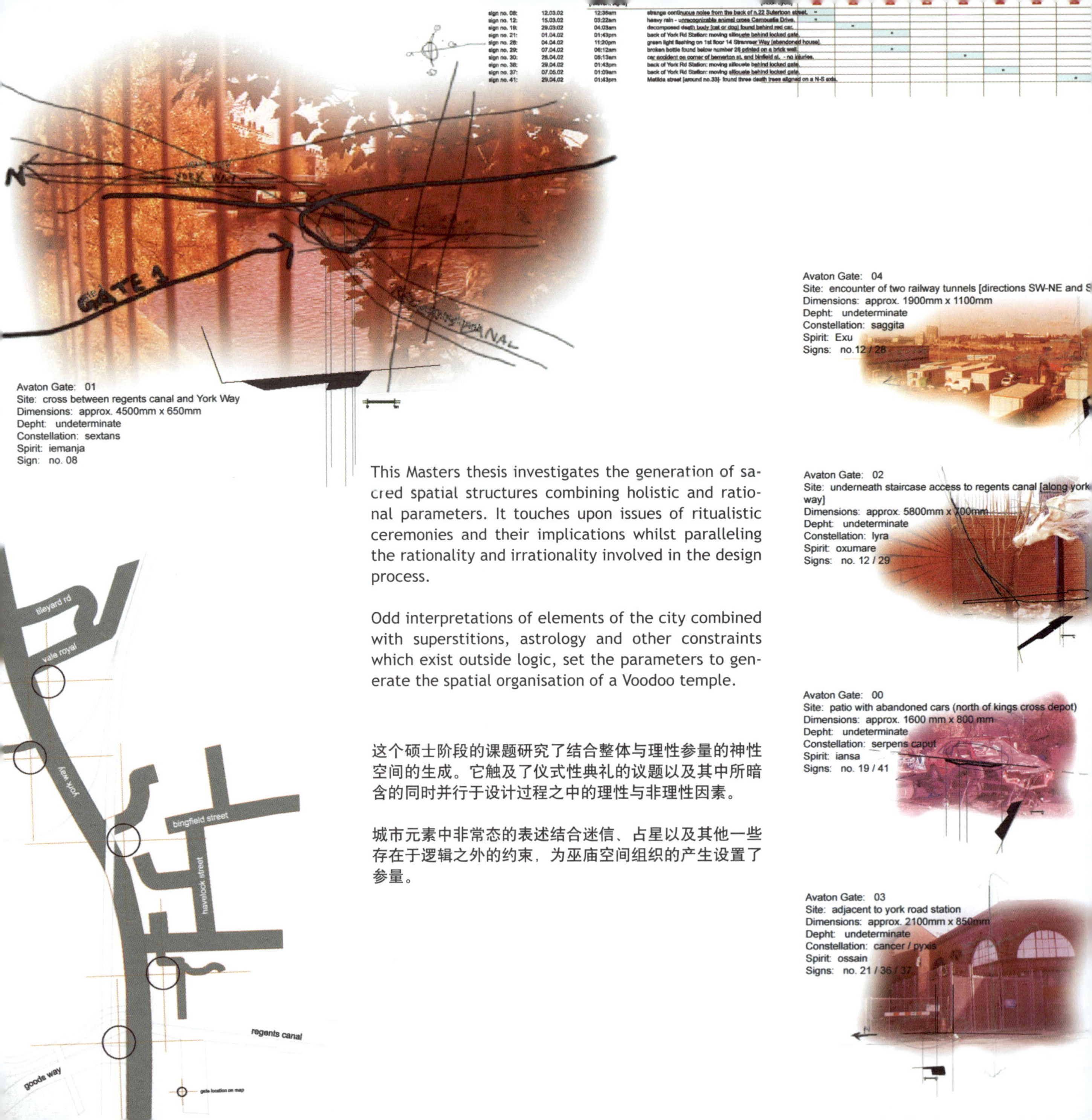

sign no. 08:	12.03.02	12:36am	strange continuous noise from the back of n.22 Sutertoon street.
sign no. 12:	15.03.02	03:22am	heavy rain - unrecognizable animal cross Carnoustie Drive.
sign no. 19:	29.03.02	04:03am	decomposed death body [cat or dog] found behind red car.
sign no. 21:	01.04.02	01:43pm	back of York Rd Station: moving sillouete behind locked gate.
sign no. 28:	04.04.02	11:20pm	green light flashing on 1st floor 14 Stranraer Way [abandoned house].
sign no. 29:	07.04.02	06:12am	broken bottle found below number 28 printed on a brick wall.
sign no. 30:	28.04.02	05:13am	car accident on corner of bemerton st. and binfield st. - no injuries.
sign no. 36:	29.04.02	01:43pm	back of York Rd Station: moving sillouete behind locked gate.
sign no. 37:	07.05.02	01:09am	back of York Rd Station: moving sillouete behind locked gate.
sign no. 41:	29.04.02	01:43pm	Matilda street [around no.33]- found three death trees aligned on a N-S axis.

Avaton Gate: 01
Site: cross between regents canal and York Way
Dimensions: approx. 4500mm x 650mm
Depht: undeterminate
Constellation: sextans
Spirit: iemanja
Sign: no. 08

Avaton Gate: 04
Site: encounter of two railway tunnels [directions SW-NE and S
Dimensions: approx. 1900mm x 1100mm
Depht: undeterminate
Constellation: saggita
Spirit: Exu
Signs: no.12 / 28

This Masters thesis investigates the generation of sacred spatial structures combining holistic and rational parameters. It touches upon issues of ritualistic ceremonies and their implications whilst paralleling the rationality and irrationality involved in the design process.

Odd interpretations of elements of the city combined with superstitions, astrology and other constraints which exist outside logic, set the parameters to generate the spatial organisation of a Voodoo temple.

这个硕士阶段的课题研究了结合整体与理性参量的神性空间的生成。它触及了仪式性典礼的议题以及其中所暗含的同时并行于设计过程之中的理性与非理性因素。

城市元素中非常态的表述结合迷信、占星以及其他一些存在于逻辑之外的约束，为巫庙空间组织的产生设置了参量。

Avaton Gate: 02
Site: underneath staircase access to regents canal [along york way]
Dimensions: approx. 5800mm x 700mm
Depht: undeterminate
Constellation: lyra
Spirit: oxumare
Signs: no. 12 / 29

Avaton Gate: 00
Site: patio with abandoned cars (north of kings cross depot)
Dimensions: approx. 1600 mm x 800 mm
Depht: undeterminate
Constellation: serpens caput
Spirit: iansa
Signs: no. 19 / 41

Avaton Gate: 03
Site: adjacent to york road station
Dimensions: approx. 2100mm x 850mm
Depht: undeterminate
Constellation: cancer / pyxis
Spirit: ossain
Signs: no. 21 / 36 / 37

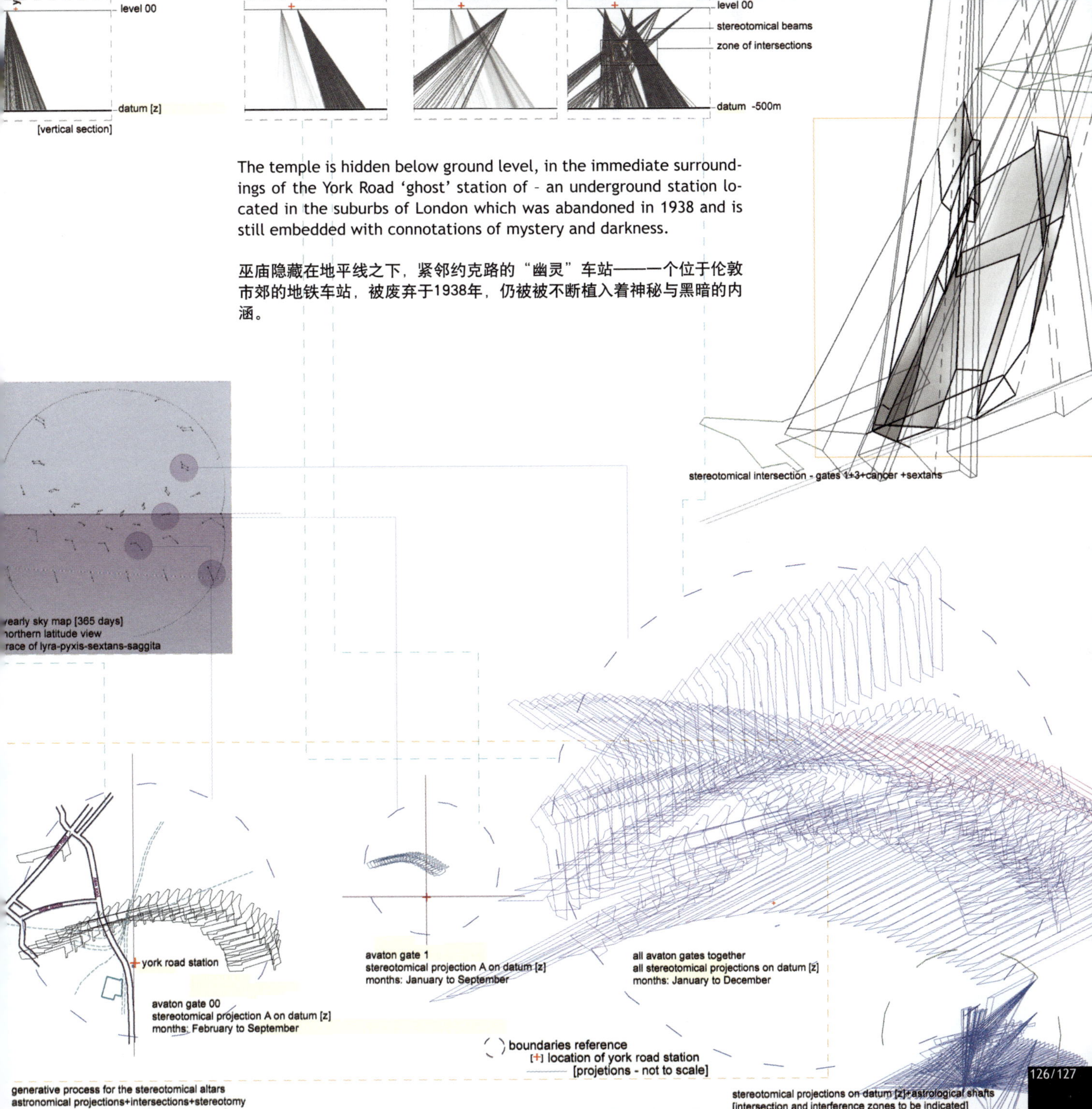

The temple is hidden below ground level, in the immediate surroundings of the York Road 'ghost' station of - an underground station located in the suburbs of London which was abandoned in 1938 and is still embedded with connotations of mystery and darkness.

巫庙隐藏在地平线之下，紧邻约克路的“幽灵”车站——一个位于伦敦市郊的地铁车站，被废弃于1938年，仍被被不断植入着神秘与黑暗的内涵。

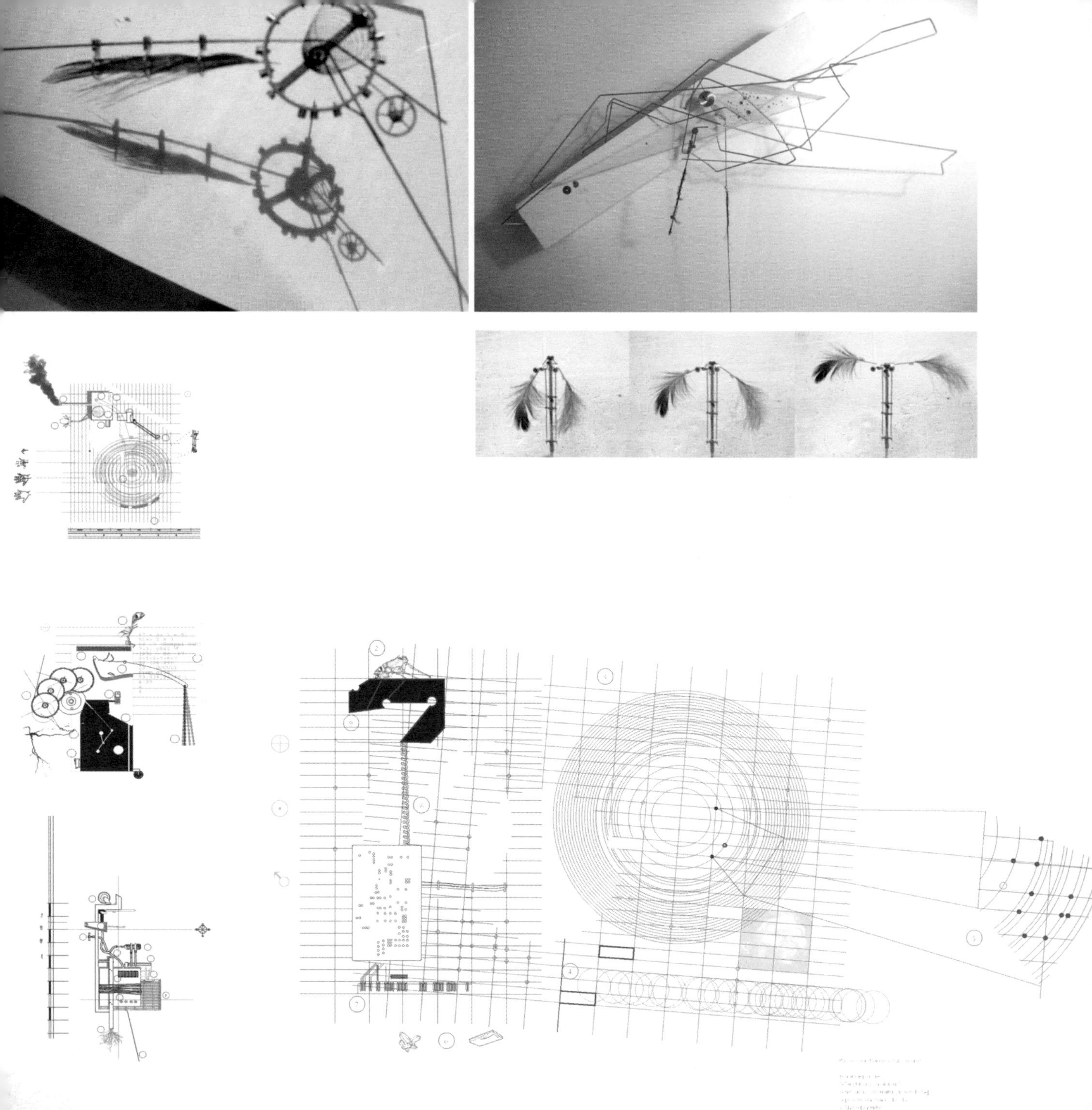

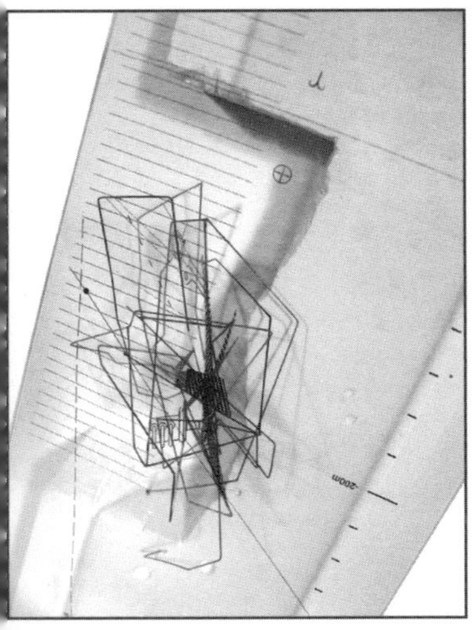

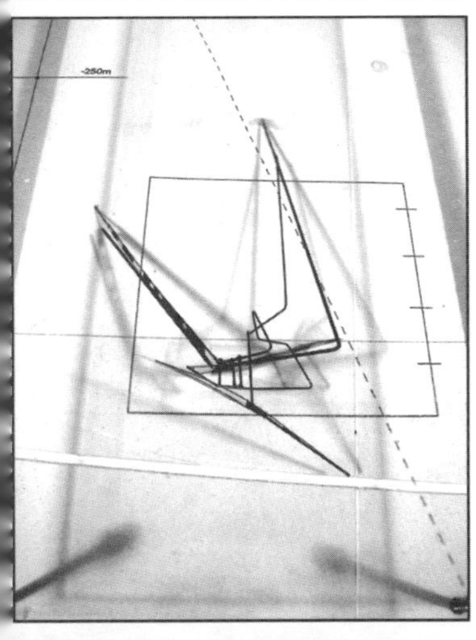

Analogous to the abandoned station, where the access is forbidden, the various gates to the subterranean temple are marginal, unclear. Inside the sacred space, machines, amulets, technological devices, objects and living creatures are re-interpreted and assembled together into a 'mechanical voodoo ritual' where ceremonies, sacrifices and offers to 'god-spirits' happen automatically - irrespective of human intervention.

The design process embraces the idealist philosophical doctrine that reality is somehow mind-correlated or mind coordinated - where the real objects comprising the 'external world' are not independent of cognising minds, but only exist as in some way correlative to the mental operations. It centres on the conception that reality as we understand it reflects the workings of mind. And it construes this as meaning that the inquiring mind itself makes a formative contribution not merely to our understanding of the nature of the real but even to the resulting character we attribute to it.

相较于被禁止进入的废弃车站，通往地下庙宇的各式入口是临界的、含混的。内在的神性空间，机械、神符、技术设备、物品和生物在一起被重新诠释与组合成“机器化的巫术盛典”，仪式、牺牲与对“上帝之灵”的供奉在这里自主发生——无关于人的介入。

设计的过程拥抱理想主义的哲学教条，那就是真实是与思维相关或相合的构成外部世界的真实物体并不独立于认知思维，而是以某种与思想运行相协作的方式存在。这将我们理解中的真实反映思维的运作的概念推向中心。它建构了这样的意义：探究性思维自身不仅对我们认识真实中的自然作出构成性的贡献，而且甚至对于我们得出的结论属性做出影响。

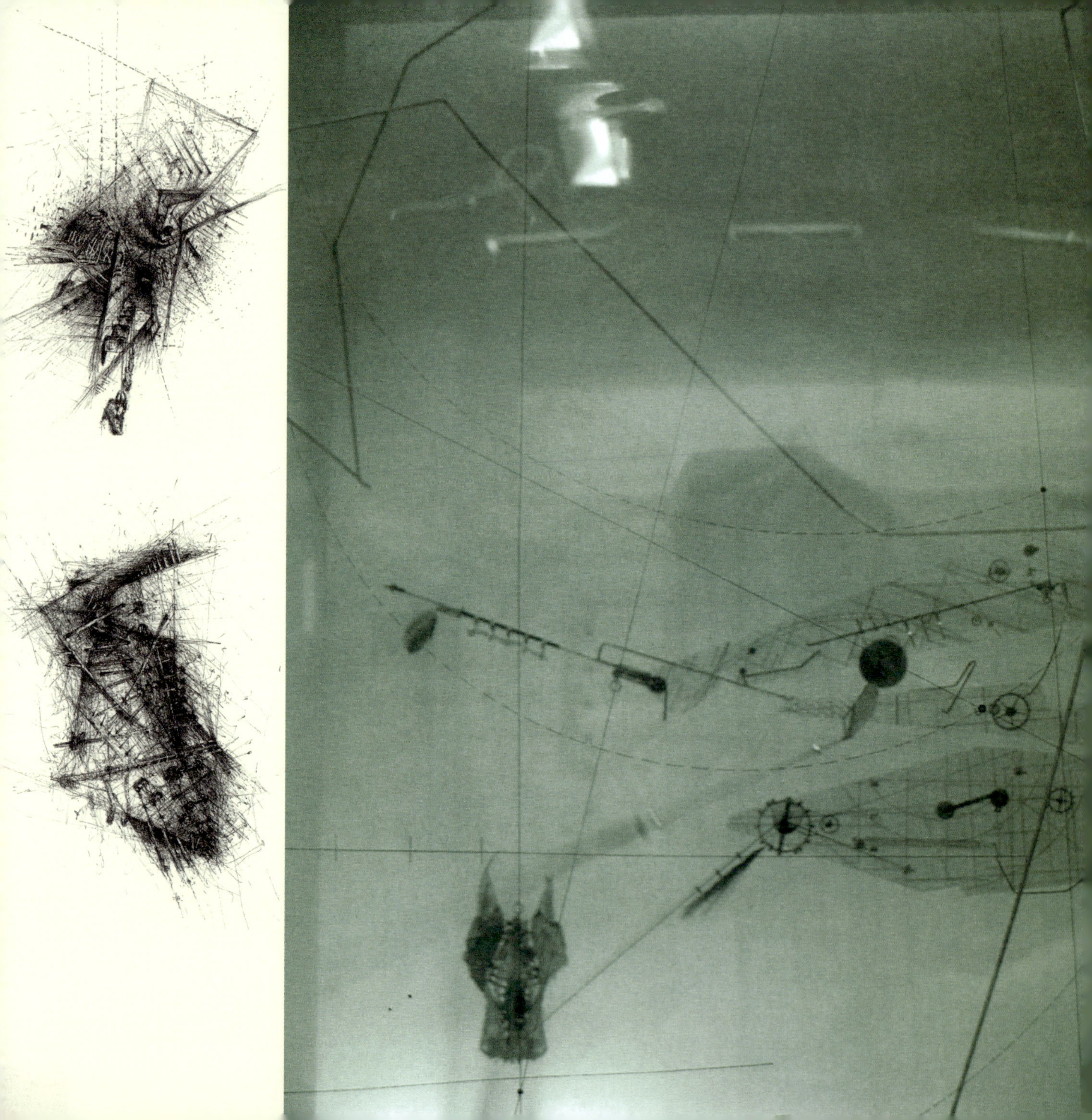

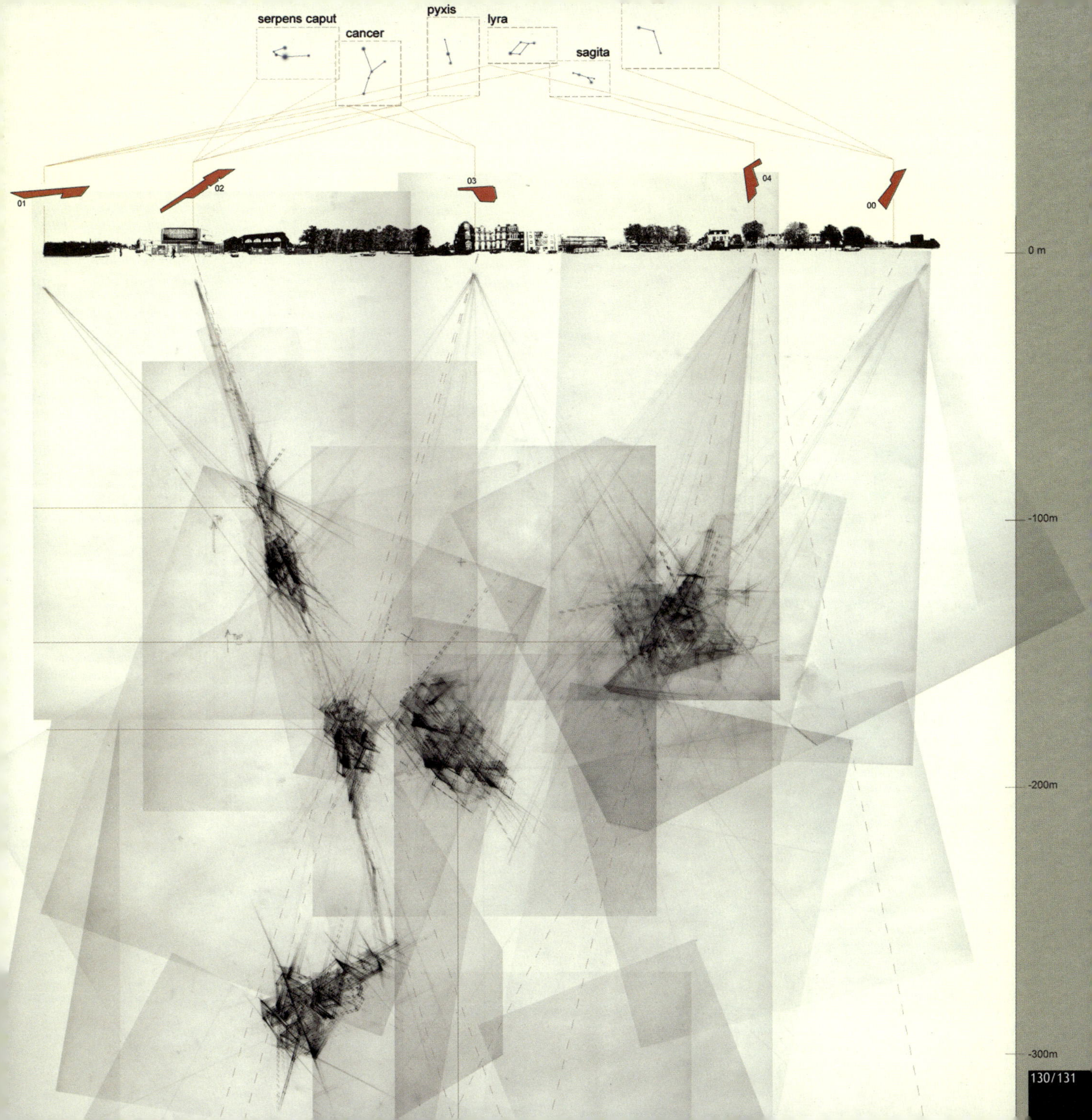
serpens caput
cancer
pyxis
lyra
sagita
01
02
03
04
00
0 m
-100m
-200m
-300m

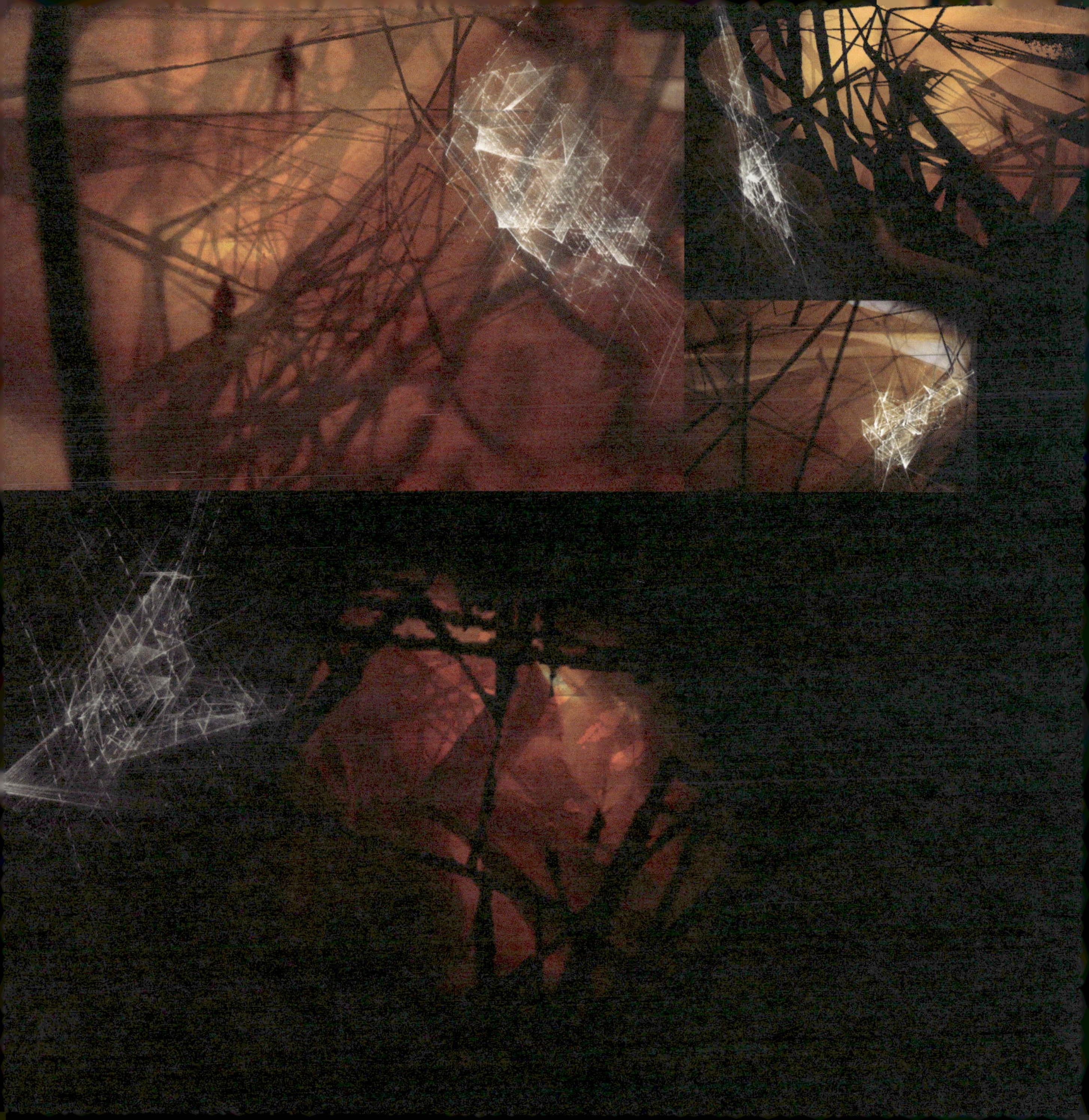

Chameleon Constellation

This is a shortlisted competition entry for a memorial for the 1999 earthquake in Chi Chi, Taiwan produced in collaboration with Amy Leung and Adrian Thompson...

Our vision for the Chi Chi Earthquake Memorial uses no architectural object or sculptural form, instead it creates a living landscape that recalls nature's gestures, a garden where the natural and artificial co-exist in seamless continuity.

We regard the site as one sweeping grass canvas 'planted' with 2034 buried uplighting luminaries across the entire expanse of the ground, circling around the trees and continuing from the land into the proposed pond, where they appear to float lightly at the water's surface.

Each of the 2034 lights is a tribute to a life lost in the 1999 earthquake in Taiwan. The diversity and individuality of each person remembered through this memorial is captured by the myriad of colours and sizes of the luminaries, each of which have their own character and rhythm.

变色的群星

本设计入围1999台湾吉吉大地震纪念设计竞赛，与艾米·梁，雅德利安·托马森合作。

我们对于吉吉地震纪念物的愿景是不采用建筑物或纪念碑形态，而是创建了一个活着的景观来唤起自然的姿态，实质是一个人工与自然浑然一体的花园。

我们将基地看作植入2034个向上发光物的一整片草质的画布。发光物散布于整个地面，环绕树林，一直延续到设计中的池塘，在那里它们在水面上轻轻飘拂。

2034盏灯中的每一盏都是对在1999年台湾大地震中丧生的人的追悼。不同颜色和大小的发光体，每一个都有自身的特征和韵律。

These ground luminaries emulate the ebb and flow of changing seasons and highlighting nature's cyclic and seasonal expression of the passage of time in a continuous choreography of light controlled by computer. In the winter all the lights are white. These gradually gather tones of blue and green with the arrival of spring. By summer the landscape blossoms to be at its most colourful with the addition of yellows, oranges, reds and purples. These vibrant colours fade to more earthy tones for the duration of autumn only to pale softly to white again in completion of the years' cycle.

Not only will the garden be a remembrance of the forces of nature, but also a garden of celebrations. The flexibility and changing aspect of the lights can allow the landscape to re-appropriate itself to different festivals and occasions. Depending on the event or celebration, the lights can be set up to blink, to dim, to interact with peoples' movements, to be set up to be all blue or all yellow, etc.

地面发光物效仿季节的潮起潮落，突显自然的循环和时光逝去的表达，是由计算机控制的光之舞蹈。在冬季所有的灯都是白色的，随着春天的到来它们逐渐获得蓝绿的色调。在夏天大地花朵遍开，以增添的黄色、橙色、红色和紫色成为其最富色彩时刻。在秋季这些色彩减退为土地的色调并渐渐再次淡化为白色，这些律动的颜色完成了一年的轮回。

花园不仅成为自然力量的纪念，同时也是庆典的花园。灯光的灵活性和变化允许地景在不同的节庆和时刻重新调整。根据时间或庆典的不同，灯光可被设置成或闪光、或朦胧、或与人们的运动互动、或被设置成全蓝或全黄，等等。

It is hoped that the memorial will grow to become more than a simple tribute to the victims of the earthquake, that over time it will find itself to be a social epicentre in the province of Chi Chi as it becomes a living, vibrant aspect of everyday life in this community.

希望此纪念物不仅仅是对地震死难者的简单致敬，更希望历经时日，它能够作为吉吉地区的社会震中，成为该社区日常生活的活跃力量。

summer
autumn

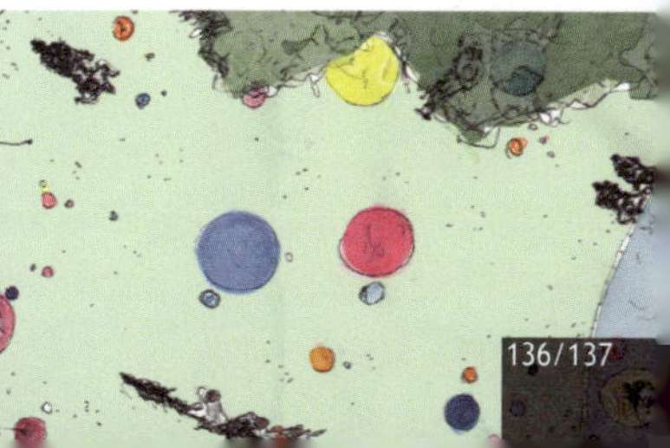

6
½ m
(M1, M20)
M 25
D'ford Crossing
Watford
London M11

迈克
米切尔
Michael
Mitchell

1998 - Illinois Institute of Technology
Bachelor of Architecture

2003 - The Bartlett
Master of Architecture

1998 - Stl Architects - Chicago
2000 - SOM - Chicago
2001 - Truex Cullins - Vermont
2002 - OHKW - Vermont

2003 - Future Systems - London
2007 - KPF (London) - London

As architects I think we need, more than the other professionals, to be aware of the narative we' re developing as our careers unfold before us. We' re trained to have a critical eye and to suggest our reflection of the world through a line on a page or the tilt of a column. We should learn this as second nature because it' s fundamental to everything else we do.

It' s another thing entirely to turn this third eye on ourselves and to become the object of our own reflection. Yet until we do this we are blind to our own motivations, and adrift with no point of reference. This is the challenge set out for the student of architecture. Know yourself, know what you' re from and tell us through your work. So that we may dream along with you.

1998 - 伊利诺伊工学院
建筑学学士

2003 - 巴特雷特学院
建筑学硕士

1998 - Stl 建筑师事务所 - Chicago
2000 - SOM 建筑师事务所 - Chicago
2001 - Truex Cullins 建筑师事务所 - Vermont
2002 - OHKW 建筑师事务所 - Vermont

2003 - 未来系统 - 伦敦
2007 - KPF（伦敦分部） - 伦敦

作为建筑师，我认为当职业生涯在我们的面前展开时，我需要比其他专业人士更多地知晓在其中发展着的叙事。我被训练得拥有一只批判的眼睛，纸上的一条线或是一根柱的斜度就能表达出我们对世界的反映。我们应该学习将这作为我们的第二天性，因为这是我们做其他事的基础。

还有另一件事彻底地将这第三只眼投向我们自身，并成为们自身映像的客体。直到我们这样做时，我们仍不明白我自己的动机，我们没有坐标点地随波逐流着。这是为建筑学生设置的挑战：了解你自己，了解你从何而来，并通过的作品告诉我们。让我们和你一起做梦。

New York City, 16 September 2001

The votive shrines went up apparently overnight in the parks throughout Manhattan. By the time I arrived with my sister, the two of us pulled to the city as much to be together as to bear witness, the layers of candle wax had fused across the asphalt of Union Square Park red, blue and white.

Spilled out across the ground that day were dreams of the American utopia, while out of the air the ash of 3000 souls still settled.

In one month I would lose my job. In one year I would be in London. The rest of life may not be charted from September to September...

but from Union Square Park.

9.11.01
NYC

Tankercity A polemical construction of salvaged oil tankers along the River Thames.

Inspiration This project begins with a series of photographs of the shipbreaking beaches of Chittagong, Bangladesh. There is a beauty implicit in the wasted scenery they depict, and it is this contradiction that Tankercity explores.

The images tell a story. Briefly, the common practice in the oil industry is to retire old oil tankers by selling them as scrap to the shipbreaking industries of China, Bangladesh and India, where they are run aground on beaches and cut apart by hand in huge ship scrap yards.

There are no environmental regulations governing these industries and because of this, international pressure has developed to halt the sale of contaminated oil tankers to the third world. The point being the first world should clean up its own mess.

The U.S. Navy has stopped sending its retired tankers and recently an English company won a contract to dispose of a contaminated Navy tanker. This didn't go down so well in the U.K. It did however provide a hundred jobs in a community that needed them, jobs that would originally have gone to the people in Chittagong, Bangladesh, who arguably need them even more.

Shipbreaking #49 – Edward Burtynski

Nickel Tailings #30 – Edward Burtynski

The world is full of economies of desperation and at every scale. This is the (perhaps absurd) point that Tankercity is born. Imagine we did scrap our own their tankers. Imagine with me, us taking on our own mess...

Environmental regulations would assure an expensive process that would create highly paid jobs for any man or woman trained to operate a cutting torch in a full contamination suit. Armies of engineers would insure state of the art decontamination and disposal facilities, while environmental scientists would scrutinize the process at every step along the way.

A site with deep-water access in a politically weak area would first have to be found to accommodate a state of the art shipbreaking yard, and luckily for Tankercity the English have the deep-water docks at Tilbury. The River Thames at Tilbury is still wide enough and deep enough to accommodate tankers, and since the English now support themselves through a service economy based widely on sales of café lattes, the docks have fallen into disuse.

All of this would raise gas prices at the pump while the industry finds its

油轮之城　　泰晤士河沿岸一个由打捞油轮构成的争议性建筑。

灵感　　项目开始于一系列位于孟加拉吉大港拆船海滩的照片。它们
一种暗藏在由其描绘的废物景观中的美，而这正是油轮之城所要探
盾。

这些图片讲述了一个故事。简而言之，在石油工业中，按一般的惯
业会把旧油轮当作废物卖给中国、孟加拉和印度的拆船工厂作退役
在那里，它们被搁浅在海滩上，在巨型的轮船废物场里被切割成散件。

由于没有环保规定来管理这些工业，因此，要
向第三世界国家出售污染油轮的国际压力不断
其要点是第一世界国家应该清除他们自己的垃圾。

美国海军已经停止出售退役油轮。最近，一家
司赢得了处理一艘有污染的海军油轮的合同。
事在英国并未消亡。它确实为需要它的社群提
百份工作，这样的工作原先会流向孟加拉吉大
然更需要它们的人们。

这个世界充满了绝望的经济，大小不等。这就
许是荒谬的）油轮之城诞生的原因。设想我们
我们自己的油轮。跟随我一起设想，我们来承
自己的混乱……

环保规定将会确保这个花费昂贵的过程会为任
受训过、能操作切割炬、穿着布满污染物的工
人或女人创造高工资的就业机会。工程师大军
艺术级别的净化和机械处理，而环境科学家会
查过程中每一步。

首先必须在政治弱势地区找到能够盛装一个艺术级别的拆船厂的基
运的是，英格兰的油轮城市拥有在提尔堡的深水码头。在提尔堡一
晤士河依然足够的宽和深，可以容纳油轮。英格兰现在靠卖拿铁咖
务业支撑，码头沉沦不用了。

当这一产业找到出路并开始盈利时，所有的这一切都会提高泵出的
格。这样我们就会达到这一议案的终点，正像每一个对改选感兴趣
家会告诉你的……

假如不是为了“油轮之城”。
将这些工人转而投入到酒吧里去是荒谬的。因此，这些油轮，这些

s and becomes profitable. And with that we've reached the end of line for this proposal, as any politician interested in reelection will you...

t weren't for Tankercity.

turn the investment of all of this labor into reinforcing bars would be icrous. And so these tankers, these brutes, might better be carefully dered for their now precious cuttings of curving steel.

er decontamination, the hulls of tankers would be cut into pieces be graded by quality and degree of curvature. Large straight flanks, rns, pointy bows; the remnants of ship after ship would be catego-ed and stockpiled - ultimate destination: Tankercity.

Tankercity I imagine is in London, England because I am here. It's visioned as an extension of public space into and along the River ames.

e Thames is a post-industrial landscape of anachronistic references a commercial heyday now long extinct. The physical separations that e kept a polluted river at arm's length from the city it serviced were propriate when the river was an industrial sewer but are now irrel-ant and hold the river back from becoming what it naturally wants to come: London's largest park.

最好就是像现在这样表现在切割精致的弯曲钢铁之中。

在净化之后，油轮的外壳将会被切割成片状并根据其弯曲的品质和程度进行分级。大而直的侧腹板，船尾，尖尖的船首；船的残件被分类与储存，在这之后——到达终级目标：油轮之城。

我构想的油轮之城是在英格兰的伦敦，因为我就在这里。它被设想成为泰晤士河的内河沿岸公共空间的延伸。

对于绝迹已久的商业全盛期，泰晤士河是一个不合时代参照的后工业景观。当它还是工业下水道时，使这条被污染的河流与它所服务的城市保持一臂之隔的物理分离是适当的。但现在的分离已与此毫不相关，还阻止着泰晤士河回归到了它原本想要的样子：伦敦最大的公园。

Tankercity would be a boardwalk extending as far upstream as the dams that mark the divide of the fresh and salt water Thames. At some points the boardwalk might be only a few meters wide, a delicate addition floating along the tides. At more urban locations, Tankercity would widen into a pedestrian mall and grow ammenities such as kiosks, restaurants and boathouses. And at Blackfriars in central London, Tankercity will rise up out of the water and expand into a massive urban appendage plugging city and river inexorably together.

Here trains, buses, and water taxis would bring rivergoers to a floating island swimming hole where Londoners could bathe in filtered river water (London has few swimming pools) or sun themselves on undulating steel beaches. Urban cabanas for river vacationers would survey the scene, treehouse-like, from atop spindly masts. At points along the promenade, intimate promontories would cantilever delicately out over the river, allowing lovers and children to dangle a toe into its waters.

One super promontory would extend a loop of the promenade across the river at Blackfriars to a sky-high viewing position in the middle of the Thames before wending its way back to the north bank and on down the river.

油轮之城将会成为一个延伸至标示泰晤士河海水和淡水分野的大坝上游的散步道。在某些点上，散步道也许只是几米宽，一个精巧漂浮在潮上的附加物。在更加城市化的地点，油轮之城将会被拓宽成步行商业街、亭子、餐厅和船库等康乐设施。并且，在伦敦中心区的布莱克法莱尔一带，油轮之城会生长出水面，延伸为一个巨大的城市附属物，将城市与河流无情地插接在一起。

这里火车、公共汽车、水上出租车将会承载着河上的游客到一个浮岛上的深水潭中。在那儿，伦敦人可以在过滤过的河水中洗浴（伦敦几乎没有游泳池）或是在起伏的钢制河滩上晒太阳。

为泰晤士河度假者准备的简易浴室可以从纺锤形桅杆顶上俯瞰风景，像树屋似的。在沿散步道的某些地点，私密性的岬角会优雅地悬挑在河上，可以让情侣和孩子们在水中晃动着脚趾。

一个超级岬角会在布莱克法莱尔延伸成跨过河岸的散步环道，在它绕回北岸降下河面之前在泰晤士河中达到一个天空高度的观察点。

Explore all the tiny places, everything under your nose.
Train yourself to see what the others do not.
Do not live without beauty and wonder.
It is not enough to merely be alive, you must know it.

You must be reminded of the magic of it all.

----- - unkown

When you have all the answers about a building before you start building it, your answers are not true.
The building gives you answers as it grows and becomes itself.

----- - Louis I. Kahn

探索一切细微的所在，万物都在你的鼻子之下。
训练你自己去看那些其他人看不到的东西。
生活中一定不能没有美与奇迹。
仅仅活着是不够的，你必须知道这些。
你必须被唤起它所有的魔力。

——佚名

如果在你开始建造一个建筑之前，
就有了关于它的所有答案，你的答案是不正确的。
建筑会在成长中给你答案并成为它自己

——路易斯 • 康

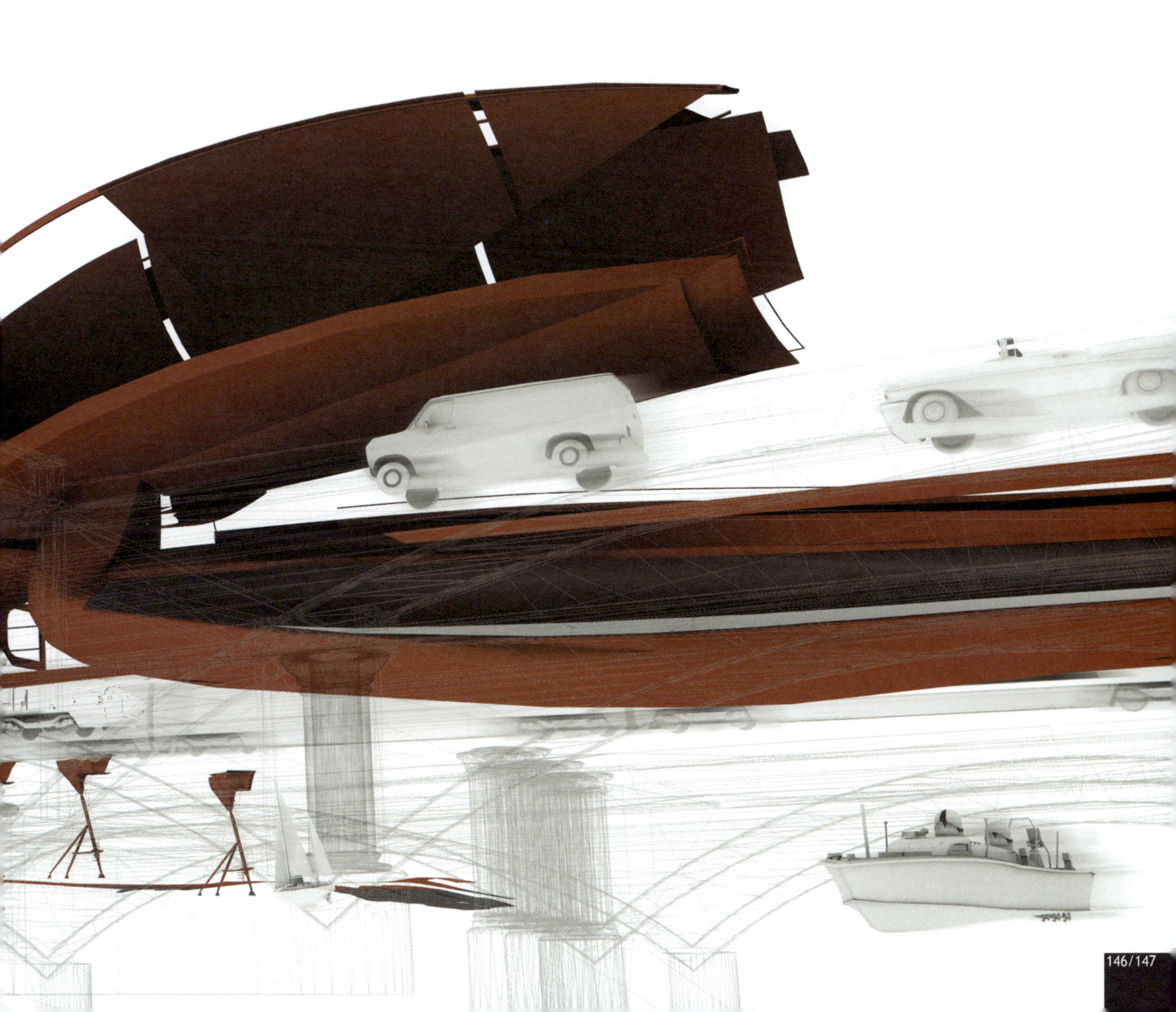

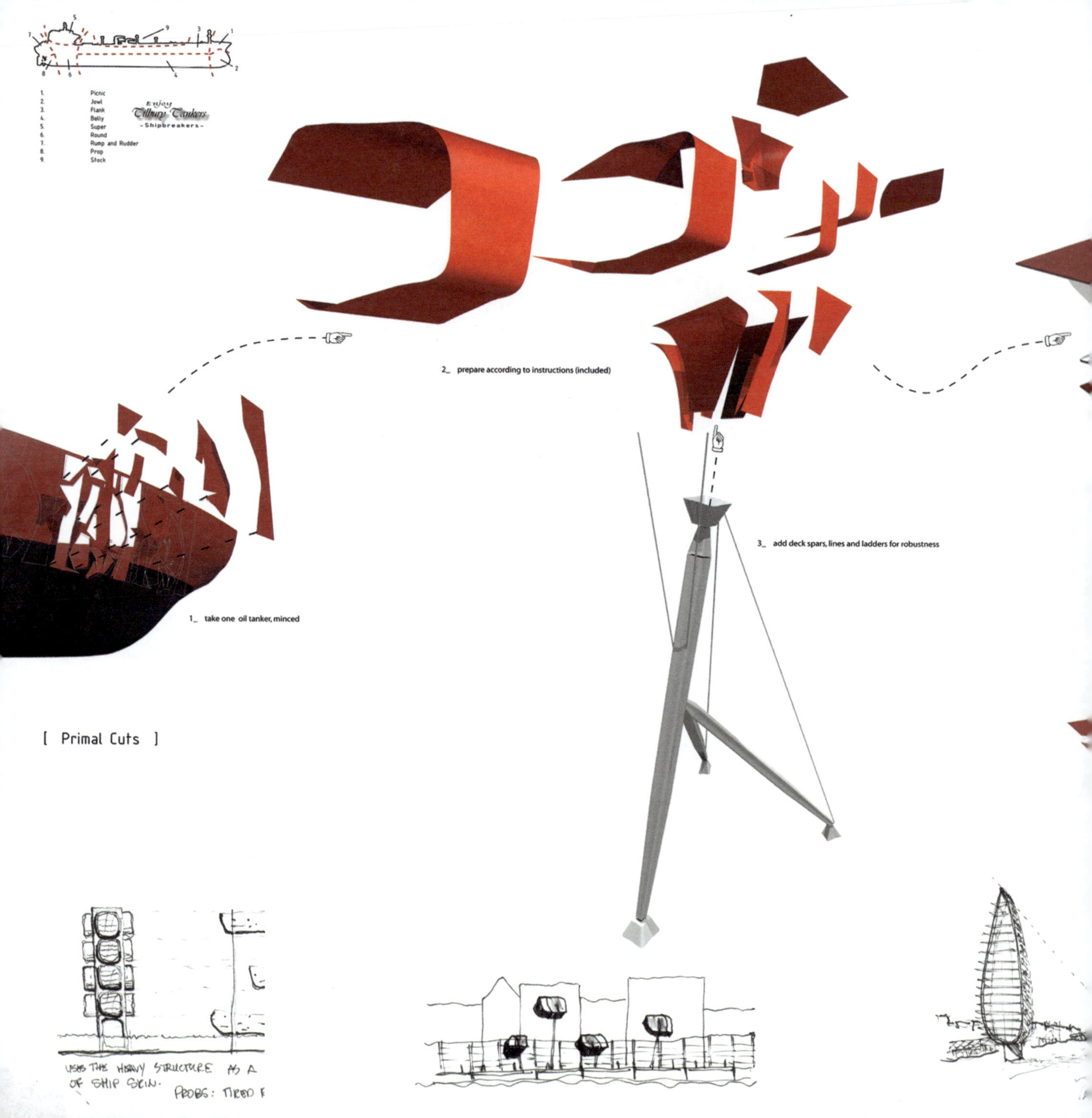
1. Picnic
2. Jowl
3. Flank
4. Belly
5. Super
6. Round
7. Rump and Rudder
8. Prop
9. Stock
Enjoy Tilbury Tankers
- Shipbreakers -
2_ prepare according to instructions (included)
3_ add deck spars, lines and ladders for robustness
1_ take one oil tanker, minced
[Primal Cuts]
USES THE HEAVY STRUCTURE AS A
OF SHIP SKIN.
PROBS: TIRED F

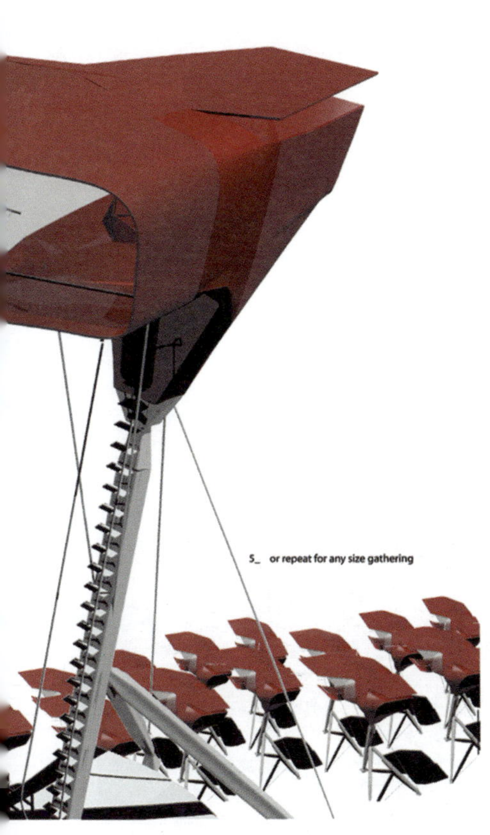

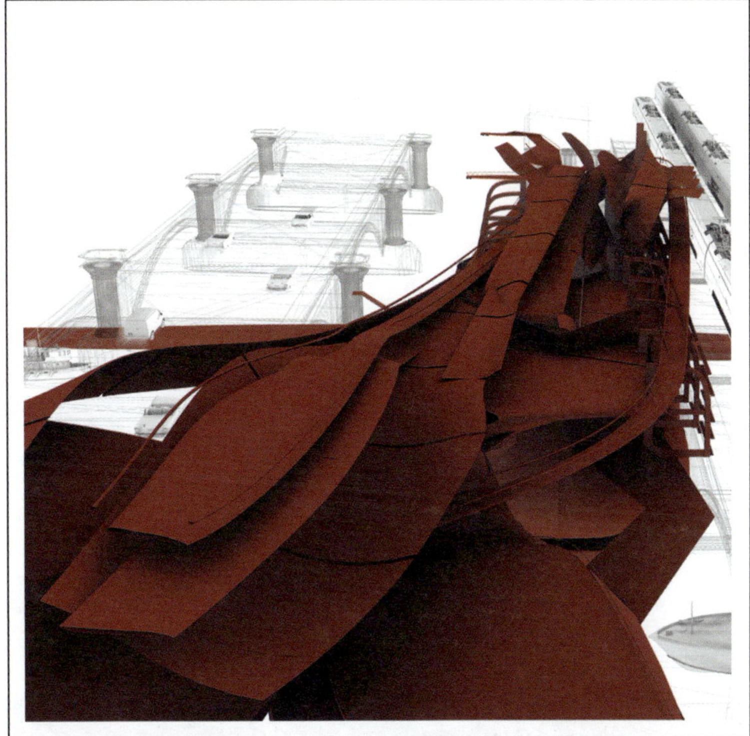

Tankercity would exist at the edge of city and river, water and land, sense and nonsense. It is a promontory, a promenade, a bridge or a building. It floats, it spans, it rises and falls with the tide. It began self-consciously with a sense of beauty defended by satire and has ended up as so much more...and innocently, much much less.

油轮之城将会存在于城市与河流，水与陆，意味与无聊之间的边界中。它是一个岬角，一个散步场，一座桥梁或是一栋建筑。它漂浮着，延伸着，随着潮水起起落落。它自觉地开始于以讽刺辩护的美感，如此丰富地结束……纯洁地，越来越少。

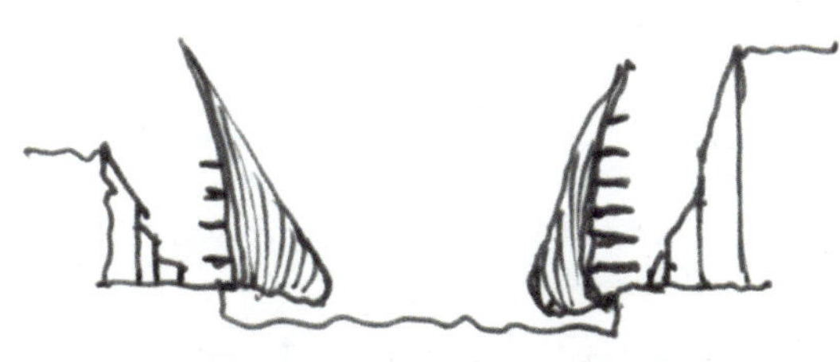

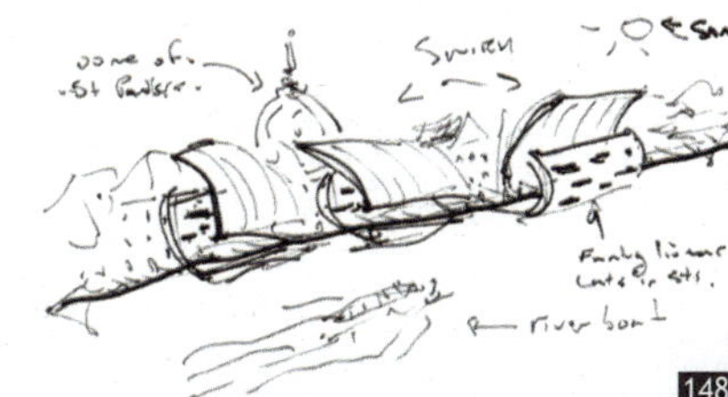

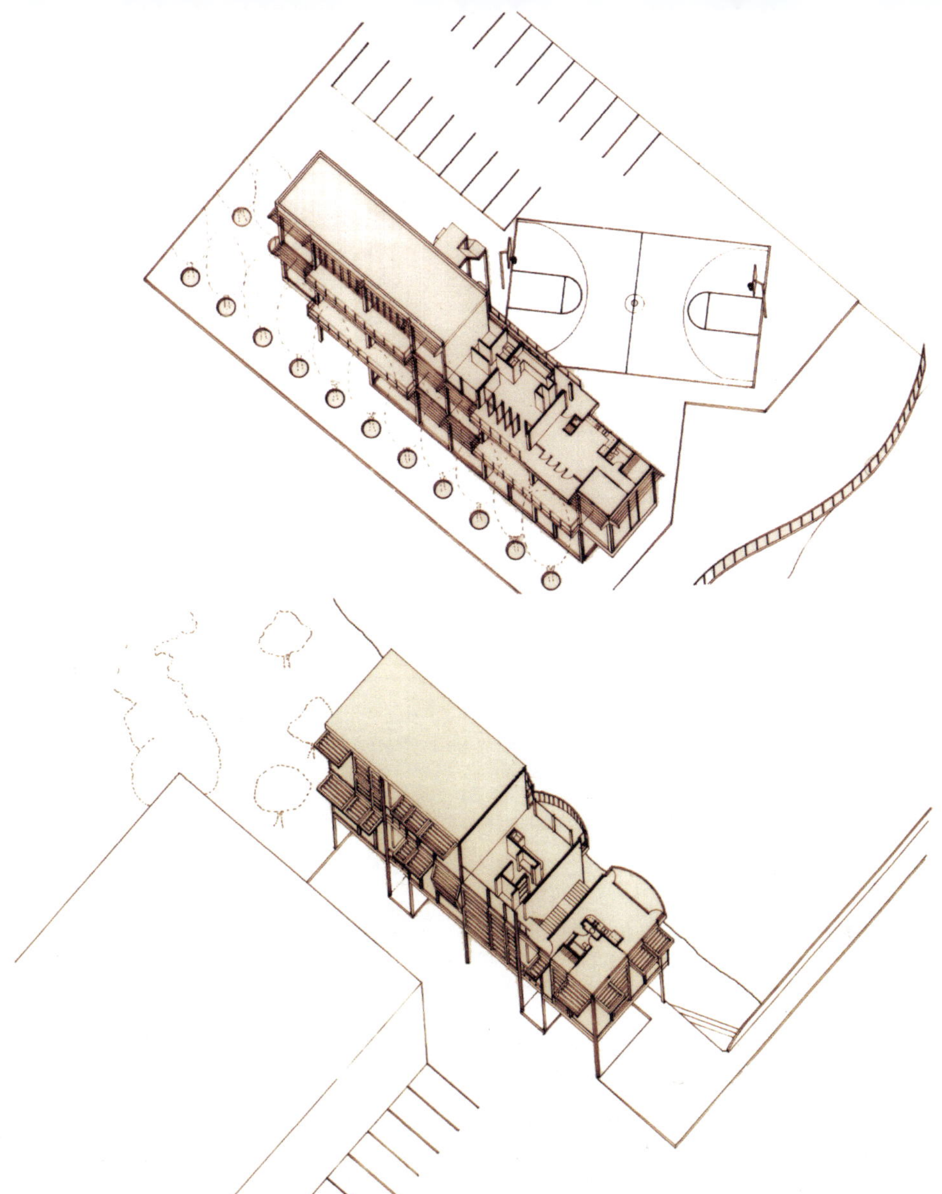

Undergraduate work - 1992 to 1997 Illinois Institute of Technology, Chicago and Tulane University, New Orleans

Clockwise from bottom left.
House for Two on Lake Michigan, Chicago. A House for a Flying Car and collage. Bayou Michou Housing, Louisianna.

Old Ford B 142
C. London
Stratford
(A11)
Docklands
Blackwall Tunnel
A12

潘
岩
Yan
Pan

在高高低低的平台上，从低处向高处攀爬，混凝土的粗糙表面摩擦着我的皮肤；从高处往低处跳下，带着一点点恐惧，脚底震得生疼。一路跑回家，在一大片“火柴盒”中找到一栋，扶着生锈的楼梯铁扶手爬到三楼，敲开门喝着祖母做的绿豆汤，跑到阳台上，我家的阳台是改造过的，三面都用玻璃包起来，我得去看看阳台上养着的一只鸡，我知道我的邻居家阳台上有其他更好玩的动物，一楼那个家伙养了一只狐狸，五楼的养了一只鹰。当然更多人还是把阳台变成了储藏室。那年我9岁，1980年代末，中国。

学习建筑是一件有意思的事。其研究对象极至日常，如此司空见惯，你每日在其中生活而不自知。直到有一天，这成为你的职业，“前建筑”时期的经验因未受所谓专业训练的影响而成为建筑师空间体验的极其特殊的部分。越是学习建筑时间长久，这些关于空间的处女记忆就越频繁地浮现于脑海。这段记忆与从体验中抽离出的专业知识对峙，为对职业的反思提供了基点。

建筑师教育作为一种职业教育，仿佛要把你同这一切割裂开来，你知道那些高高低低的台子是苏联式设计的观礼台，为了某些仪式；你知道那些整齐划一的“火柴盒”是苏联式规划下为了批量建造和适应北方寒冷气候而设计的集合住宅，你知道它们的合理深度和长度。然而在我以前受到的教育中，体验并不重要，你学会的是如何把生动丰富的空间变成有可能并不说明任何问题的平立剖，专业知识在这里意味着割裂，拒绝这种割裂，我来到伦敦，陷入了另一场斗争。在五光十色的巴特雷特学院的建筑光谱中，我近乎失明。如果用我以前对于建筑学的定义，这里一半以上的人很难说是在做建筑。当我看到一个人把几只金属棍从不知是什么的黄色油状物里提出来，并且向别人介绍这如何联系到他对于时间的理解时，简直绝望了。好吧，做我感兴趣的，做我能理解的，慢慢试着了解其他人的思维（关于我的巴特雷特作品，在作品介绍里有详细叙述）。

巴特雷特的课程是一个导入，当我完成学业时，我知道一场真正的旅行才刚刚开始。吹动着风帆的是我的兴趣，这些兴趣凝结着人生的记忆，我不知它将把我带向何处，但我感觉到我相信自己的力量。在此，如果有些武断地说，我的记忆汇聚成这三个方向：人工世界中的现代景观；当代文化中的视觉体验；被科技扁平化的智息社会中的材料物性。

Platforms, high and low, I was climbing from the lower one to the higher one. Skin scratches the rough concrete surfaces; Jumping from the higher to the lower, with a little bit fear, convulsed the feet aching. Running back home, navigating towards one of the residential unit among many "matchboxes", I climbed up to the third floor following the rusty iron handrails of staircase, knocked the door open, and gulped a large bowl of green bean soup made by grandmother, then went to the balcony, said hello to my dear hen raised in the balcony, which was enveloped by glass on three sides. I knew that my neighbors had more interesting animals on their balconies: a fox on the first floor, an eagle on the fifth. But most people converted their balconies into storage rooms. That year, I was nine, end of the 1980s, China.

Studying architecture is an interesting thing, the objects is so commonplace that you are living in them everyday almost without noticing, until one day the interest became your career. The experiences in pre-professional period became a special part of an architect's spatial experience. The longer architecture training, the virgin memories of space more frequently surge in your mind. These memories are confronted with the professional knowledge, which are extracted from the experiences, offering the base point of rethinking your profession.

As a professional education, it seems the architects' education tends to depart you from all of these. You know that the low and high platform are Soviet-style rostrum for ritual; you know those uniformed "matchboxes" are collective residential units in Soviet-style planning for massive manufacturing and adapting to the cold climate in north China, you know the reasonable depth and length of these buildings. However, in my previous education, the experiences were not important, what you leant is how to translate vivid colorful space into plans, elevations, sections, which may not mean anything. Professional knowledge meant split, I rejected this split, so I came to London, felling into another was. In the architecture spectrum of Bartlett School, I was almost blind. In my previous definition of architecture, more than half of people here are not doing architecture. When I saw a guy lifting several pieces of metal sticks out from some unknown yellow oil-like thing, and introduced to others that it was about his understanding of time, I was just desperate. Well, I do what I am interested in, do what I can understand now, and try to understand others' ideas gradually. (About my Bartlett works, there are detailed descriptions in the works part.)

Bartlett course is just an introduction. When I completed the study, I knew that a real voyage just started, my interest, which condensed my life memories, blows the sail. I don't know where it would take me to, but I can feel I believe my strength. Here, if I say it arbitrarily, my memories converge into these three directions: modern Landscape in the artificial world, visional experience in contemporary culture, materiality in the fleeting society which has been flattened by technology.

布莱克浦轨道娱乐

城市

布莱克浦是一座城市，在海的背景下展开，海，海滩，海岸散步大道，海堤，有轨电车轨道，滨海大道，一层一层从海面推向城市。与海岸线相对是公共建筑集群，经济，社会价值的富集地带。再伸向内陆就是漫无边际的居住区。

布莱克浦是两座城市，居民的布莱克浦和游客的布莱克浦。它们彼此参差交界却又壁垒分明。日常生活的忙碌和梦幻世界的沉醉定义了各自的居民。

元素

1．旅馆

在两座城市的交界处，旅游经济驱动游客的城市扩张它的边界，在居民的城市刚刚退却的地方，游客城市的经济模式与居民城市的建筑形态在这里达到平衡，形成这样的场景：数十个一个模子倒出来般的二层小住宅像装配线上的产品一样沿道路整齐排列。小房子底层的玻璃凸窗里挂着“Vacancies”“No vacancies”这样的显示有房或者客满的牌子。牌子是整体发光或者霓虹灯的，以使夜间投宿的旅客不用进入旅馆询问就能得到信息。

这样的牌子，在如此的“工业化”布局的街区中，使互相类似的布莱克浦的家庭旅馆完成了一个变化：从提供人与人交流场所的建筑空间转化为直接与人交流的机器。在这样的意义上，布莱克浦家庭旅馆就是居住的机器：“住舱”。机器智能探知是否有空位，以屏幕加以显示，达到信息的反馈，提供给人以判断选择的可能性，由此产生人机交流，这也是布莱克浦的家庭旅馆的运行机理，所不同的是家庭旅馆的机械化程度不够，还需要人来辅助进行开门，收银，这样一个机械臂，一台自动刷卡机就能完成的简单动作。旅馆，一个典型儒勒·凡尔纳式的未来机械。

2．轨道

两侧是高高的电线杆，地面上是深深凹陷的四条铁轨。海堤与海边大道就此隔离开来，布莱克浦的有轨电车轨道就是这样在大海与城市之间的绵延十数公里的条形地带。

布莱克浦的有轨电车轨道是一种自然现象，它贯穿整个城市，从看不到的城市外的起点到看不到的城市外的终点。

布莱克浦的有轨电车轨道是一项城市景观计划。它以地面和空中的钢铁线昭示着这个城市的方向感。它在城市与海洋的交界处增加了一个层次，给这样的过渡关系更丰富的可能性。

布莱克浦的有轨电车轨道当然是一个大的机械系统的一部分。当电车隆隆开过的时候，这样作为轨道的原意被一丝不苟的履行着。

3．运动体

布莱克浦有轨电车的塔式接触器擦在电缆上偶尔迸出火化；布莱克浦塔里笼子样的电梯被钢缆拽着嘎吱的上下；大大小小的娱乐机器蠢笨的蠕动，震得上面的游客屁股麻木。

作为一座娱乐城市，布莱克浦具有一种独特的运动的品质。往复式的机械运动显然与乐趣无缘，作为娱乐的机器试图在物理的规范下寻求突破，一种非规则性。在这里，实质完全控制，表面运动状态疯狂而无法预知。娱乐机器的设计制造与追求效率与精确控制的普通机械，在诉求与手段上产生强烈的分野。

城市

"比现实更疯狂"才具有娱乐的噱头，而作为一座普通的城市自然没有试验建筑的野心。娱乐的目的和营建的思维惯性使布莱克浦产生奇妙的混合。意义与外表发生多重的错位，于是布莱克浦成为一位不自觉却绝对激进的探索者。探索的不自觉保证了这样的过程的真诚性，没有一丝的矫情——这样的探索完全根据需要。

建筑与机械的边界被打破，既有的概念不足以描述事实。在老工业时代建筑统治的布莱克浦，城市却透露出一种戏虐的未来性。

设计

将整个城市强力拉入的舞台，自然与城市的分界，可供戏弄的老工业遗脉，业已铺就的运动平台和供能系统。设计在电车轨道上展开。

这是一组娱乐的机器，秉承布莱克浦机械娱乐的传统。简单笨拙的齿轮，轴承模拟着自然界生物的运动，粒子的运动，并把机械的运动方式极端化。他们是机械的杂耍，娱乐的空间。当数十上百个这样的物体在轨道上运行，整个布莱克浦将成为机器与人类的乐园。

这也是一项可发展、可生长的工程，新的机器不断地设计出来。

当机器的寿命终结，会有一部分被选择出来，以其自身的可动结构适应海岸的悬崖，成为海边眺望与休息的平台。机械转生为建筑。

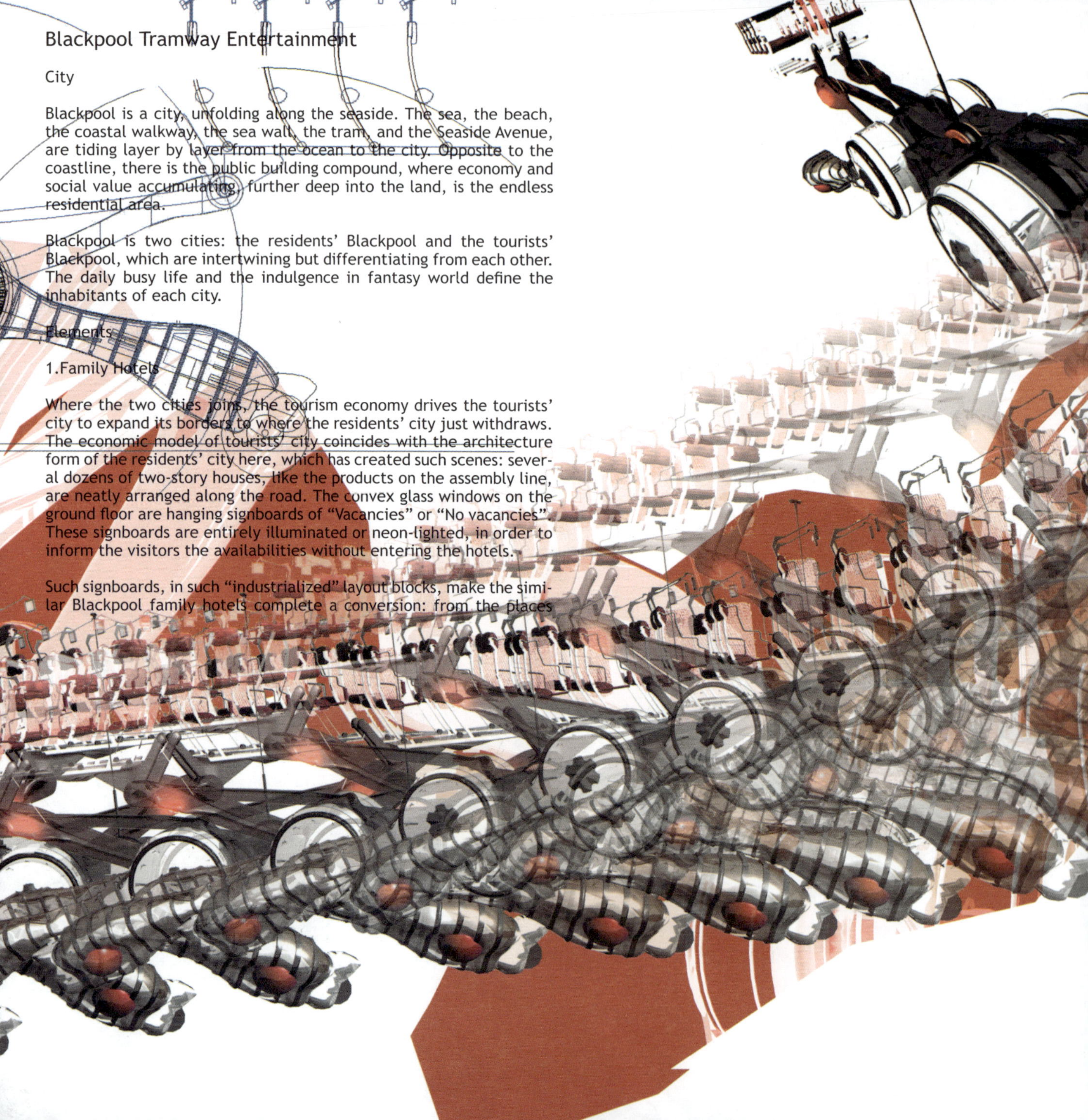

Blackpool Tramway Entertainment

City

Blackpool is a city, unfolding along the seaside. The sea, the beach, the coastal walkway, the sea wall, the tram, and the Seaside Avenue, are tiding layer by layer from the ocean to the city. Opposite to the coastline, there is the public building compound, where economy and social value accumulating, further deep into the land, is the endless residential area.

Blackpool is two cities: the residents' Blackpool and the tourists' Blackpool, which are intertwining but differentiating from each other. The daily busy life and the indulgence in fantasy world define the inhabitants of each city.

Elements

1.Family Hotels

Where the two cities joins, the tourism economy drives the tourists' city to expand its borders to where the residents' city just withdraws. The economic model of tourists' city coincides with the architecture form of the residents' city here, which has created such scenes: several dozens of two-story houses, like the products on the assembly line, are neatly arranged along the road. The convex glass windows on the ground floor are hanging signboards of "Vacancies" or "No vacancies". These signboards are entirely illuminated or neon-lighted, in order to inform the visitors the availabilities without entering the hotels.

Such signboards, in such "industrialized" layout blocks, make the similar Blackpool family hotels complete a conversion: from the places

accommodating people's communication into the machines directly communicate with people. In this sense, Blackpool family hotels are machines for living: "living capsules". The machines are intelligent to address vacancies, display the information by screens, in order to let people make choices, which produce an interaction between machines and men. This is also the operation model of Blackpool family hotels, the only difference is, the mechanization level of family hotels are not sufficient enough, which needs men for opening the doors, reception and cash dealing. In the mechanical version, a mechanical arm, an automatic

card swipe machine will be able to complete all these works as simple actions. Mechanical hotel, a typical Jules Verne's futuristic machinery.

2. Tracks

Four tracks depressed deeply on the ground; tall electricity poles on both sides of the tramway, The sea wall and the seaside road separates here. In this way, the tramway in Blackpool forms a 10-km-long strip between city and the sea.

The Blackpool tramway is a nature phenomenon, they runs through the city, from the unseen starting point outside of the city to the unseen end beyond the city.

The tramway in Blackpool is an urban landscape project. It illustrates

the orientation of the city by the steel lines on the ground and in the air, which adds a layer to the juncture of the city and the sea, offering richer possibilities for the transition, and a venue for life stories.

Blackpool tramway is certainly a part of a mechanical system. When the thundering trams pass by, the original intention of the tracks is implemented precisely.

3. Moving Objects

The tower-shape contactors of Blackpool trams occasionally spark on the cables; the cage-like lift cabins in Blackpool Tower are stretched by the steel cables, creaking, up and down; the big and small entertainment machines clumsily squirm, shaking tourists' asses numb.

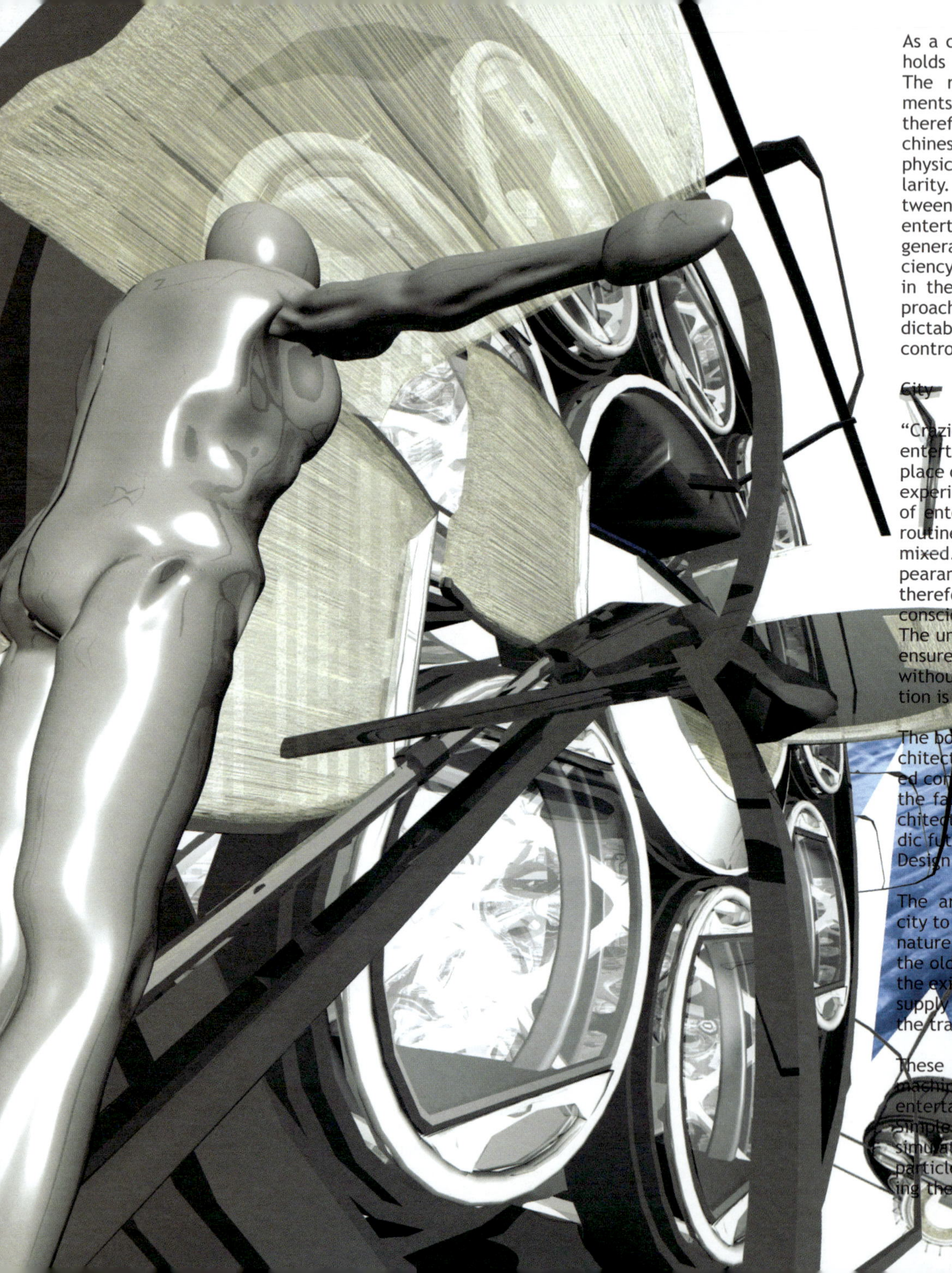

As a city for entertainment, Blackpool holds a unique quality of movement. The reciprocating mechanical movements have nothing to do with fun, therefore, these entertainment machines have to break rules within their physical tolerances, a kind of non-regularity. There are great distinctions between the design and manufacture of entertainment machines and those of general machines which pursues efficiency and predictable precise control, in the respects of intentions and approaches. Here, underneath the unpredictable crazy moving superficies, a full control dominates.

City

"Crazier than reality" produces more entertainment tricks. As a commonplace city, Blackpool has no ambition of experimental architecture. The purpose of entertainment and the construction routines make Blackpool wonderfully mixed. The significance and the appearance are multiply interlaced, therefore, Blackpool has become a non-conscious but absolute radical explorer. The unconsciousness of the exploration ensures the faith during this process, without any factitiousness- the exploration is completely triggered by needs.

The boundary between machine and architecture has been broken. The existed concepts are insufficient to describe the facts. In the old industrial age architecture dominated Blackpool, a ludic futuristic smell flouring in the air.

Design

The arena which hijacked the whole city to join, the boundary separates the nature and the city, the residual vein of the old industrial for entertainment, the existed moving platform and energy supply system. Design is launched on the tram tracks.

These are a group of entertainment machines, carrying on the mechanical entertainment tradition of Blackpool. Simple clumsy gears and shafts are simulating movements of organism and particle in the nature, and even making these mechanical movements more

extreme. They are mechanical acrobatics and places of entertainment. When tens of hundreds of such objects are moving on the tracks, the entire Blackpool will be a paradise for machine and human.

The project is developing and growing, new machines will be designed constantly.

When these machines end their service life, some of them will be selected to be seaside observation platforms for sightseeing and rest, by adapting the coastal cliffs with their flexible structures.

Machines are reincarnated to be architecture in their eternal life.

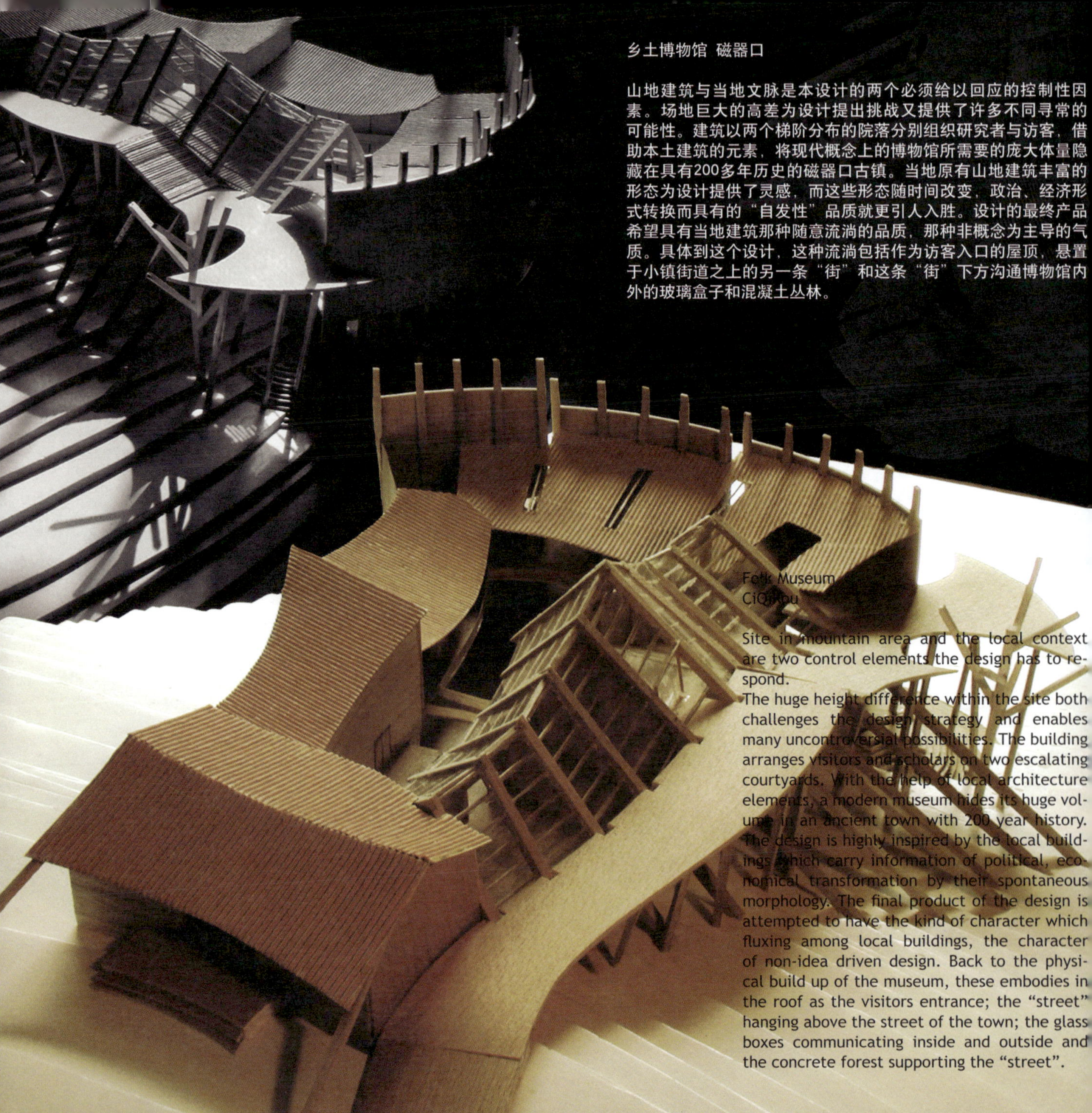

乡土博物馆 磁器口

山地建筑与当地文脉是本设计的两个必须给以回应的控制性因素。场地巨大的高差为设计提出挑战又提供了许多不同寻常的可能性。建筑以两个梯阶分布的院落分别组织研究者与访客，借助本土建筑的元素，将现代概念上的博物馆所需要的庞大体量隐藏在具有200多年历史的磁器口古镇。当地原有山地建筑丰富的形态为设计提供了灵感，而这些形态随时间改变，政治、经济形式转换而具有的“自发性”品质就更引人入胜。设计的最终产品希望具有当地建筑那种随意流淌的品质，那种非概念为主导的气质。具体到这个设计，这种流淌包括作为访客入口的屋顶，悬置于小镇街道之上的另一条“街”和这条“街”下方沟通博物馆内外的玻璃盒子和混凝土丛林。

Folk Museum CiQiKou

Site in mountain area and the local context are two control elements the design has to respond.

The huge height difference within the site both challenges the design strategy and enables many uncontroversial possibilities. The building arranges visitors and scholars on two escalating courtyards. With the help of local architecture elements, a modern museum hides its huge volume in an ancient town with 200 year history. The design is highly inspired by the local buildings which carry information of political, economical transformation by their spontaneous morphology. The final product of the design is attempted to have the kind of character which fluxing among local buildings, the character of non-idea driven design. Back to the physical build up of the museum, these embodies in the roof as the visitors entrance; the “street” hanging above the street of the town; the glass boxes communicating inside and outside and the concrete forest supporting the “street”.

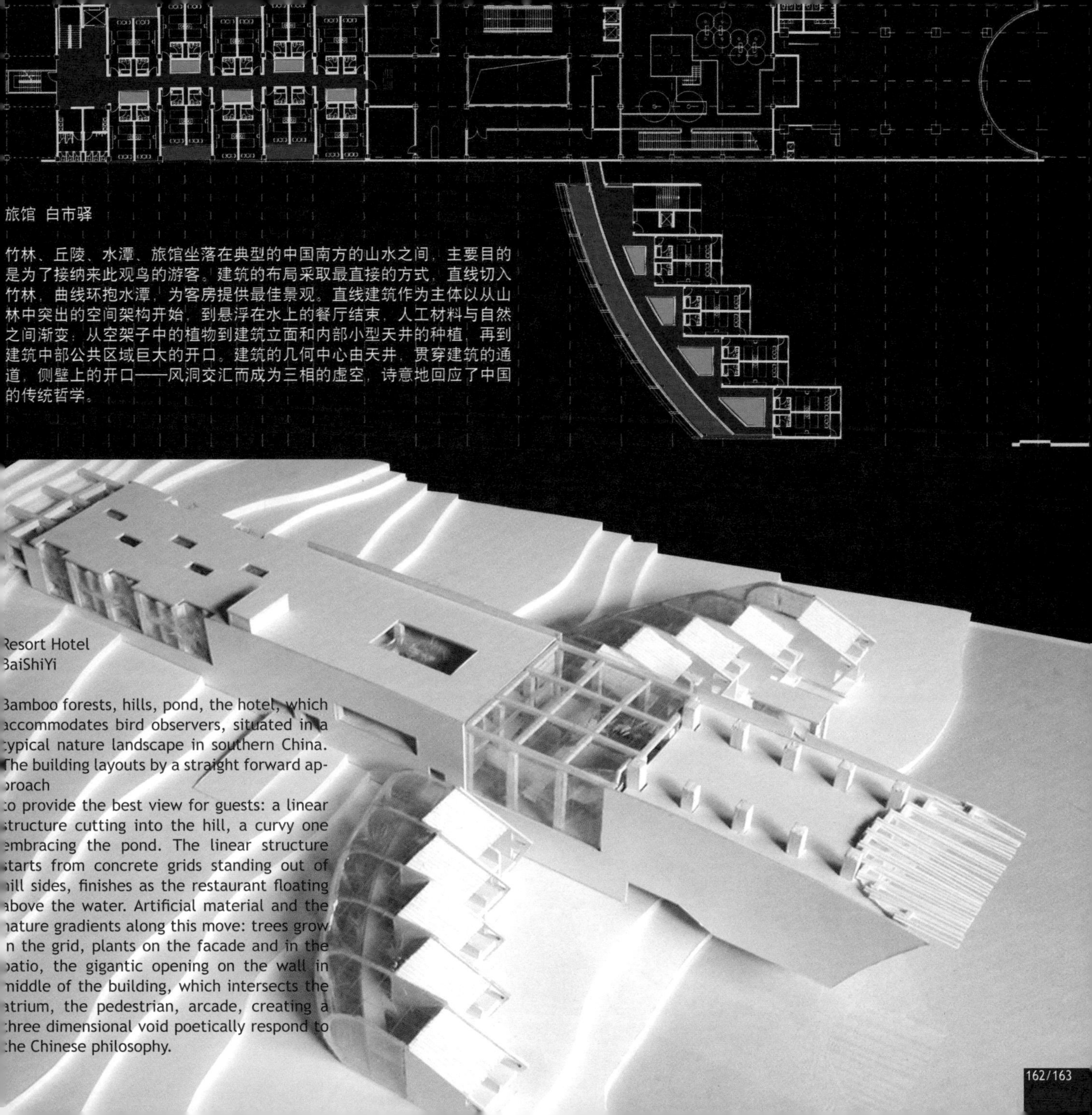

旅馆 白市驿

竹林、丘陵、水潭、旅馆坐落在典型的中国南方的山水之间，主要目的是为了接纳来此观鸟的游客。建筑的布局采取最直接的方式，直线切入竹林，曲线环抱水潭，为客房提供最佳景观。直线建筑作为主体以从山林中突出的空间架构开始，到悬浮在水上的餐厅结束，人工材料与自然之间渐变：从空架子中的植物到建筑立面和内部小型天井的种植，再到建筑中部公共区域巨大的开口。建筑的几何中心由天井，贯穿建筑的通道，侧壁上的开口——风洞交汇而成为三相的虚空，诗意地回应了中国的传统哲学。

Resort Hotel
BaiShiYi

Bamboo forests, hills, pond, the hotel, which accommodates bird observers, situated in a typical nature landscape in southern China. The building layouts by a straight forward approach
to provide the best view for guests: a linear structure cutting into the hill, a curvy one embracing the pond. The linear structure starts from concrete grids standing out of hill sides, finishes as the restaurant floating above the water. Artificial material and the nature gradients along this move: trees grow in the grid, plants on the facade and in the patio, the gigantic opening on the wall in middle of the building, which intersects the atrium, the pedestrian, arcade, creating a three dimensional void poetically respond to the Chinese philosophy.

当代艺术博物馆 重庆南山

现代人造景观同时作为人造物与自然场景的内在张力令人着迷，作为对于地景建筑的持久兴趣，重庆艺术家村画廊试图以地形被塑造的方式创造建筑，完成从地质学语言到建筑学语言之间的一次转译。山地即作为场地，也作为建构方式上的比照，既成为建筑的背景，也成为建筑嬉戏的伙伴。

Museum of Contemporary Art South Mountain Chongqing

The internal tense of landscape, both as a part of nature and as an artificial product in the modern world, is intriguing. Responds to the constant interests in landscape, Museum of Contemporary Art in Chongqing attempts to generate building in the way the contour line being generated, thus to accomplish a translation from geological language to architectural language. Mountain is both the site and a reference point tectonically. Both back ground and playmate.

旋转塔楼 都市

作为另一种自然介入建筑的方式，旋转塔楼以建筑几何形与观看点之间的互动模糊了其几何形态。建筑在垂直方向上分为六个单元，每个单元各自包含十个互相扭转一定角度的楼层，各单元相互镜像，叠置，这样构成旋转一回转的秩序。建筑的边界呈现出不确定的，不稳定的状态。因观察角度、位置的不同，边界线表现为直线、折线、弧线、曲面。

Twisting tower Metropolis

As another way nature been evolved, the twisting tower blurs its geometric boundary by the interaction between building geometry and the viewing point. It divided into 6 blocks in which each floor contains 10 twisting floors. Blocks mirrored in pairs and stacked onto each other, thus, the edge of the building came into an uncertain, unfixable status. By different angle people stand, the edge of building appears as a straight line, zigzags and curves.

P Short stay
15 minutes free
Arrivals
Zone D
ick up area
and
Disabled
parking
Cars
Coaches
At any time
No loading
At any time

帝诺斯
帕帕蒂米卓
Dinos
Papadimitriou

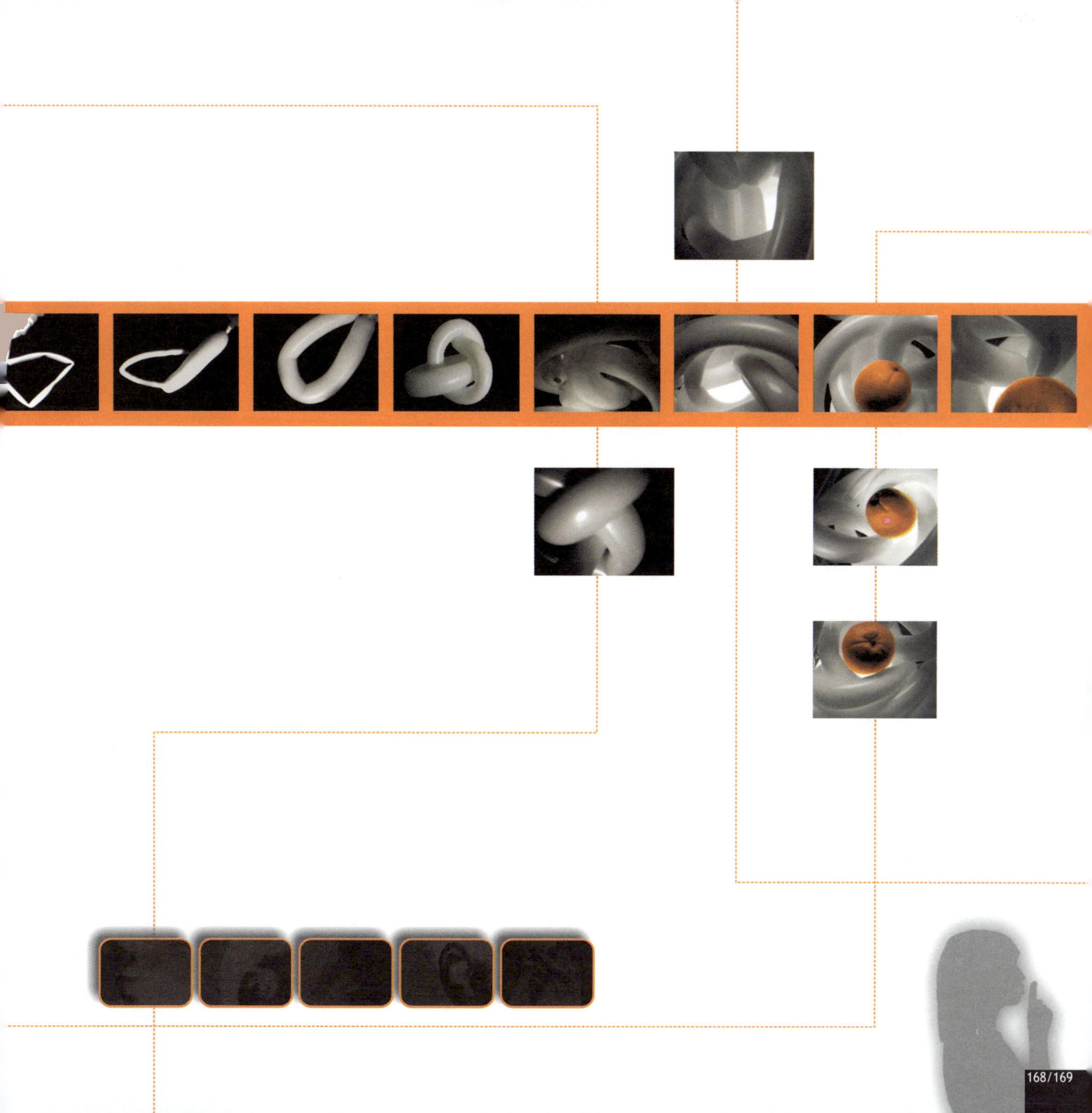

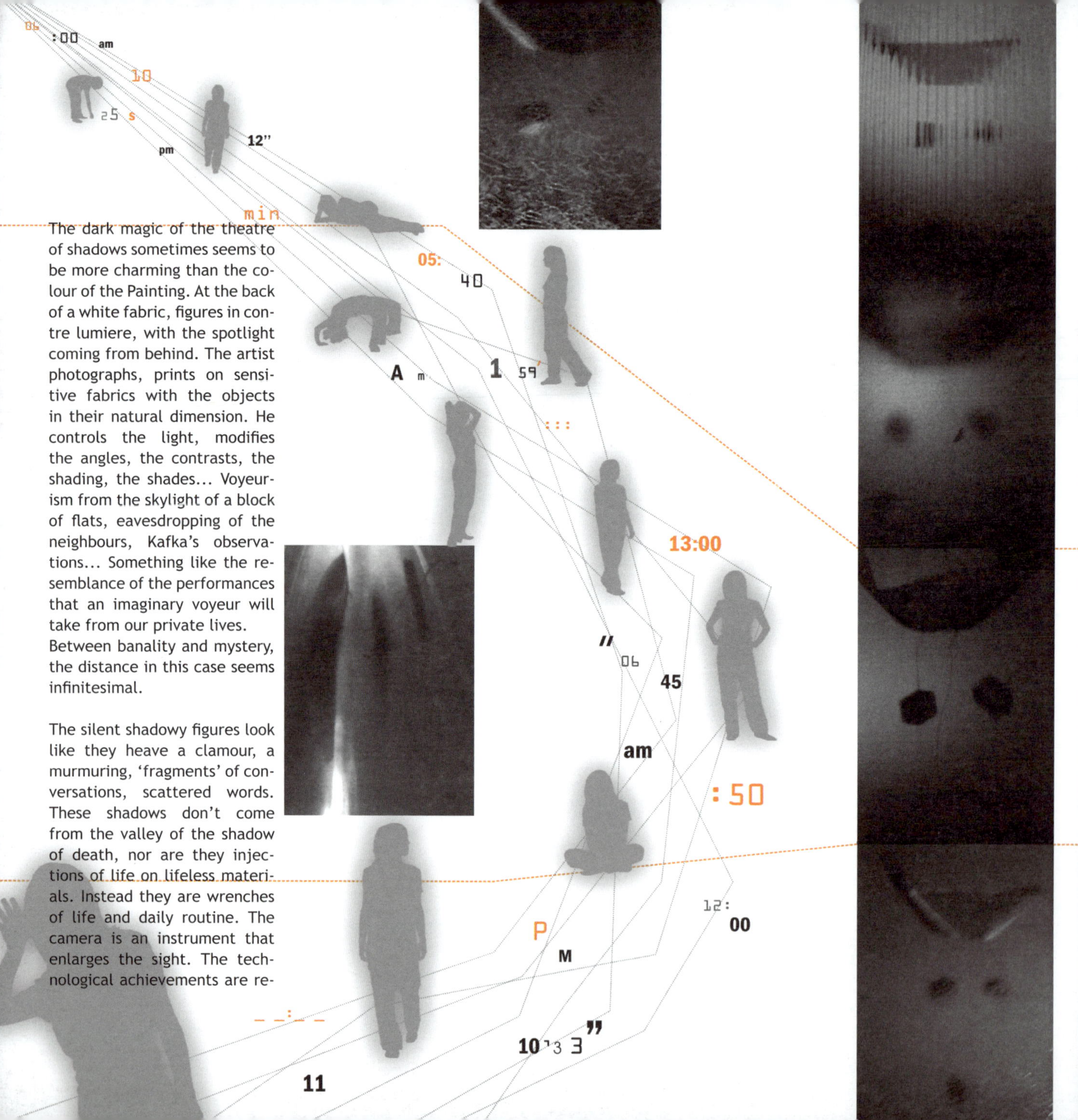

The dark magic of the theatre of shadows sometimes seems to be more charming than the colour of the Painting. At the back of a white fabric, figures in contre lumiere, with the spotlight coming from behind. The artist photographs, prints on sensitive fabrics with the objects in their natural dimension. He controls the light, modifies the angles, the contrasts, the shading, the shades... Voyeurism from the skylight of a block of flats, eavesdropping of the neighbours, Kafka's observations... Something like the resemblance of the performances that an imaginary voyeur will take from our private lives. Between banality and mystery, the distance in this case seems infinitesimal.

The silent shadowy figures look like they heave a clamour, a murmuring, 'fragments' of conversations, scattered words. These shadows don't come from the valley of the shadow of death, nor are they injections of life on lifeless materials. Instead they are wrenches of life and daily routine. The camera is an instrument that enlarges the sight. The technological achievements are re-

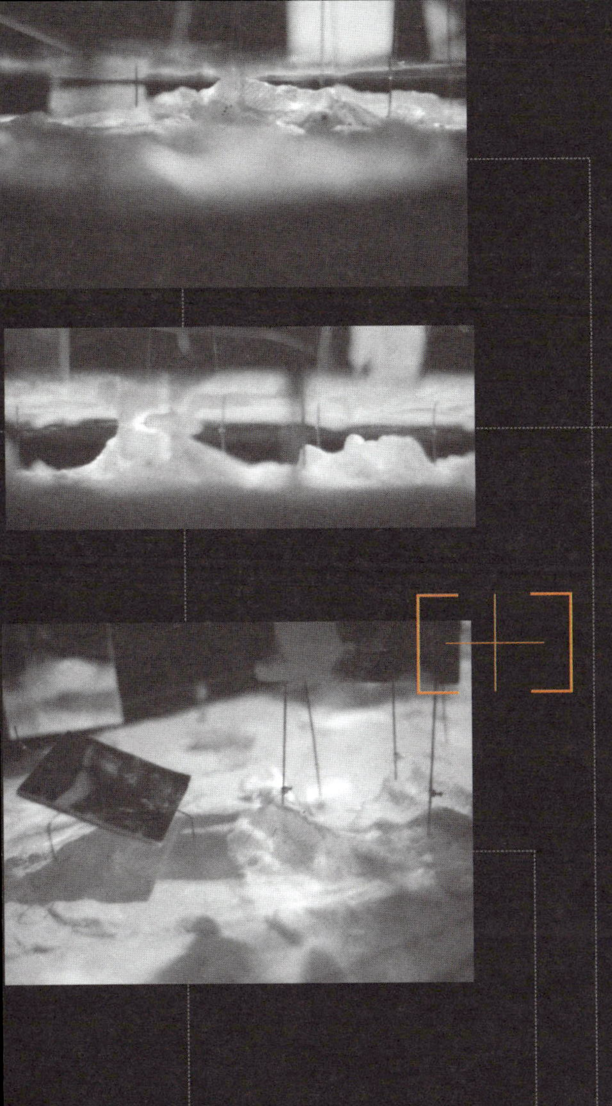

garded as "additional parts" used to reinforce the power of the individual. Benjamin sees the camera as a surgical instrument that penetrates the human body; therefore by looking at the human body in pieces, it can penetrate even deeper than its reality. Moreover, the credibility of the snapshot, the mechanical means, no longer exists.

In fact, it is about a theatre of shadows that is slaughtered elliptical, semi-complete. The spectator's brain complements what is missing. There is a short-sighted impression of the picture.

剧场阴影造成的黑暗魔力有时看起来比油画的色彩更加迷人。在白色织物的背景中，人物身处从后部射来的聚光灯追光中。艺术家的照片以真人大小的天然尺度印在感光的织物上。他控制灯光，调整角度、对比度、阴影、遮光物……来自于公寓天窗的偷窥者，偷听着邻居的对话，带着卡夫卡式的观察……这类似于表演的虚构偷窥者将削弱我们的私生活。

平庸与神秘之间的距离在这儿似乎无限小了。

沉寂的影子像是在低语、喃呢、像是对话的"片断"，只言片语。这些影子并非来自死亡之影的深谷，他们也非无生命材料的生命注射剂。相反地，他们是生活和日常惯例的扭曲。照相机是一个可以放大视野的工具。技术成就被视为增强个人力量的"附加部分"。本杰明把照相机视为一个穿透人类身体的外科工具；通过对人类身体碎片式的观察，它可以洞穿到超越其本体的更深处。此外，快照的可信度，机械式的手段，不复存在。

事实上，这是一个形状残破的、半完成的椭圆形影子剧院。观众的大脑补充了丢失的部分。这是一种近视的景象。

Penetration
Distorted time
Shrink
Real time
On/
S t r et c h

Spontaneous
Perceptive
10:09:45
Funny
06:19:37
06:36:14
Fragile

City Campings refer to those people who are keen on camping and nomadic life. However, up until now, they only had the chance to put up their tents in the country and not in the big urban centres, which they would like to visit, bringing along their personal space. Therefore, the idea of the collapsible rooms came up as a modern and urban version of the tent. In reality, it is a mobilized accommodation system that one can carry with him by adjusting it to the back of his car, to the back of a train, to the luggage space of an airplane. The way in which they unfold, demands the existence of special installations. These can be found in the City Campings which can also supply the collapsible rooms with electricity, water, even internet access. In general, City Campings can provide to their visitors a variety of facilities as well as places to exercise any leisure time activities. Specifically, City Campings London is accompanied by a parallel unit of baths.

“城市营地”来源于那些热衷于宿营和游牧生活的人。然而，迄今为止，他们只能在农村，而不是在城市中心搭建帐篷，带上他们的私人空间造访城市。因此，可折叠空间的概念作为一个现代和城市化版本的帐篷被提出。实际上，一个可移动性的居住体系可以被随身携带，通过调整可以适应汽车的后部，火车的后部，以及飞机上的行李空间。由于它们展开的方式，要求一些特殊装置的存在。这些可以在同时提供拥有电力，水源，甚至网络接口的可折叠空间的城市营地找到。一般来说，城市营地可以为他们的游客提供各种各样的设施和进行休闲活动的场所。特别的是，伦敦的城市营地甚至配备了洗浴单元。

1. restaurant
2. storage space
3. Wc's
4. reception
5. city-campings
5.
4.
2.
1.
3.
Right
Front
Back
Index

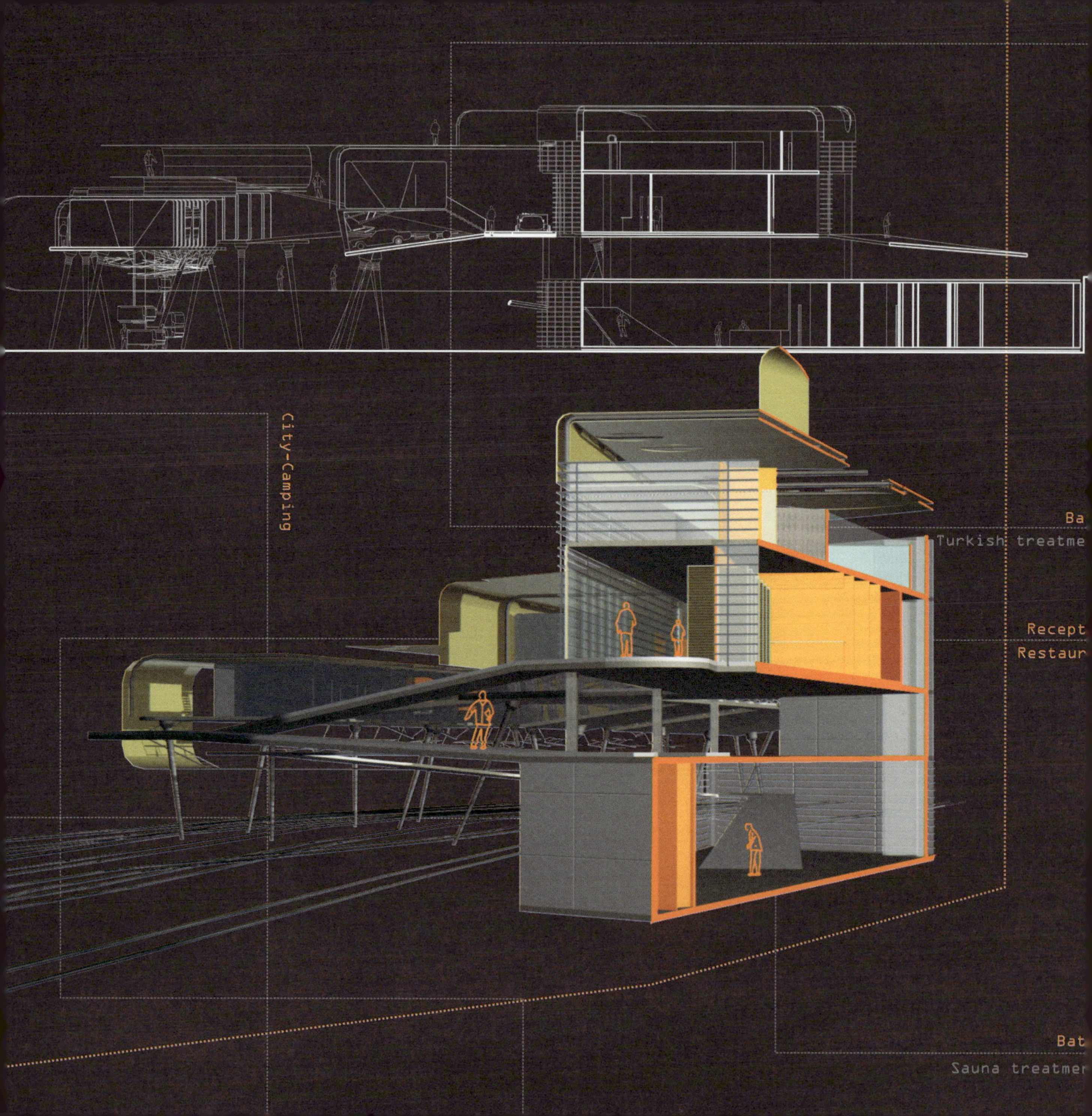
City-Camping
Ba
Turkish treatme
Recept
Restaur
Bat
Sauna treatmer

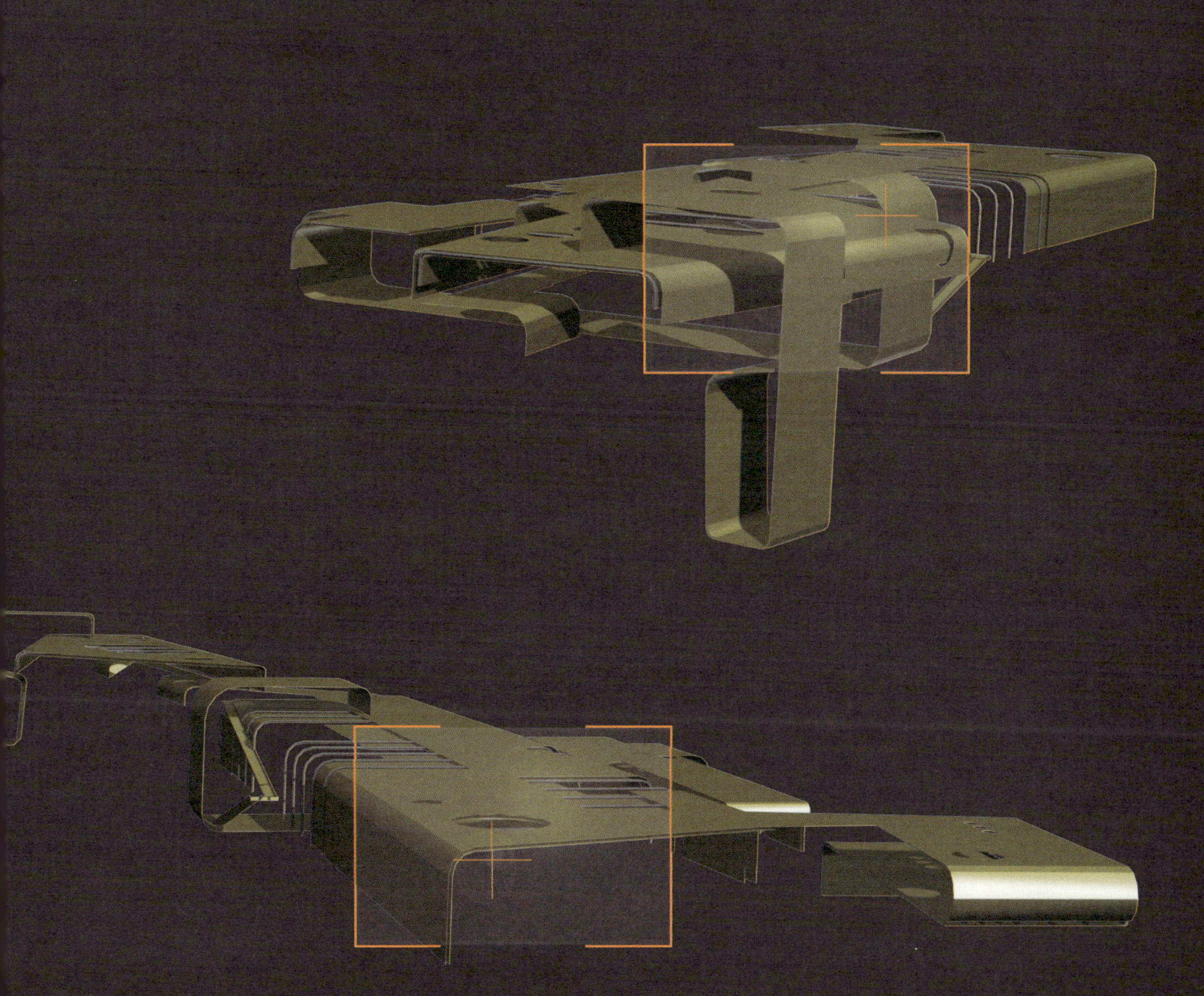

he cover not only protects city-campings, but also creates another one layer where activities can take place...

页盖不仅仅保护了城市营地，还创造了各种活动可以发生的另一个层次。

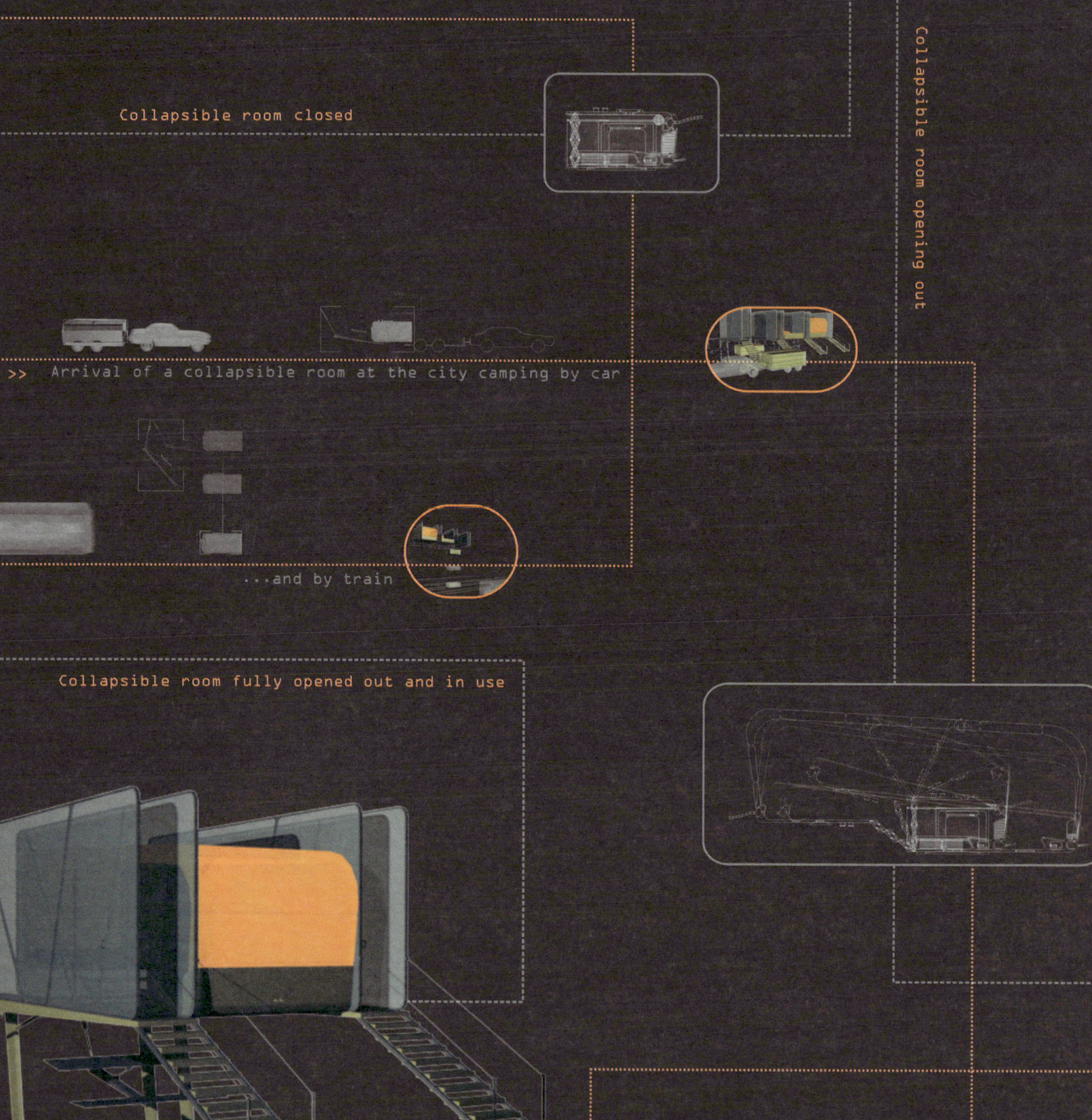

Collapsible room closed
Collapsible room opening out
>> Arrival of a collapsible room at the city camping by car
...and by train
Collapsible room fully opened out and in use

● Standard
○ Extra
▬ Not available

		Models	
		CR-S	CR-X
Dimensions	Width/Depth/Heigth (Folded)*	300/600/40	600/600/40
	Width/Depth/Height (Unfolded)*	300/300/12	300/300/24
	Volume (folded)**	72	144
	Volume (unfolded)**	10.8	21.6
	Set up time***	60	90
	Weight****	35	70
Equipment	Bedroom	●	●
	Bathroom	●	●
	Seating area	●	●
	Kitchen	●	●
	Laundry/clothing iron	○	○
	Spa	▬	○
	Gym	▬	○
	Reading area	○	○
	Internet access	○	○
	Mini bar	○	○
	T.V/DVD/Hi-Fi	○	○
	Disabled	○	○

*cm **m^3 ***sec ****kgr

opposed to the impersonal character of the hotel rooms, the Collapsible rooms be adjusted to the demands of the user transforming it into a ‘home away m home’. It provides the convenience that someone may need: from bed and hroom to spa and gym...

对立于酒店房间的非个人化特征，可折叠的房间可以根据使用者的需要被加以调整，变形成“远离家的家”。它提供了人们可能需求的便利：从床和卫生间到温泉疗养和健身 ……

6
½ m
(M1, M20)
M 25 D'ford Crossing
Watford
London M 11

提奥
萨隆托格鲁
Theo
Sarantoglou

THEO SARANTOGLOU
232 av. W Churchill, 1180 Brussels, Belgium E-MAIL: Theosarantoglou@gmail.com

Academic Background
2002-2003 Post professionnal Masters in Architectural Design, The Bartlett School of Architecture, Prof. P. Cook, London.
1994-1999 Diploma in Architecture, with High Distinction, ISAI-Victor Horta, Brussels, Belgium.
Professional Experience
2008- ASY. EU : Director
2006-2007 ASYMPTOTE, New York, USA: Senior Project Architect
2004-2006 FUTURE SYSTEMS, London, UK: Project Architect
1999-2002 PDDA, Brussels, Belgium: Architectural registration - Project Architect
Workshops and Teaching
2007-2009 ARCHITECTURAL ASSOCIATION / Research cluster organiser with A. Dempsey and Y. Obuchi.
2007 >GSD HARVARD / Masters studio / with Hani Rashid and Chris Johnson
2007 >GSAPP - COLUMBIA / Advanced studio 6 / Spring semester with Hani Rashid and Alex Pincus
2004-2005 >ISAI V. HORTA / Workshop leader, 2 weeks workshop, 2000, 2001, 2002, 2005.
2004 >BARTLETT SCHOOL OF ARCHITECTURE / external Critic
2003 >ARCHITECTURAL ASSOCIATION / External examiner, intermediate unit 8, Y. Obuchi with A. Dempsey.
Personal Projects
2004 >Kuwait Bowling Club, Hotel and Office Towers / Kuwait
2004 Finalist / with Q. Alshaheen, Mike Mitchell, Federico Celoni, Volkan Alkanoglu, Yan Pan.
2003 >Hellinikon Metropolitan Park / Athens, Greece.
2001 Competition / with Mike Mitchell, Dora Sweijd, Volkan Alkanoglu, Yan Pan, Ricardo Ostos.
>Urban Design and Nursery / Brussels, Belgium
Public competition: 1st prize
>Case study for an Asphalt Square / Brussels, Belgium

提奥・萨隆托格鲁

地址：232 av. W Churchill, 1180 布鲁塞尔，比利时 E-MAIL：Theosarantoglou@gmail.com

学术背景

2002–2003 建筑设计硕士，伦敦巴特雷特建筑学院，在彼得•库克的督导之下

1994–1999 比利时，布鲁赛尔， ISAI–Victor Horta大学优异毕业的职业学位执业文凭，优秀毕业

职业经验：

2008至今 渐近线，欧洲：负责人

2006–2007 渐近线，美国，纽约：高级项目建筑师

2004–2006 未来系统，英国，伦敦：项目建筑师

1999–2002 PDDA，比利时，布鲁塞尔：注册项目建筑师

执教经验

2007–2009 伦敦建筑联盟，聚落研究，与A•丹普赛与Y•奥布什一同组织

2007 >哈佛GDS，硕士工作室，与哈尼•莱希德，克瑞斯•约翰逊合作

2007 >哥伦比亚GSAPP，高级工作室6，春季学期，与哈尼•莱希德和艾里克斯•平克斯合作

2004–2005 >ISAI V. HORTA，为期2周的工作室，2000，2001，2002，2005

2004 >巴特雷特建筑学院，外部评审

2003 >伦敦建筑联盟，外部主审，三年级第8组，Y•奥布什与A•丹普赛合作

个人项目

2004 >科威特保龄球俱乐部、旅馆与办公塔楼，科威特
进入最后一轮（与大卫•塔基曼，麦克•米歇尔，沃肯•奥卡纳格鲁，潘岩合作）

2004 >海棱尼孔公园都市，雅典，希腊
UIA 国际创意竞赛（与大卫•塔基曼，麦克•米歇尔，沃肯•奥卡纳格鲁，潘岩，里卡多•奥斯图斯合作）

2003 >比利时，布鲁塞尔，城市设计方案和幼儿园设计，
公开竞赛一等奖

2001 >比利时，布鲁塞尔，沥青广场的景观设计，个案研究和设计

URBAN PATCHWORKS explores the idea of the individual and his representation as an essential and enriching part of a collective representation such as a collective housing scheme. It is a collage of prefabricated inhabitable units of variable size and program (single storied, double storied, three storied, longitudinal, offices spaces and shops). The housing block is enriched in its evolution by the diversity of family types, cultures, generations and social background. The collective housing is not anymore composed by the repetition of a normative inhabitable unit (Unite d' habitation of Le Corbusier in Marseille) with a living space optimised for an average proportioned man (modulor). The patchwork façade gives a clarified reading of the different social entities that compose the scheme (single-parent, single, couple, student, family, office or shopping)

“城市拼贴工程”探索了个体的概念，并将其表现作为整体表征的首要和丰富性要素，例如集合住宅方案。这是不同大小和计划安排的预制居住单元的拼贴（单层、双层、三层、纵向的、办公空间和商店）。住宅街区被家庭种类、文化、世代和社会背景的多样性带来的演进所丰富。集合住宅的构成不再重复标准化居住单元（勒·柯布西耶在马赛的居住单元）和常规比例人（模数）所优化的起居空间。拼贴立面对构成方案的不同社会实体（单亲、单身、伴侣、学生、家庭、办公或购物）给出了一个清晰的解读。

Elevation

东立面

平面

nsion process

生成过程

different Units can expand through time to fit changes in occupa- There are three expansion modules: 7m, 14m, 21m. The interior e is adjustable to peoples needs.

不同的单元可以随着时间的发展不断扩展以适应在居住过程中的变化，这里有3种扩展模式：$7m^2$，$14m^2$，$21m^2$。室内空间可以根据人们的需要进行调节。

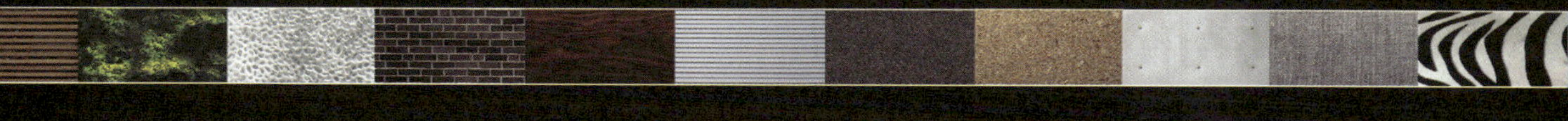

inhabitants can also choose the degree of openness of their flats ards public space. The façade is composed by a glazed skin and in- able shutters that can be rearranged periodically by the inhabitants rding to the seasons or their personal tastes. The inhabitants can ose between a set of materials to customize their façade. A trans- nt module allows an unlimited choice of facade wallpapers. The is a collective garden where people can meet for a promenade or nise picnics and barbecues.

居住者还可以选择其公寓面向公共空间的开敞程度。立面是由光滑的表皮和可倾斜的百叶窗组成，可以由居住者根据季节或个人口味进行周期性的再布置。住户可以在一系列材料中进行选择，来定制他们的立面。半透明的模式允许立面墙纸有无穷的选择。屋顶是一个集合式的花园，人们可以在这儿碰头去散步，或是组织野餐和烧烤。

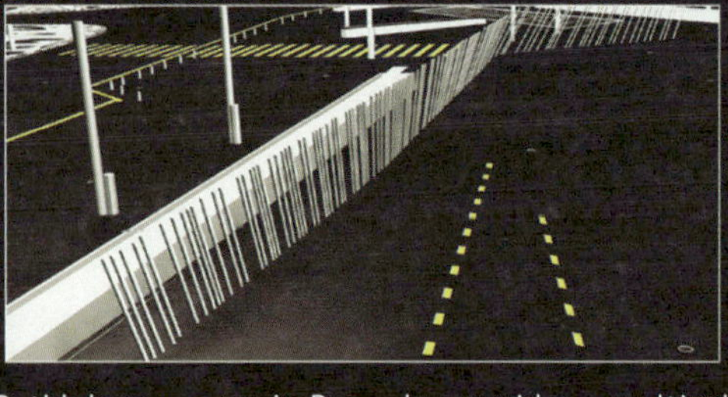

Bethlehem square in Brussels, provides a multicultural specificity, principally inhabitants of Mediterranean origin with a very intense outdoors lifestyle which the current arrangement does not take into account. A specificity of the square is its geographic position in Brussels between the summit and the hollow of its valley. Its space is generated by the convergence of several streets linking high and low points, giving it a particular topography, an urban clearing. Our position was to merge the sidewalk, the street, the parking and the square into a continuous asphalt surface to support a Mediterranean lifestyle that doesn't recognise these distinctions (i.e. in Sienna in Italy, roads and sidewalks and squares are continuous). This smooth surface is topographically informed to optimise the different occupations by providing slopes for cycling promenade and skateboarding, horizontal surfaces for football and picnic, higher points for resting and viewing and lower points for gathering. The fluctuations of the ground surface are meant to stimulate the body, with or without the use of a vehicle (bicycle, skate, roller-skate). There are places of stability and place where the body is more exposed to gravity. New limits with different degrees of permissivity can be defined by topographical means, physical barriers, vegetation and sometimes sign. The vegetation concept is to replace existing trees which have leaves only three months a year, with a variety of pine trees, similar to those that can be found in the Mediterranean, with persistent leaves "aiguilles".

位于布鲁塞尔的伯利恒广场，提供了多元文化的特征。尤其是地海裔的居民，基本上来说地中海裔的居民有非常强烈的户外生活式，这是现有的布局所没有考虑的。广场的一个特性是它在布鲁尔的地理位置在其所处谷地的顶峰和低谷之间。它的空间由几条接高低点的街道收束而成，赋予一块城市空地特殊的地形。我们职责是将人行道、街道、停车场和广场结合到一个连续的沥青表中，来支持对这种原有空间区分并不认同的地中海式生活方式（如在意大利的锡耶纳，道路、人行道和广场之间是连续的）。这平滑的表面如地形般地被感知，为自行车漫步和滑板提供坡道，足球和野餐提供场地，将高点留给休息和眺望，将低点留给聚会使种种不同功能得以优化。地表的起伏意在刺激人体，无论是否于交通工具（自行车、溜冰鞋、滚轴冰鞋）。这里有颇具稳定性场所和使身体更加暴露于重力作用之下的场所。具有不同通行度新界限可以被地形的形态，物理性的栅栏，植被和间或出现的标所界定。关于植被的想法，是去掉现存的那些一年中只有三个月叶子的树，代之以各式各样的松树，与可以在地中海地区找到的相同，它们带着不凋的“针叶”。

ephemeral vegetation strip is to be laid out annually by local school
ldren, in a pedagogical project. The square can host events such as:
arket, concerts, sports competitions, and school holidays. A lighting
eme is integrated in the ground surface to intensify nightlife.

一个暂时性的植被带在一个教育活动中被当地的学童排布出来。广场可以举办诸如集市、音乐会、体育比赛和学校节庆之类的活动。一套照明方案被整合进地表以强化夜生活。

Airports have very authoritarian layouts that often generate devaluated peripheral areas. They occupy immense territories. For example Hong-Kong's Airport surface is equivalent to a third of Manhattan. Schiphol, in Amsterdam, is considered to be the biggest city in the Netherlands, in terms of surface.

According to a recent report by the EU, each of the top-20 airports will be congested or saturated by 2010. The development of 430 regional airports is going to be encouraged during the next decade to help decongestion the main ones. If we added up all their surfaces the resulting territory would have absurd proportions sometimes bigger than some European states. A horrifying perspective if we look at how airports are usually laid out.

Airports materialise the myths and ambitions of a global world. Air-travel used to be the privilege of wealthy people living in the capital. During the last decade it has gradually become accessible to most people.

The project is a speculative infiltration scheme for Birmingham Airport's forbidden territory. The scheme will engage with the site on a basis that repairs the separation between city and airport by exploring possible ways of revitalizing the landscape. The construction of an airport implies the evacuation of history, topography, vegetation diversity, economic activity, social life and even law system. The project proposes ways of inhabiting this space by reintroducing these values. It is a site of expectation that could become the expression of cultural diversity and interaction by overlaying programmatic elements to stimulate encounters by different categories of people. It is a 24hour city accessible for passengers in transit, people using the highway and people from the neighbourhood. The forbidden territory is a question ground .It is an opportunity to address issues of globalisation and to question established hierarchies in the way airports are currently laid out and experienced.

机场具有非常独裁的设计，通常会使外围区域贬值。它们占据了广阔的地域。举例来说，香港机场的平面相当于三分之一个曼顿。阿姆斯特丹的Schiphol机场，以平面面积来说，可以被认为是兰最大的城市。

根据欧盟最新的报告显示，在2010年之前，每一个排名前20位的市机场都会变得拥挤或饱和。430个地区性机场将被鼓励发展，用分散主机场的拥塞。如果我们合计所有这些机场的平面，因之而的地域将占有荒谬的面积比例，有的大过一些欧洲国家。假如我看看机场通常是怎么被设计的，这将是个可怕的前景。

机场物化了全球化世界的神话与野心。空中旅行曾经是居住在首的富人的特权。在过去的十年中，这已经逐渐成为大多数人可以及的了。

本设计项目是对伯明翰机场禁地的投机性渗入。方案基于修复城与机场之间的分隔，以探索重塑景观的可能方式来干预基地。一机场的建造意味着历史、地形学、植被多样性、经济活动、社会活甚至法律体系的退出。方案提出重新引入这些价值以重新激活一空间。这将是一个充满期待的场地，可以变成对文化多元性的达和以项目性元素的重叠来刺激不同人群所带来的互动。这是一对转机的乘客、使用公路的人和近邻的人24小时开放的城市。禁是有问题的场地。这是一个标定全球化问题和质疑已确定的等级系的机会，而当下的机场就是以这样的等级制度被规划出来和被验的。

e are completely cut off from the territory we cross. This prevents us from perceiving it as a place, from being fully in it. A condition of “deja-vu” d disorientation. We could be anywhere: Paris, Athens, or Brussels.

们被我们穿越的领域所彻底隔离。这妨碍了我们把它作为一个地点去感知，妨碍了完全投入其中。这是一种“幻觉记忆”和方向感迷失的状态。我们可在任何地方：巴黎、雅典或布鲁塞尔。

mingham’s International Airport is situated approximately 12Km south east of the city centre.

明翰的国际机场位于距城市中心12 Km远的东南方。

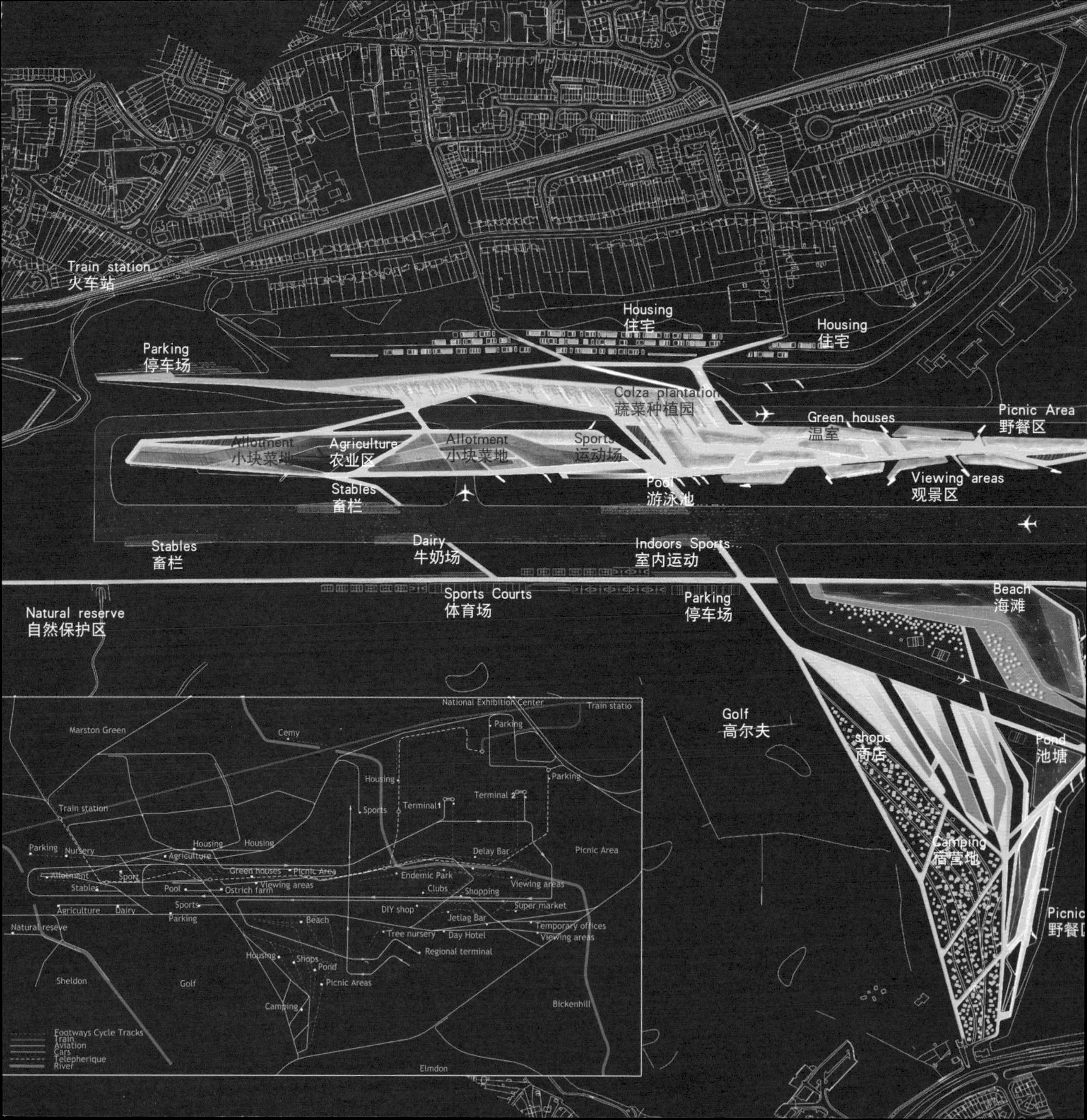

Train station
火车站
Housing
住宅
Housing
住宅
Parking
停车场
Colza plantation
蔬菜种植园
Picnic Area
野餐区
Green houses
温室
Allotment
小块菜地
Agriculture
农业区
Allotment
小块菜地
Sports
运动场
Stables
畜栏
Pool
游泳池
Viewing areas
观景区
Stables
畜栏
Dairy
牛奶场
Indoors Sports
室内运动
Sports Courts
体育场
Parking
停车场
Beach
海滩
Natural reserve
自然保护区
Golf
高尔夫
shops
商店
Pond
池塘
Camping
宿营地
Picnic
野餐区
National Exhibition Center
Train statio
Marston Green
Cemy
Parking
Housing
Parking
Terminal 1
Terminal 2
Sports
Train station
Housing
Housing
Delay Bar
Picnic Area
Parking
Nursery
Agriculture
Green houses
Picnic Area
Endemic Park
Allotment
Sport
Viewing areas
Stables
Pool
Ostrich farm
Clubs
Shopping
Viewing areas
Sports
Super market
Agriculture
Dairy
DIY shop
Parking
Jetlag Bar
Natural reseve
Beach
Temporary offices
Tree nursery
Day Hotel
Viewing areas
Regional terminal
Housing
Shops
Pond
Sheldon
Golf
Picnic Areas
Camping
Bickenhill
Footways Cycle Tracks
Train
Aviation
Cars
Telepherique
River
Elmdon

Housing
住宅
Terminal 1
1号航站楼
Terminal 2
2号航站楼
Parking
停车场
Picnic Area
野餐区
Endemic Park
风土公园
Delay Bar
航班延迟酒吧
Clubs
俱乐部
Viewing areas
观景区
shopping
购物区
DIY shop
自助商店
Super market
超市
Jetlag Bar
长途旅行吧
Temporary offices
临时办公室
Tree nursery
育苗站
Viewing areas
观景区
Day Hotel
日间酒店
Regional terminal
区域性交通港

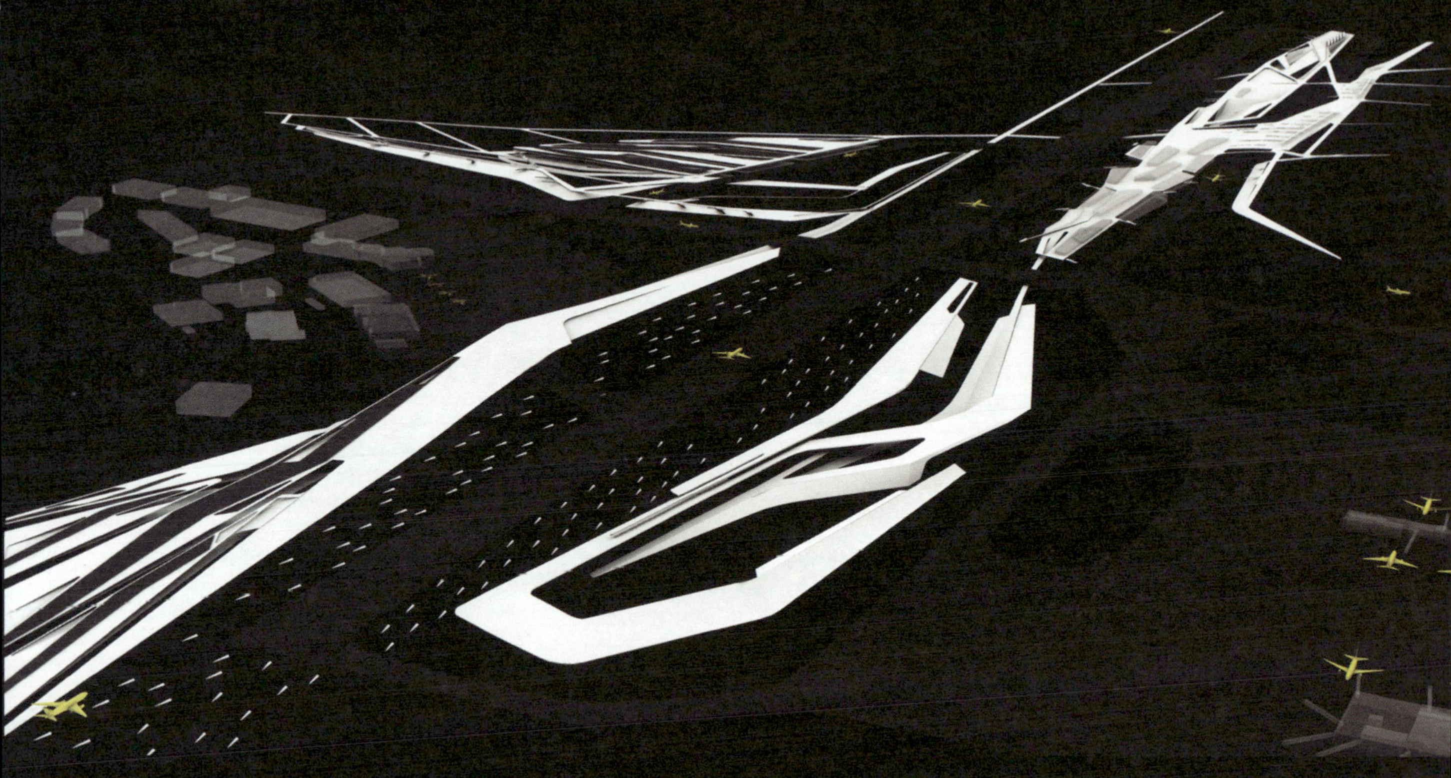

A Passengers view of the Jetlag city. The Green houses are lit at night
乘客视点的长途旅行综合症城市。温室在夜间被点亮了。

View of a bridging platform over the taxiway giving access to the pa
Colza plantations supply the energy for the green houses and some of t
indoors sports and cultural programs.
一个跨越滑行道的连接公园的桥状平台视图。育种场为温室和一些室内

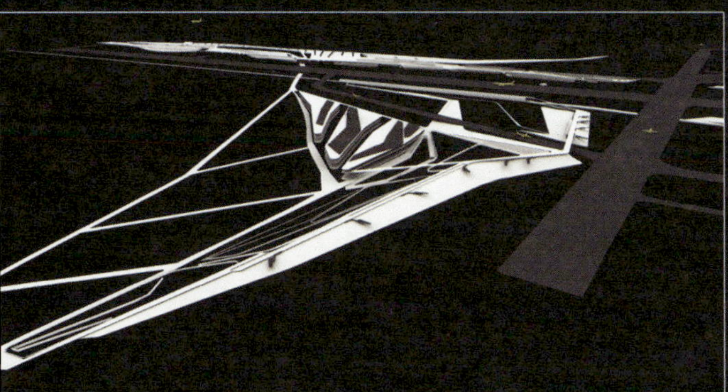

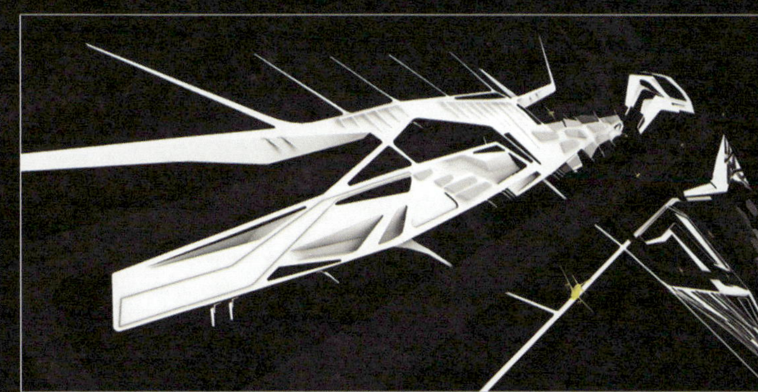

e scheme is reconnecting the isolated surfaces that are situated be-een the runway and the taxiways to the surrounding neighbourhoods, e highway and the terminal building. Connections are created by bridg-; platforms and underground links, thus allowing the airfield to be infil-ated by the surrounding city. The surface is a park made of a sequence leisure and gardening areas followed by an agricultural park with green uses and colza plantations, and an endemic park which is directly con-cted to the terminal building. The underground spaces are occupied various programs such as local sports facilities, community centres, dustrial spaces, shopping malls, temporary offices and night clubs ...

这一方案重新建立起跑道和滑行道间被隔离的表面与周边近邻、公路和候机楼的联系。这种联系由桥状平台和地下联结建立起来，这样使得机场被周边城市所渗透。表面是由一系列休闲和花园区域所构成的公园，紧接着花园的是有着温室和育种场的农业公园和一个直接与候机楼相连的地域性植物公园。地下空间被各种各样的项目占据着，例如当地的体育设施、社区中心、工业空间、购物中心、临时办公区和夜总会……

ew from inside the endemic park. Recycled turbines in the trees are ing converted into sleeping capsules for passengers in transit. A "take-Bar" provides viewing platforms for watching the planes take off and d.

或性植物公园内的视图。循环利用的树间的引擎被转化成提供给转机乘客

View from the tunnel linking the endemic park to the green house area.

将地域性植物公园连向温室的隧道视图。

Cambridge
(M 11 (N))
B. Stortford
A 120
Non-motorway
traffic
London
Harlow
M 11 (M 25)
Y584
FEV
POLICE

丽莎
修娃
Lisa
Silver

Lisa Silver was born in London, UK in 1978. She studied for a BA in Architecture at the University of Nottingham. She completed the Diploma at the Bartlett School of Architecture gaining a Distinction and Sir Banister Fletcher Prize for the highest marks in the Diploma in Architecture final examination and a runner up prize for the RIBA Presidents Silver Medal. Lisa has worked for TP Bennett Architects, Sheppard Robson Architects, and is currently working for Ian Ritchie Architects in London.

One of the most important things I have learnt during my architectural career is to tell the truth. "No one likes a liar" my mother often told me and she wasn't wrong. Tell lies in architecture and you won't be popular in the playground. Pull the wool over their eyes, begin a life of fraud and deceit, be mutton dressed up as lamb, but they'll eventually see you for what you are; a big fat liar!

In developing my architectural ideas I have steered further and further away from the notion of a uniform skin which encloses and hides space, is untrue to the ideas and concepts it contains, lacks relation to the context that surrounds it and effectively excludes the users it aims to attract. My work tries to find ways for the users and the function of the space to become the so called "enclosure" or at least clearly represent them at the building boundary instead of wrapping them up on the inside. In recent years architects have been increasingly conscious of expressing the building structure and allowing the user to enjoy the space through the physical description of its construction. In taking this concept further, I would like to believe that one could not only be truthful to the structure, but to the very things that demanded the space in the first place; the objects, the people, the systems, the activities. If the space is to contain these, why can't the space be these?

The idea that activity could create space first occurred to me when I saw a performance called De La Guarda, an Argentinean group of - well basically wild people! For two hours they entertained the audience by bouncing, slamming, propelling, sliding and throwing their bodies against surfaces, materials and objects as well as each other and the audience, making a symphony of sounds in the process and culminating in a sensual spatial experience far beyond that of your wildest dreams. In effect the sound and movement created by the physical interaction between the animate and the inanimate, created an energy which was effectively an enclosure.

In fact, tactility is a perfect way to be truthful to space. A building is a physical object and we should be encouraged to touch, taste and smell it as part of the spatial experience. Our sense of sight alone allows light, colour, pattern, depth and scale to guide one through space. Can we even begin to imagine the extent of stimulation received by experiencing a space through all five senses? I'm certainly a sucker for stroking buildings- I actually ran to touch the Guggenheim in Bilbao - that delectable titanium was just too much for me to handle. Many of my favourite buildings are my adult comfort blankets and I say, lets have some more!

The traditional building skin often hides the sensual stuff of the building, enclosing the contents in a homogenous membrane to tidy up the loose ends. The architect tries her up-most to hide the "dirt" of the building elements that make the building tick. How would the user react if everything was on display and how would it affect the separation of public and private space. What if the architect said; "take me as I am- warts and all." My first Diploma project was inspired by a wicker chair that was comfortably parked in the toilet of a brick lane cafe. This alien object existed as a tactile element in a typically sterile, hygienic environment. The existence of such a sociable object as a chair within the space of the toilet insinuated the presence of more than one user, suggesting that the space was not as private as it first appeared. Thus the meaning of both the chair object and the toilet space were subverted or changed by their unusual and unnatural coexistence.

My diploma work used the notion of the alien object to question attitudes towards traditional building materials. It began with the pure fascination of pocessions. The things people own and have around them tell a story of their experiences and interests. These items that populate our living environments are assorted into cupboards, shelves and other bland surfaces that society provides for our most treasured articles. Why should our objects, whether meticulously designed or tastelessly irrelevant, bare this sordid fate? Instead of our pocessions fitting into the boundaries of our living space, can't our living space be designed around our objects. Or should our living space be made of objects?

To extract an object from its original meaning by making it a building material both changes one's attitude toward the object and the building. It breaks the building down into its components and allows us to understand and interact with the physical fabric that surrounds us. So many potentially useful materials are discarded because they do not fit into the bracket of bricks and mortar. The car for example is filled with many highly engineered components that could have many uses within our built environment. The limited lifespan of the car, and its adverse effects on the environment as waste, make it a suitable candidate for our city of objects.

In the following work, items that hold value to the individual inhabitant are continually recomposed into spatial relationships and a dynamic environment is created which is entirely responsive to the individual creator. The inhabitant gives their own personality to the space through the objects they hold dear. The objects become intriguing by both expressing their original meaning and their new meaning as a building component. Could it be possible that spaces made of specifically selected collections of objects could lead to an entirely personalised environment perhaps- and dare I say it- negating the need for the so called architect?

丽莎·修娃，1978年生于英国伦敦。在伯明翰大学学习取得建筑学学士学位。她在巴特雷特建筑学院修习建筑专业文凭时取得优秀成绩，获得了表彰建筑系学生毕业设计最高成绩的巴尼斯特·弗莱彻爵士奖，并在英国皇家建筑师学会主席评选中获得银奖。
丽莎曾就职于TP Bennett建筑师事务所，Sheppard Robson建筑师事务所，目前就职于伦敦Ian Ritchie建筑师事务所。

在我的建筑职业生涯中，我所学到的最重要的事之一就是说实话。我妈妈经常告诉我："没人喜欢撒谎者"，她说得没错。在建筑上说谎，你就不可能受人欢迎。障人眼目，开始欺骗和谎言的人生，披上羊皮，但最终会被人识破：大话王！

在我发展我的建筑理念的时候，我越来越远离用统一表皮包裹和隐藏空间的观念，那对于它包含的概念和想法是不真实，缺乏与周围文脉的关系，并隔绝了它本来应该吸引的使用者。我的工作试图寻找一种方式让使用者和功能成为所谓的"围合"，或者至少在建筑的边界上清晰地表达它们而不是将它们包裹在里面。在近些年中，建筑师越来越意识到建筑结构的表达，并且允许使用者从结构的物理性表达中享受空间。为了将这个概念带得更远，我愿意相信不仅仅能对结构真实，而且能对任何对空间有第一位需求的东西的真实：物体、人、系统、活动。如果空间包含这许多，为什么空间不能就是这许多？

我第一次产生活动可以创造空间的想法是在我看一个所谓的德·拉·戈达的表演的时候。那是一个阿根廷的基本上是野人的表演团！在两个小时里，他们将他们的身体跳、甩、推、滑和扔向不同表面、材料、物体以及他们和观众之间来愉悦观众，制造出序列的声音交响曲，达到肉欲的空间体验高潮，远超出你最狂野的梦。实际上声响和运动生成自运动者和不动者之间的物质互动，制造出有效的围合空间的能量。

事实上，触感是通向空间真实性的完美道路。建筑是物质的，我们应该被鼓励将触摸，品尝和嗅闻作为空间体验的一部分。仅我们的视觉本身就允许光、色、图样、深度和比例来引导人穿过空间。我们能够甚至开始想像通过所有五官感觉接受的空间体验带来的刺激延伸吗？我一定是抚摸表面的痴迷者——实际上我跑着去触摸毕尔巴鄂的古根海姆——那让人愉悦的钛金属的丰富性而使我不能把握。很多我喜欢的建筑都是我成年的毯子，我说：让我们拥有更多吧！

传统的建筑表皮经常掩盖建筑的感观因素，用巨大的膜围合其内容，整合起松散的肢端。建筑师尽他的最大可能隐藏起建筑的"污垢"，那些令建筑滴答作响的元素。如果一切都展现出来，使用者将会如何回应，并且它会怎样地影响公共和私密空间的划分？如果建筑师说，"以我本来面目待我——不加掩饰的"会怎么样呢？我的第一个毕业设计，受到的启发来自于一把被舒适地安置在一个咖啡馆厕所里的柳条椅。这样的异端物体，作为触觉元素存在于一个典型的消过毒的卫生的环境中。椅子，这样一个社会性物体的存在，影射出超过一个使用者的在场，暗示出这个空间已不像它开始看起来那样私密。以这样的方式，椅子和厕所空间的意义都被它们的不同寻常的和非自然的共存所改变了。

我的毕业设计用外来物体的概念质疑了对传统建筑材料的态度。它以对财产的纯粹迷恋开始。通过围绕在人们身边的物品讲述关于他们的经历和兴趣的故事。这些在我们的生存环境中堆积繁殖物品被装进橱柜、架子和其他乏味的外壳中，社会为我们最珍贵的物品提供的就是这种东西。为什么我们的物品，不论多么的精心设计和远离乏味，都要承受这样可悲的命运？相对于将我们的财产放进我们居住空间的边界，难道我们的居住空间不能围绕我们的物品设计吗？或者我们的生活空间就是由物体组成的？

将一个物品从它的源义中抽取出来，将其变为同时改变人对物体和建筑的态度的建筑材料。它把建筑打碎为组件让我们理解围绕在我们身边的物质机理，并与之互动。许多潜在的可用材料被抛弃了，因为它们不适合砖石和抹灰结构。举例来说，内部充填的高技术组成材料，可以在我们的建筑环境中得到多重的应用。汽车使用寿命的限制，以及它作为废物对环境的负面影响使之成为我们物质城市中合适的候选对象。

在接下来的工作中，对个体有价值的物品被重组进空间关系中，一个完全响应于个体创造者的动态环境被创造出来。居住者通过他们呵护的物品将他们的自我个性给予空间。这些物品由于同时表达它们的源义和作为建筑组成物的新意义而变得媚人。由选出的物体集粹而成的空间有可能带来整个环境的完全个人化吗？或许，我是否敢说，探讨了所谓建筑师的必要性呢？

FORBIDDEN FRUIT

FROZEN SUCCULENCE

CONSTRUCTED FRAGILITY

Spaces that take their form from objects rather than the traditional enclosure or boundary suggest transformation, tactility and subversion. The human qualities of objects provides a medium that can be easily handled and recomposed making the creation of the space as important as the space itself. Object architecture promotes a transfiguring context where the user can be entirely interactive with their environment and allow the narratives of their possessions and found items to characterise the space. The variation of forms, functions, textures and life - cycles that different objects embody, provides an endless and complex palette of materials that far exceeds the limitations of the typical wall, roof and floor. The resulting spaces may maximise these embodied qualities and/or subvert the meanings of the objects to suit the new environment. Therefore, when objects are implanted into spaces to which they do not naturally belong and to which they appear alien, new rhythms and conditions are established as the new and existing adapt and transform in order to co-exist.

由物品而不是传统的围合或边界形成的空间暗示了变形、触感
覆。物品的人文品质提供了一个易于打理和重组的媒介，这使得
的生成变得和空间自身一样重要。物品建筑提升了变形中的文脉
这样的文脉中，使用者可以与他们的环境完全互动，并允许他们
对所有物和找到物品的叙事来表征空间。不同物体所包含的形式
能、质地和使用寿命是千差万别的，提供了远超出典型的墙、屋
楼板限制的、无尽的、复杂的材料调色板。由此而来的空间可以
化地包含这样的品质，并/或打乱物品的意义来适应新的环境。
当物品被植入它本来并不自然从属且表现为异端存在的空间时，
物品和原有的物品互相适应和变形以达到共生，这样，新的韵律
件就建立了。

EDIMENTARY SURVIVAL

ALE OF RECODIFICATION

MARY-LOU'S BATHROOM

JIMMY-RAY'S STUDY

BILLY-BOB'S GARAGE

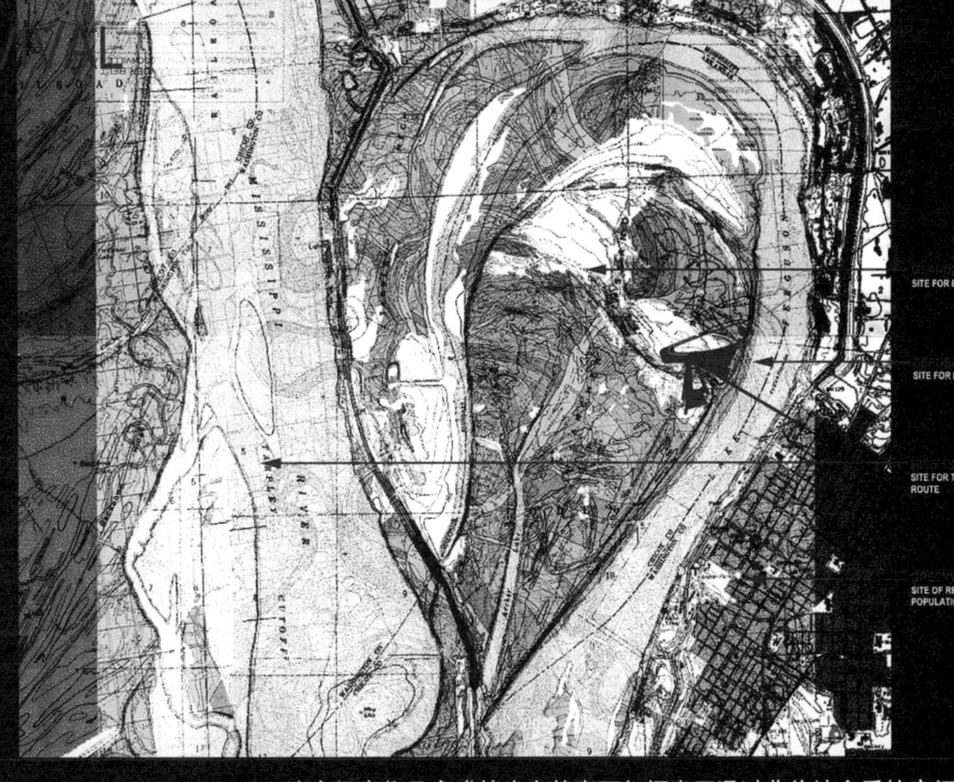

tale of survival through the composition of objects explores not only physical construction of space through an unconventional medium, also how the characteristics and capabilities of its makers allow it to its specific form. The narrative explores the psychological response of characters to confinement to a house and follows how personal and d objects are used within their environments, firstly to maintain sanity then as a way of life. Part I is set inside the house during confinement follows how the increasingly obsessive behaviour of the characters is cted in their environments through their compulsive composition and ersion of obscure objects.

II sees the characters breaking out of their confinement and colonising ite with enclosures, systems and forms constructed and composed of cts. The final part illustrates the working enclave where dwelling and ct based enterprises established by each character form interconnecting s between which people and objects move in balanced rhythms.

tory narrates an alternative living condition where people, objects and e are entirely interdependent. This "enclave of objects" edges on the orary in its ability to be continually transfigured and recomposed though presentation of the personalities of its users through their inventive position of personal and found objects suggests an immense permanence. r way, through an architecture of objects, the characters have the lom to create a physical and spatial interpretation of themselves.

这个经由物品合成的生存故事不仅探索了通过非传统介质的空间物质构成，而且探索了空间的创造者的特性和能力怎样允许空间采取特定的形式。这一叙事探索了被限定在一座房子中的三个角色的心理反应，接下来是个人的和找到的物品怎么在他们的环境中使用，首先保证合乎情理，然后成为一种生活方式。

第一部分被限制在房子之内，紧跟着的是角色的递增的必然行为怎样通过他们的必然组成物和不明物体的颠覆反映在他们的环境中。第二部分可以看作是角色打破对他们的限定并且以构建围合，系统和形式的物体对场地进行殖民。 最后一部分图解了每个角色所基于的作业地块，在其中，每个角色将构架其对于栖居和物体所具有的企盼，角色之间形成互相关联的地带，在这此间，人和物体节奏和谐地运动于其中。

这个故事叙述了一个另类的人、物和空间完全相互依赖的生存环境。通过对其使用者个性的表达，对个人和找寻到的物体的创造性组成，这片“物的领地”将其本身置于连续变形和重组的能力所带来的暂时性的边锋之上；并发现了物体暗含着一种浩瀚的永恒。另一方面，通过一种物体建筑学，角色们拥有了创造他们自己的身体与空间阐述的自由。

CONFINEMENT 禁闭

Our story begins on a meander cut-off on the Mississippi river near Greenville, where three characters, Mary-Lou, Jimmy-Ray and Billy-Bob are imprisoned inside the bathroom, the study and the garage of a dilapidated farmhouse. Over time, the siblings develop various types of spatial dyslexia and obsessive behaviour which is imparted onto their confinement through the objects contained within each room. Mundane objects are inventively and compulsively subverted and recomposed to provide a more stimulating environment until these inventions can no longer be contained by the conventional confines of the house.

我们的故事开始于靠近格林村的密西西比河的一个转弯点。这里有三个人物：玛丽•卢、吉米•雷和比利•鲍勃。他们分别被禁闭在一个荒废农舍中的浴室、书房和车库中。随着时间的流逝，这些同胞们通过每个房间所包含的物体，发展出各式不同的被附加于他们禁闭生活的空间识别障碍类型和强迫性行为。日常物品被创造性地且强制性地被打乱和重组，以提供更具刺激性的环境，直到这些发明不再被包含在房子惯常的范畴之内。

MARY-LOU 玛丽•卢
THE BATHROOM 浴室

Confinement to the bathroom rendered Mary-Lou claustrophobic and obsessed with cleanliness. She therefore configures sponges bottles and other bathroom items into a barrier against the dirt and a nature look out point orientated towards a tiny window. In this wayshe minimise contamination and maximise spaciousness thus satisfying her idiosyncrasies and need for mental stimulation.

被囚禁在浴室中使得玛丽•卢患上了幽闭恐惧症，并为洁癖所困扰。因此她重新布置了海绵浴块、瓶瓶罐罐和其他的洗浴用品，构成一个抵御污垢的壁垒和一个天然的朝向一个小窗的外窥点。通过这种方式，她最小化了污垢，并最大化了宽敞，以此来满足她的癖好和对精神刺激的需要。

JIMMY-RAY 吉米•雷
THE STUDY 书房

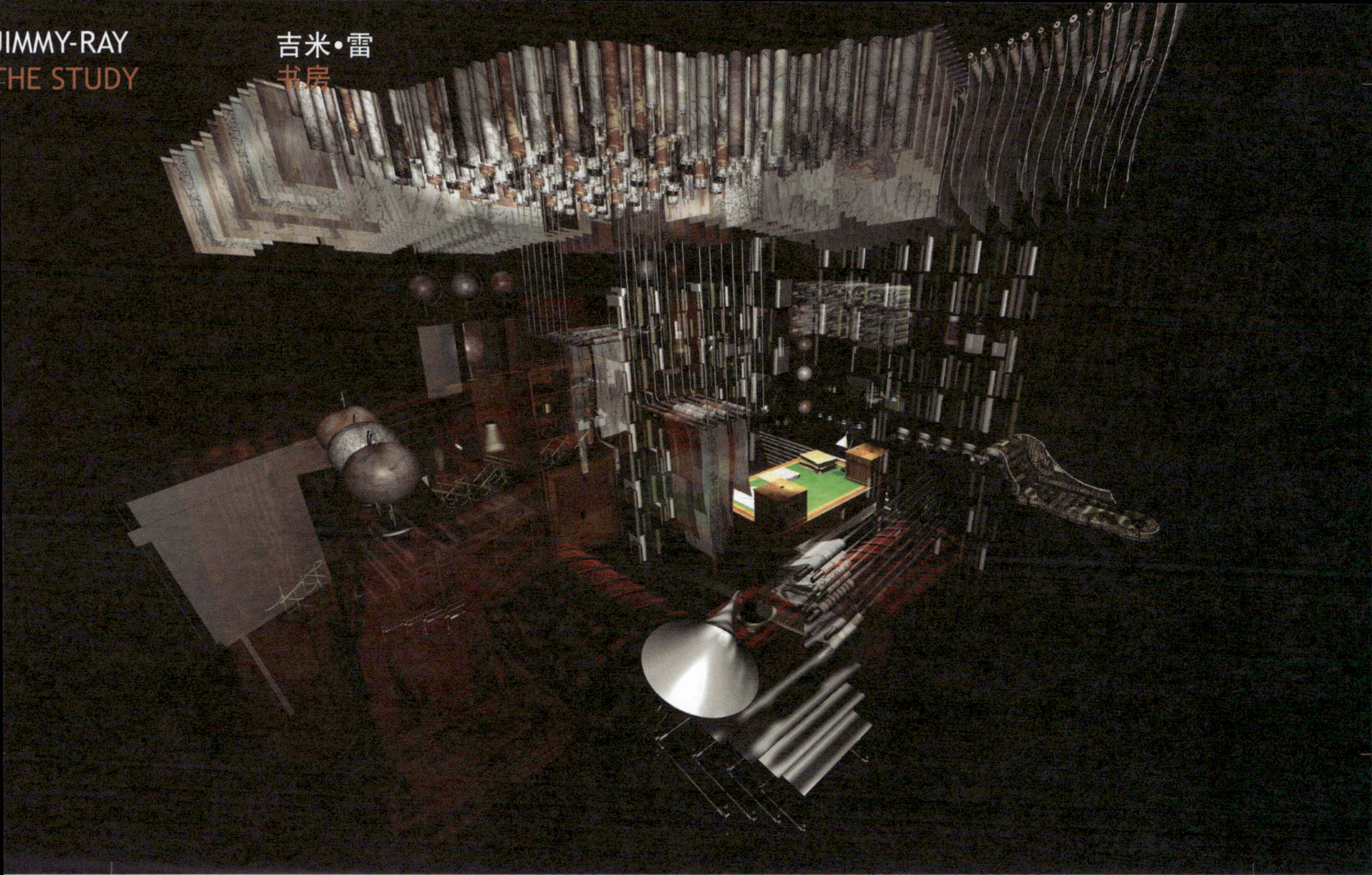

study containing an endless array of written and graphical information on the Mississippi river, provides jimmy-ray with an obsession for studying, chiving and arranging. as newspapers, books, maps and photographs are neurotically reconfigured, spatial flows that can be likened to the eroding d depositing power of the Mississippi itself, gradually shift the boundaries of the room until they cease to exist. And still he continues to relentlessly ravel, fold and order this papery haven.

房容纳着无穷的的关密西西比河的文字与画信息的阵列，使吉•雷执迷于研究、归档安置。当报纸、书籍、图和照片被神经质式地新安排，空间的流动就密西西比河锚固和沉积力量，逐渐地改变着房的疆界直到它们消失。且，他仍然继续着对这纸质天堂无情的拆解、叠和排序。

BILLY-BOB 比利•鲍勃
THE GARAGE 车库

In the dim light of the garage, Billy-bob deteriorating eye sight leads to a heightening of the other senses and a spatial conception governed by texture and sound. He composes discarded, disused and dismantled electrical and garden equipment into functional sculptures and a living space within a broken car carcus.

在车库暗淡的光线下，比利•鲍勃不断恶化的视力导致了他其他知觉的提高。而且，空间的概念是由肌理和声音支配的。他把被废弃的、无用的和被拆卸的电器和园艺设备组合成一个功能性的雕塑和处在一个损坏的汽车构架中的起居空间。

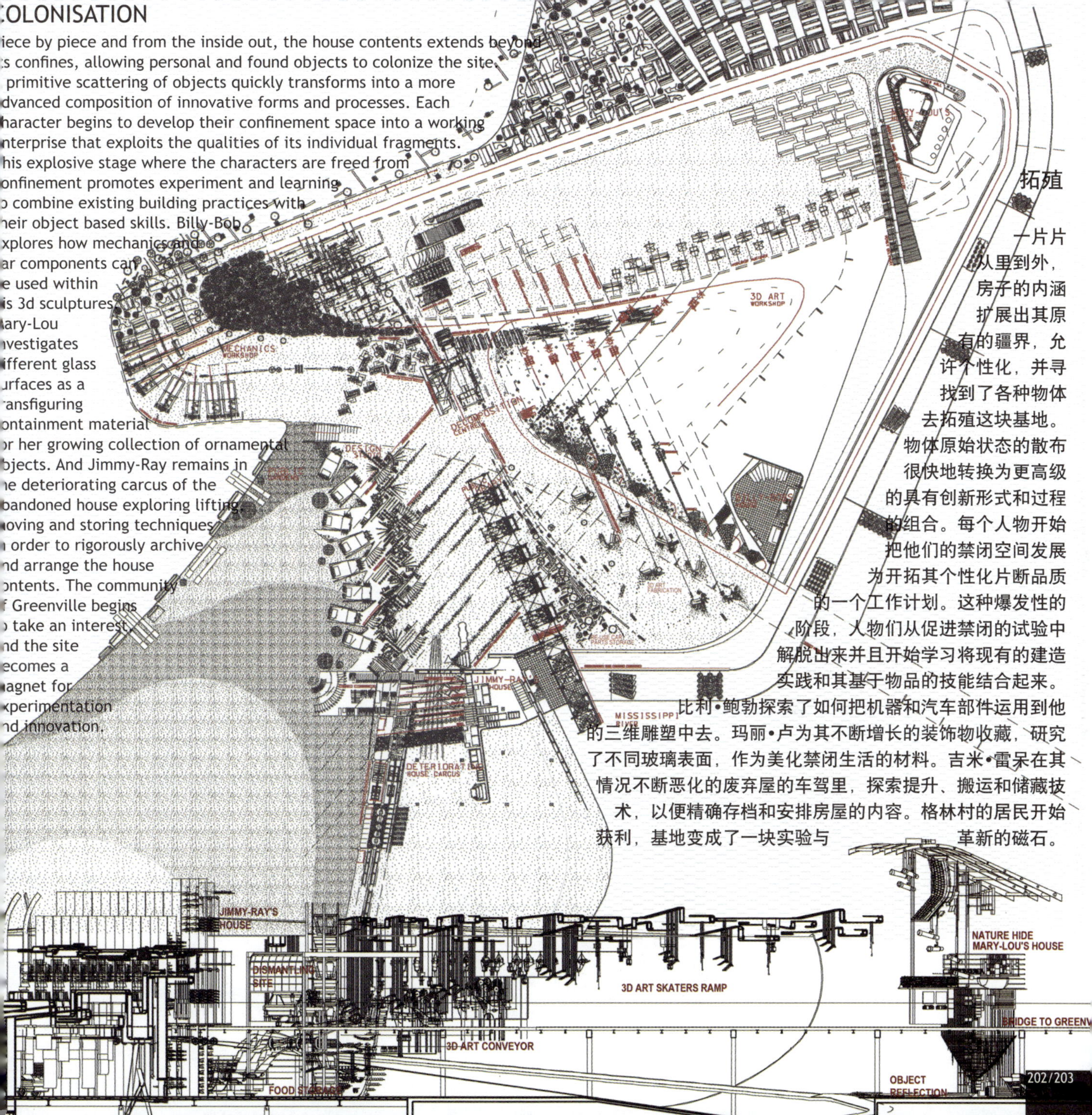

:OLONISATION

iece by piece and from the inside out, the house contents extends beyond :s confines, allowing personal and found objects to colonize the site. primitive scattering of objects quickly transforms into a more dvanced composition of innovative forms and processes. Each haracter begins to develop their confinement space into a working nterprise that exploits the qualities of its individual fragments. his explosive stage where the characters are freed from onfinement promotes experiment and learning o combine existing building practices with heir object based skills. Billy-Bob xplores how mechanics and ar components can e used within is 3d sculptures. ary-Lou ivestigates ifferent glass urfaces as a ansfiguring ontainment material or her growing collection of ornamental bjects. And Jimmy-Ray remains in he deteriorating carcus of the bandoned house exploring lifting, ioving and storing techniques i order to rigorously archive nd arrange the house ontents. The community f Greenville begins take an interest, nd the site ecomes a agnet for xperimentation nd innovation.

拓殖

一片片从里到外，房子的内涵扩展出其原有的疆界，允许个性化，并寻找到了各种物体去拓殖这块基地。物体原始状态的散布很快地转换为更高级的具有创新形式和过程的组合。每个人物开始把他们的禁闭空间发展为开拓其个性化片断品质的一个工作计划。这种爆发性的阶段，人物们从促进禁闭的试验中解脱出来并且开始学习将现有的建造实践和其基于物品的技能结合起来。比利•鲍勃探索了如何把机器和汽车部件运用到他的三维雕塑中去。玛丽•卢为其不断增长的装饰物收藏，研究了不同玻璃表面，作为美化禁闭生活的材料。吉米•雷呆在其情况不断恶化的废弃屋的车驾里，探索提升、搬运和储藏技术，以便精确存档和安排房屋的内容。格林村的居民开始获利，基地变成了一块实验与革新的磁石。

GLENVILLE 格林村

MISSISSIPPI 密西西比

y Lou's bathroom objects weave upward into a narrow tower that houses a panoramic look-out point for nature enthusiasts and voyeurs. her personal ects are contained within sufaces and meshes of varied transparency that slide and pivot to vary levels of light, reflection and opacity within the er.

丽•卢的浴室用品向上交织进入一个狭窄的塔，替一个给天生的狂热者和偷窥狂而设的全景外窥点提供空间。她的私人物品夹在不同透度的表皮与孔隙之间，在塔中的各种不同层次的光线、反射和黑暗中滑动、旋转。

ne site has matured into an environment where the three siblings and the local community innovatively combine new objects, techniques and process to an enclave for living and object based activities. Three entirely interdependent zones which have developed from the rooms of each character ncourage people and objects to move through the site in balanced rhythms. The enclave is continually reconfigured through constant gleaning, building and use. In this way, Glenville physically illustrates the idiosyncrasies and practices of its makers, by the honest and inventive composition of their ersonal and found objects.

基地已经成熟地融入了环境，在这儿，这些同胞们和当地居民创造性地把新物品、技术和程序结合进生活中的，基于物品的活动的领。三个由每个角色的房间发展而来的相互完全独立的区域激励着人和物体以平衡的韵律移动过场地。这一领地通过持续地捡拾、建造使用，被不断地改装着。通过这种方式，格林村以他们个人的和找寻的物体的忠实和创造性的构成，物理性地描绘了它的创造者的陆气质和实践。

Billy-bob's garage has transformed into a 3d art workshop, car dismantling platform and mechanics yard that is linked via a bridge to Greenville. The workshop takes the form of a skateboard ramp where art material poured from a roof suspension system is spread by the physical movements of the skaters so that people and the process of painting become the art form. The car dismantling platform allows car components to be used throughout the site both as an art medium and as a key building material.

比利•鲍勃的车库被转换成为一个三维的艺术车间，汽车分解台和机械场通过一座桥连接到格林村。车间采取滑板坡道的形式，这样艺术材料从屋顶的悬挂系统以滑板玩家般的物理运动分散下来以使人和喷涂过程成为艺术形式。汽车分解平台允许汽车零件同时作为艺术媒介和建筑的关键材料在整个基地中使用。

Jimmy-ray's study expands into a distribution site for collected and abandoned objects using the deteriorating carcus of the house as a framework for distribution and deposition. His labyrinth of memories maps and books are archived into a public space composed of dismantled car components.

吉米•雷的书房扩大为一个分配收藏品和废弃品的基地，以每况愈下的废弃屋的车驾作为分配和储存的构架。他的迷宫般的记忆地图和书籍在有分解的汽车构建组成的公共空间中存档。

GREEN MAN INTERCHANGE
Walthamstow A114
Leytonstone. Leyton
Central London. Stratford
Blackwall Tunnel (A102(M))

祖贝尔
苏提
Zubair
Surty

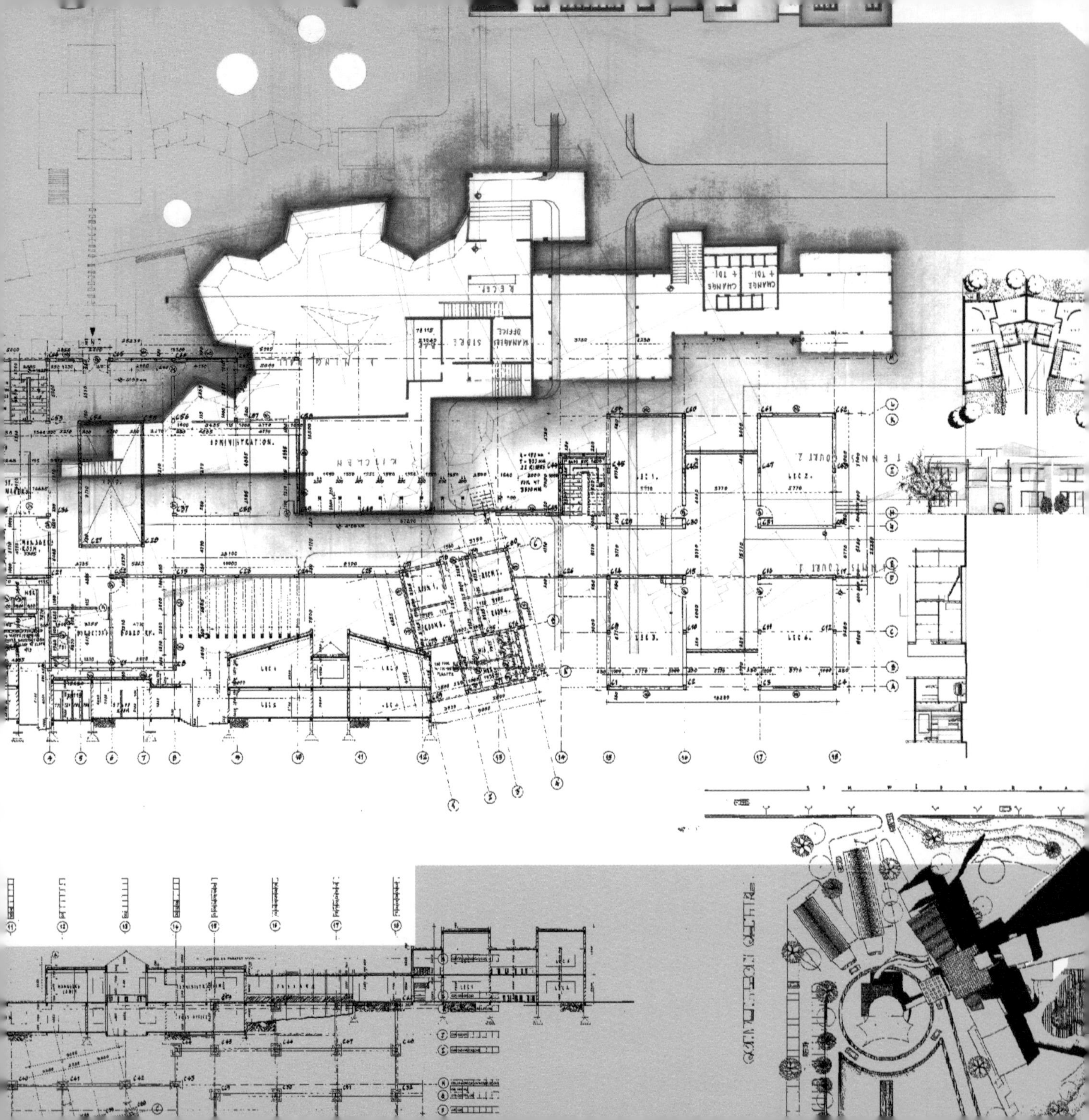

ıbair Surty
7A Canadian Avenue, London SE6 3AU, UK
ob - 0793 223 5461 Tel - 020 8314 0444
ıbairsurty@gmail.com
ationality - Indian
ducation
)02 - 03 M.Arch (Distinction - Thesis) The Bartlett School of Architecture University College London, UK
)95 - 00 B. Arch (Hons) IES College of Architecture University of Mumbai, India.
eaching Experience
)02 - 03 Exteriors Interiors - Mumbai, India, Lecturer
E.S College of Architecture - Mumbai, India, Visiting Lecturer
'ork Experience
urrent Hamiltons. - London, UK
)04 - 06 RTKL - UK Ltd. - London, UK
)04 - 06 Cazenove Architects - London, UK
)03 - 04 V + L Associates - London, UK
)03 - 04 Mobile Response Unit - London, UK, Co Founder
)00 - 02 M Zubair Surty Architects - Mumbai, India, Principal
)99 - 2000 Mungekar & Associates - Mumbai, India
orkshop
)03 Invited participant of ARCH Desant 3 (Dirmitrovgrad, Russia). Urban regeneration workshop organised
/ center of contemporary Architecture (CCA) Moscow.
ıblications
)03 "Mesh" architecture magazine, London
:ook - Inc" M.Arch magazine, London
)02 "Futures", IES Dept of Ekistics & Research, India
)01 "Inside Outside", Design magazine, Nov issue, India

l贝尔・苏提
址：17A Canadian Avenue，伦敦 SE6 3AU，英国
机：+44 793 223 5461 电话：+44 20 8314 0444 Email：zubairsurty@gmail.com
人资料
生日期：1978年2月22日
籍：印度
育背景
)02年—2003年 建筑设计硕士（优秀—论文） 巴特雷特建筑学院，伦敦大学学院，英国
)95年—2000年 建筑学学士（荣誉毕业） IES建筑学院，孟买大学，印度
教经验
)02年—2003年 "室内，室外"印度孟买， 讲师
IES建筑学院—印度孟买， 客座讲师
作经验
前 汉密尔顿建筑师事务所 — 英国伦敦
)04年—2006年 TKL英国有限公司 — 英国伦敦
)04年—2006年 Cazenove建筑师事务所 — 英国伦敦
)3年—2004年 V+L联合事务所 — 英国伦敦
)03年—2004年 "移动反应小组" — 英国伦敦，合作创办人
)00年 M 祖贝尔 苏提建筑师事务所 — 印度孟买，负责人
)99年—2000年 Mungekar及其合伙人建筑事务所— 印度孟买
讨会
)03年 受邀参加ARCH Desant 3（Dirmitrovgrad，俄罗斯）
莫斯科当代建筑中心举办的城市复兴研讨会
版
)03年 《网眼》建筑杂志，伦敦
库克—公司》建筑设计杂志，伦敦
)02年 《未来》，ISE群居研究学院杂志，印度
)01年 《内外》 设计杂志， 11月刊，印度

My dear friends, the following text is an attempt to take you through the journey from the point when I learned about the word 'ARCHITECT'. Magical word isn't it. Fascination for the word made me what I am today.Never knew my class five, 500 word essay would change my life to this extent.

I.E.S College of Architecture, Mumbai, India is where I started my architectural education. An environment where computers were a stigma, conceptual studies and investigations was waste of time and energy. Ignorance to radical and experimental architecture constantly prevailed. Everything else seemed O.K.

亲爱的朋友们，以下的文字试图带你经历从我开始知道“建筑师”这个词汇的旅程。富有魔力的词汇，不是吗？这个词汇的魔力使我成为现在的我。500字的短文在这样的程度上改变了我的人生。

印度孟买I.E.S建筑学院是我开始我的建筑教育的地方。在这个环境中，用计算机是一种耻辱。概念上的研究和调查是浪费时间和精力。长期以来对于激进和实验性建筑的忽视十分盛行。除此以外的一切似乎也还不错。

Regd. No.:- 950/8

DAI WELFARE SOCIETY

B/M/9, (10X10), Cheeta Camp, Trombay, Mumbai-88.

Date

To,
Zubair Surty,
IES ' s College Of Architecture,
Bandra Reclamation, Bandra West,
Mumbai – 400 050.

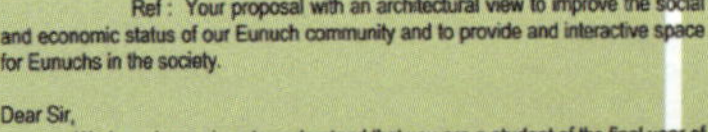

Ref : Your proposal with an architectural view to improve the social and economic status of our Eunuch community and to provide and interactive space for Eunuchs in the society.

Dear Sir,

We have been given to understand that you are a student of the final year of architecture at IES College and the topic that you have chosen for your thesis is of improving the social and economic status of the eunuch community and providing an interactive space for the eunuchs of Mumbai.

It is indeed of immense pleasure to us that the youth of our country, such as yourself, is taking such a keen interest in the life of the Eunuchs in our country. We at the 'Dai Welfare Society', are looking forward to provide the mumbai eunuchs with the amenities proposed by you and also trying our level best to make the people aware of role of this section of society on the rest of the community.

We have been planning to create a common ground between the masses and the eunuchs, by giving the latter a platform to exhibit their talents and create skills, thereby helping to educate the eunuchs and the people, and bridge the gap between these two sections. It is heartening to see that your proposal for the Mumbai eunuchs has been chalked out along the same lines.

On behalf of the 'Dai Welfare Society' , I wish you all the best in this endeavour and assure you of our co-operation at all times in this regard, as the research carried out by you would be of great importance to us. I would highly appreciate if you provide us with the copy of your thesis

Yours sincerely,

WELFARE SOCIETY

It was my fourth semester when I met my design tutor who had just joined the school after completing his masters from USA. This was the turning point of my architectural education. I was bombed with why's, what's and how's of architecture. These questions transformed my thinking towards architecture in total. I found myself using libraries and bookshops more often then I use to. Architectural design was no more an overnight tracing paper scribbling over a drawing board, and then feeling happy the next day with the satisfaction of the design tutor.

Instead, it became more of a process involving concept development, conceptual studies, sketches and working with study models. Architect is not just a designer of spaces but has a great role to play in creating and integrating society and its components. Addressing this issue was my final year dissertation 'Eunuchs an Encounter'. The thesis was more about researching and understanding the eunuch community and its functioning then designing just another building to house them in. Aim of the thesis was to make Eunuch community (community of people who are neither male nor female), which is completely shunned by the society more acceptable, and get them absorbed in the urban mainstream and how architecture becomes more apt to do so. My thesis gave me an opportunity to understand and be with a group of eunuchs belonging to a mysterious community whom the society wants to avoid. My final year in architecture didn't just contribute to my architectural personality but helped me grow as a person.

在我第四个学期时，我遇到了一位刚刚在美国完成硕士学位，新加入学校的设计老师。这是我建筑学教育的一个转折点。我被建筑中的为什么，是什么和怎么做一阵狂轰乱炸。

这些问题彻底转变了我的建筑思想。我发现自己利用图书馆和书店的次数比从前多了许多。建筑设计不再是头天晚上伏在图板上，在草图纸上乱画一气，第二天因为老师的满意而感觉陶醉。

取而代之的是一个包含概念发展、概念研究、绘制草图与做工作模型的过程。建筑师不仅仅是空间的设计者，还在创造与整合社会及其组成部分中担当重要的角色。关注这些问题开始于我最后一年的论文“阉人一种遭遇”。这个主题更多的是关于研究和理解阉人社会及其机能，然后设计一个供他们居住的建筑。

对这个主题的关注是为了使阉人（既非男也非女的社群）这个被社会回避的团体更能够被接受，使他们被纳入社会主流，以及建筑怎样能更适应这一需要。我的主题给我一个理解阉人和与他们这个属于神秘群落并被社会所回避的团体共处的机会。我的建筑学的最后一年不仅仅对我建筑个性的形成有所贡献，还帮助我长大成人。

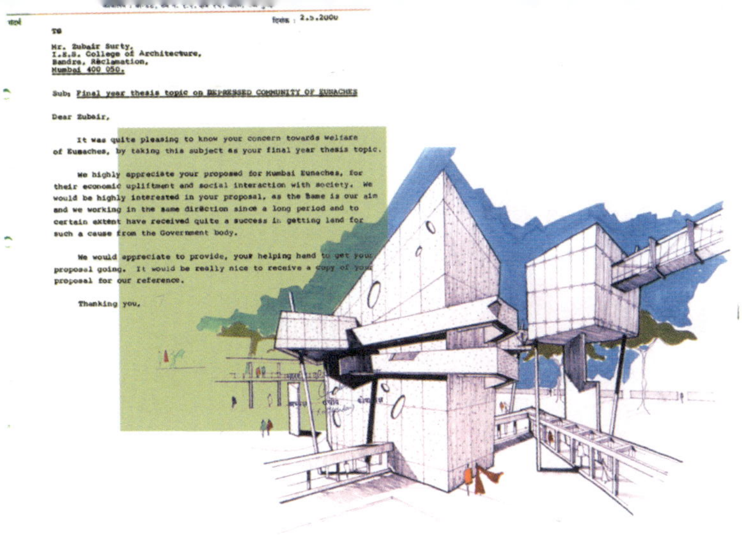

दिनांक : 2.5.2000

TO

Mr. Zubair Surty,
I.E.S. College of Architecture,
Bandra, Reclamation,
Mumbai 400 050.

Sub: Final year thesis topic on DEPRESSED COMMUNITY OF EUNACHES

Dear Zubair,

It was quite pleasing to know your concern towards welfare of Eunaches, by taking this subject as your final year thesis topic.

We highly appreciate your proposed for Mumbai Eunaches, for their economic upliftment and social interaction with society. We would be highly interested in your proposal, as the Same is our aim and we working in the same direction since a long period and to certain extent have received quite a success in getting land for such a cause from the Government body.

We would appreciate to provide, your helping hand to get your proposal going. It would be really nice to receive a copy of your proposal for our reference.

Thanking you,

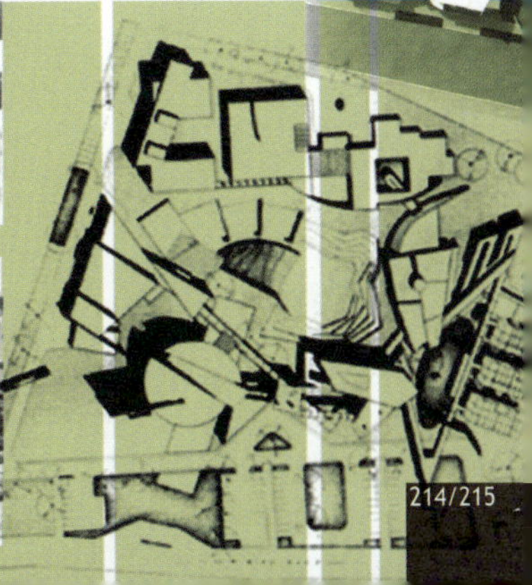

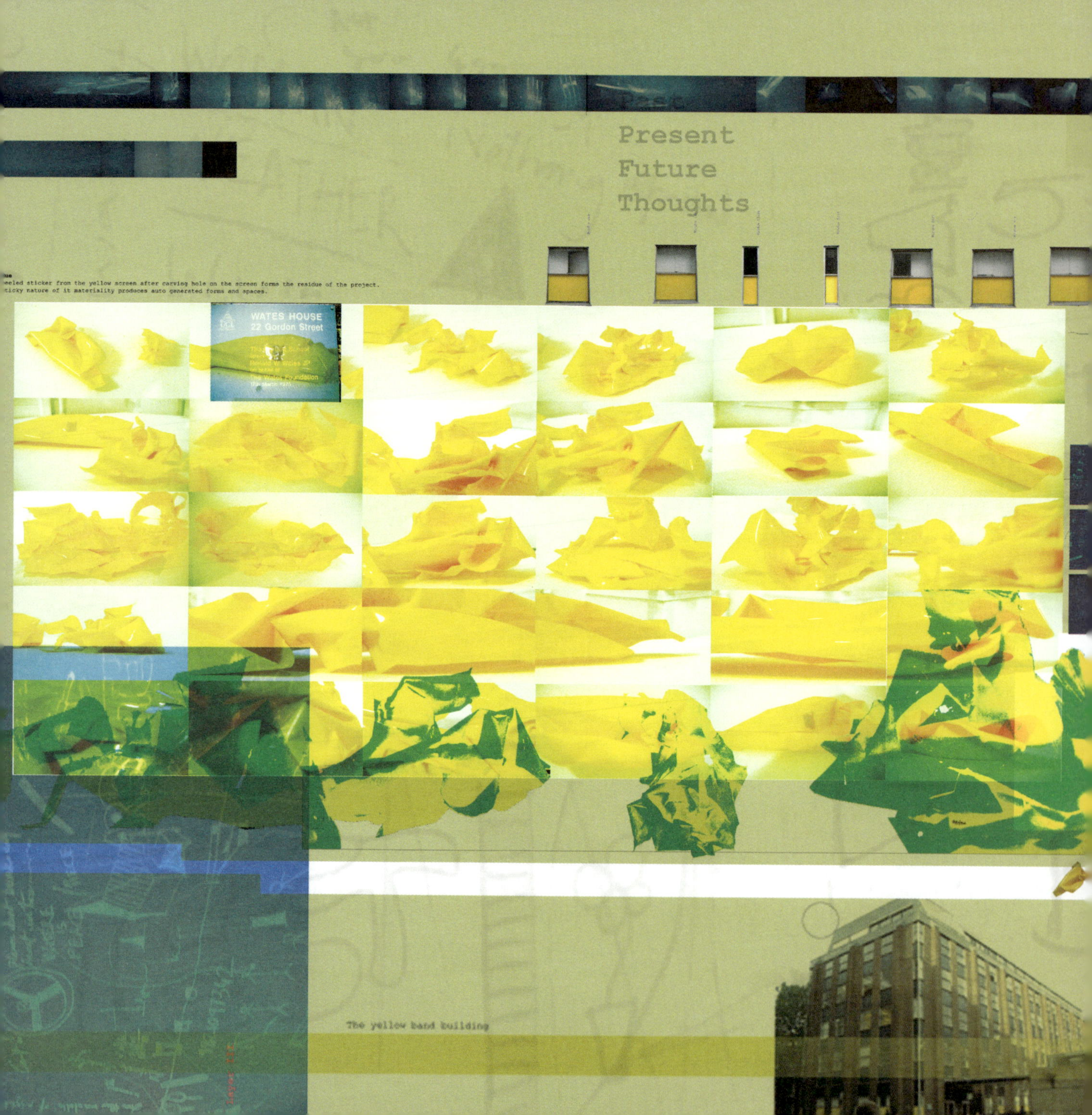
Present
Future
Thoughts
ue
eeled sticker from the yellow screen after carving hole on the screen forms the residue of the project.
ticky nature of it materiality produces auto generated forms and spaces.
WATES HOUSE
22 Gordon Street
The yellow band building

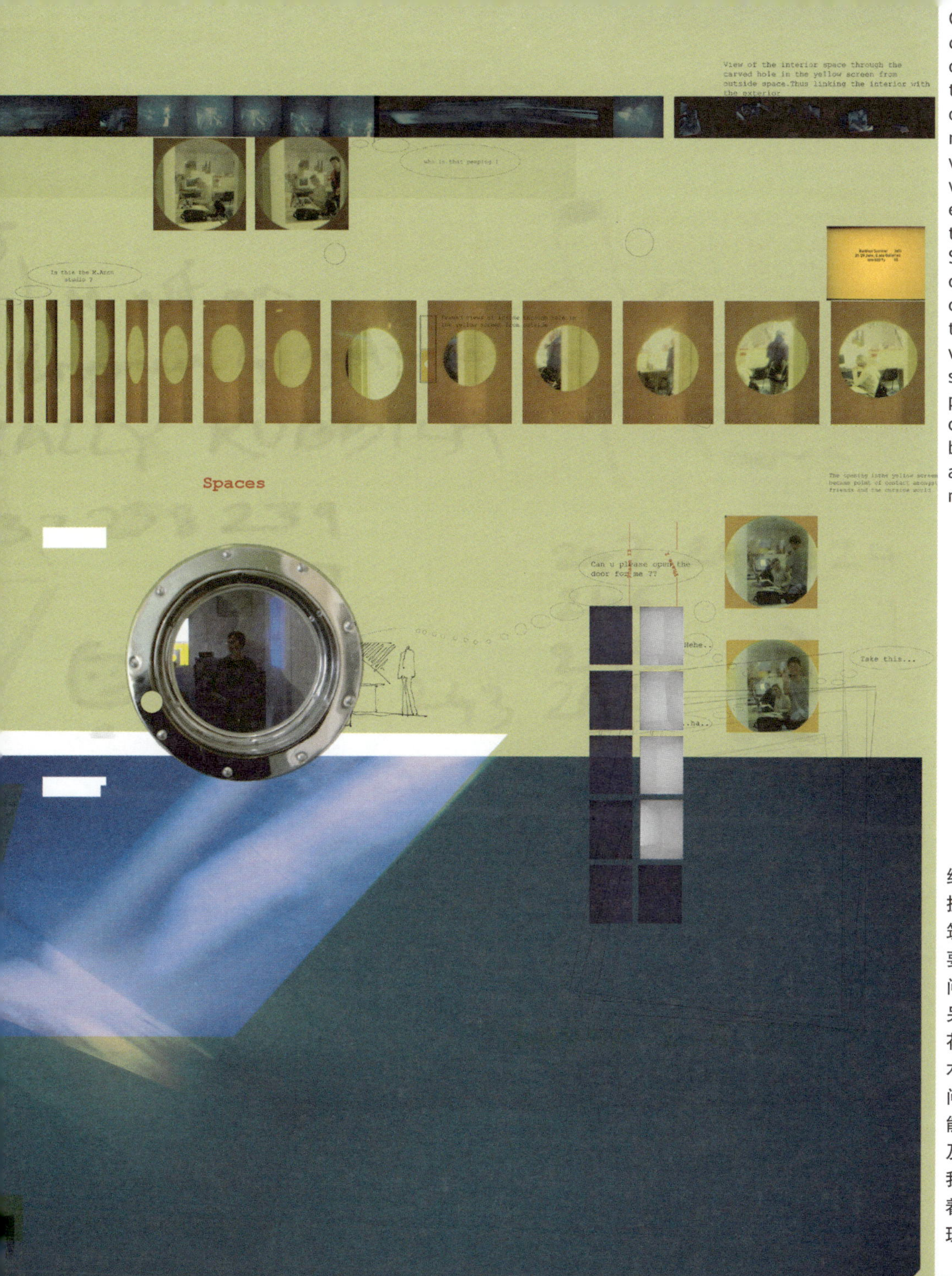

Completing my education I immediately plunged myself into my own practice. It was like welcome to reality from the dream world of architecture. Client requirements, project budgets, dealing with contractors and consultants were some of the aspects, which exposed me to the other side of the coin.

Spending all time on site with contractors, carpenters, electricians, plumbers and other consultants I learned the essential bits, which makes architecture possible. During these two years of practice and teaching in two architecture schools, I have always been occupied by the thoughts of anonymous architecture, where need is the only brief for design.

结束我的教育后，我就立刻把自己投入了实践之中。这就像欢迎从建筑的梦想世界来到现实中。客户的要求、项目的经费、与承包人和顾问们打交道是暴露在我面前的硬币另一面的一些方面。

花去所有的时间在工地与承包商、木工、电工、水工和其他的各种顾问打交道，我学到了使建筑成为可能的必要元素。在为期两年的实践及在两所建筑学校任教的过程中，我总是被对无名建筑的思考占据着，对其来说需要就是唯一的设计理念。

Such architecture lies in abundance in Mumbai ironically juxtaposed with the so called architect designed buildings, for me these so called designed buildings misses out on character, innovation, aesthetic and all other essentials of architecture.

On the other hand these small settlements and dwellings witness the very truth of the statement "Necessity is the mother of all inventions". My admiration of these user-designed aesthetics raised several unanswered questions in my mind. I found myself a great fan of the spontaneous innovation of their needs. These everyday observations and question were running parallel with commercial side of my practice. Realizing me being absorbed by the commercial stream of architecture came along a decision to take a break, Post graduation 'Master's in architectural design'. Some time out where I could dig answers for my unanswered questions and explore and stretch further the existing boundaries of architecture.

I chose London as the city, which has characteristics quite similar to that of Mumbai, where I come from. Also assuring myself of Bartlett as the school to be in I started my year with hopes and expectations. Free time being the theme for the year we started with introductory projects. The initial two projects were my immediate response to the notion of free time. First project was a two-phase light installation questioning the veracity and reality of the free time. There is nothing called free time, our mind is never free it's always occupied with something or the other.

We finish one job and start another. There exist no specific period of free time; it comes as uncontrollable fragments of time. This formed the base argument of the installation. Light played a vital role of a 'catalyst' a mediator changing the quality of space, revealing and concealing at the same time. Second project dealt with the same issue of free time but of friends. The project was an opening created in the existing window screen creating visual link between outside with inside and vice-versa. Both these projects raised several architectural issues like co-existence, incision, juxtaposing virtual against real, nothing & nothingness, the in-between and many more to be experimented during the year. From all words 'nothing' intrigued me the most. The word was unfathomable and complicated to handle. On discussion with tutors I developed a methodology to organize my series of thoughts, observations and experimentation with the word nothing.

The thoughts and observations were then transformed graphically into drawings and then into 3d models for experimentation. The result of this process gave me series of vignette projects leaving me confused with no single point or direction to hold on to. I was looking forward for the coming crits to get some advice. After the crits my position was nothing better then it was earlier. The best advice during the critic was to refer Marcel Duchamp's large glass. It seemed I got the key to the lock, and indeed it was. I needed to sew all my vignettes in to one project just like Duchamp did for his large glass. Twelve years of all his vignettes were knit together in the large glass. My hunt for a common theme combining all my vignettes was on set.

The Baloon

这样的建筑学大量存在于孟买，反讽似地与所谓的建筑师设计的建筑并置着。对我来说，这些所谓的有设计的建筑遗漏了个性、创新、美学和所有其他建筑学的要素。

在另一方面，这些小的定居点和住居见证了绝对的真理论断——“需求是发明之母”。我对这些使用者设计的美学的赞赏使几个尚未被回答的问题浮现在我的头脑中。我发现自己成为了这种根据需求而自发的创新的超级爱好者。像这样的日常观察和提问与我实践中的商业化的一面平行运转。当意识到自己正被卷入建筑的商业洪流中，我决定暂停一下，进行研究生阶段的学习——建筑设计硕士学位。一段时间的休整可以使我为那些尚无答案的问题找出解答，并探索和进一步拓展建筑学现有的疆界。

我选择了伦敦作为这个城市，它与我所来自的孟买有着极其相似的特征。也确保对巴特里特作为是我充满希望和期待开始我的学年的学院。休闲时光是我们这一年的主题，我们开始了开场项目。最初的两个项目是我们对休闲时光这一内容的直接回应。第一个项目是两段式的光装置，对休闲时光的多样性与真实性提出了疑问。没有什么能真正称得上休闲时光，我们的头脑从来没有空闲过，总是被这样或那样的事情占据着。

我们完成一项工作，又接着开始另一项。并不存在所谓休闲时光的特别时期，它就像无法控制的时间片断一样到来。这构成了这个装置最基本的内容。光扮演了重要的“催化剂”媒介的角色，改变着空间的品质，同时暴露着和隐藏着。第二个项目同样是休闲时光的主题，但却是关于朋友的。这个项目是在现存窗户屏蔽上的一个开口，这样就造成了内外之间的视觉联系和正反互逆。这样的项目都呈现出了一些诸如共生、割裂、并置虚拟替代现实，虚无与虚无性，居间等的有待这一年中实验的建筑问题。在所有的词汇中“虚无”是最能激起我兴趣的一个词。这个词是深不可测的和难以捉摸的。在和指导老师的讨论中，我发展了一种方法论来组织我对于虚无这个词汇的系列化的思考、观察和试验。

During this time we were off for our study trip to Los Angeles and Las Vegas.
Spending just one night in Las Vegas I was convinced that cynicism of the present Las Vegas situation of excess would help me collate the vignettes. I began generating my immediate response of Las Vegas in terms of drawings, models and collages. Using them as tools to criticise, reflect or highlight the existing Las Vegas scenario Bugsy 'Benjamin Siegel' a gangster whose vision for Las Vegas is held responsible for the present Las Vegas became my inspiration for the story.
I developed a storyline about the final wish of bugsy's soul visiting Vegas and admiring the extreme versions of his vision for the desert city. The storyboard formed the base of the projects guiding me towards the selection of site and development of the site programme. Selected site 'The Boneyard', a final resting ground of all neon signs from the strip. The boneyard was situated off the main strip completely secluded from the main action. These dilapidated signs lying in the boneyard were once the reason of strip glamour. It became very important to physically or programmatically connect the site with the strip. I preferred programme to do the needful... The boneyard to be developed as rhetoric Las Vegas museum fuelled by the underground casino, connecting site with the strip.

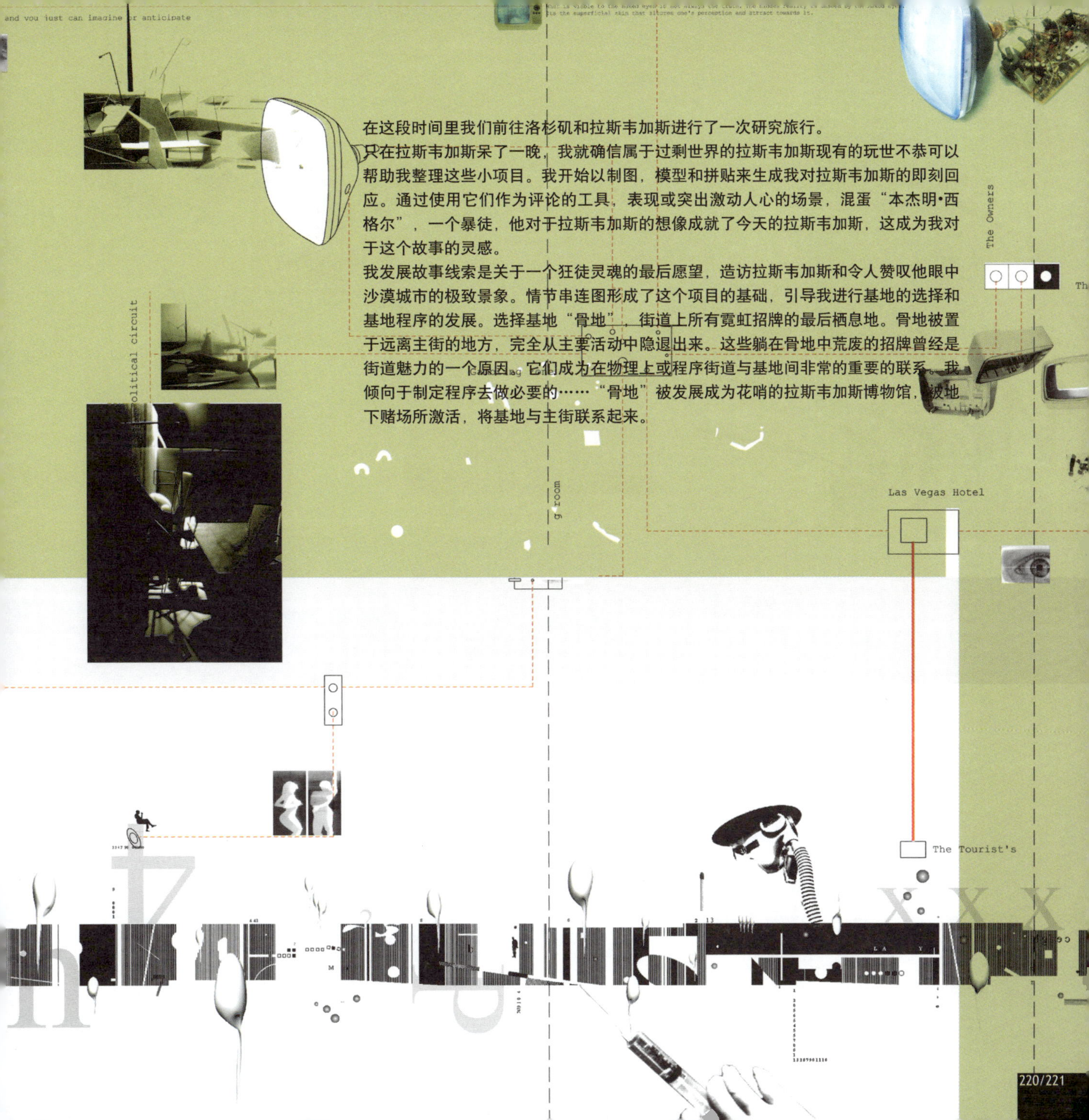

在这段时间里我们前往洛杉矶和拉斯韦加斯进行了一次研究旅行。

只在拉斯韦加斯呆了一晚，我就确信属于过剩世界的拉斯韦加斯现有的玩世不恭可以帮助我整理这些小项目。我开始以制图，模型和拼贴来生成我对拉斯韦加斯的即刻回应。通过使用它们作为评论的工具，表现或突出激动人心的场景，混蛋“本杰明•西格尔”，一个暴徒，他对于拉斯韦加斯的想像成就了今天的拉斯韦加斯，这成为我对于这个故事的灵感。

我发展故事线索是关于一个狂徒灵魂的最后愿望，造访拉斯韦加斯和令人赞叹他眼中沙漠城市的极致景象。情节串连图形成了这个项目的基础，引导我进行基地的选择和基地程序的发展。选择基地“骨地”，街道上所有霓虹招牌的最后栖息地。骨地被置于远离主街的地方，完全从主要活动中隐退出来。这些躺在骨地中荒废的招牌曾经是街道魅力的一个原因。它们成为在物理上或程序街道与基地间非常的重要的联系。我倾向于制定程序去做必要的……“骨地”被发展成为花哨的拉斯韦加斯博物馆，被地下赌场所激活，将基地与主街联系起来。

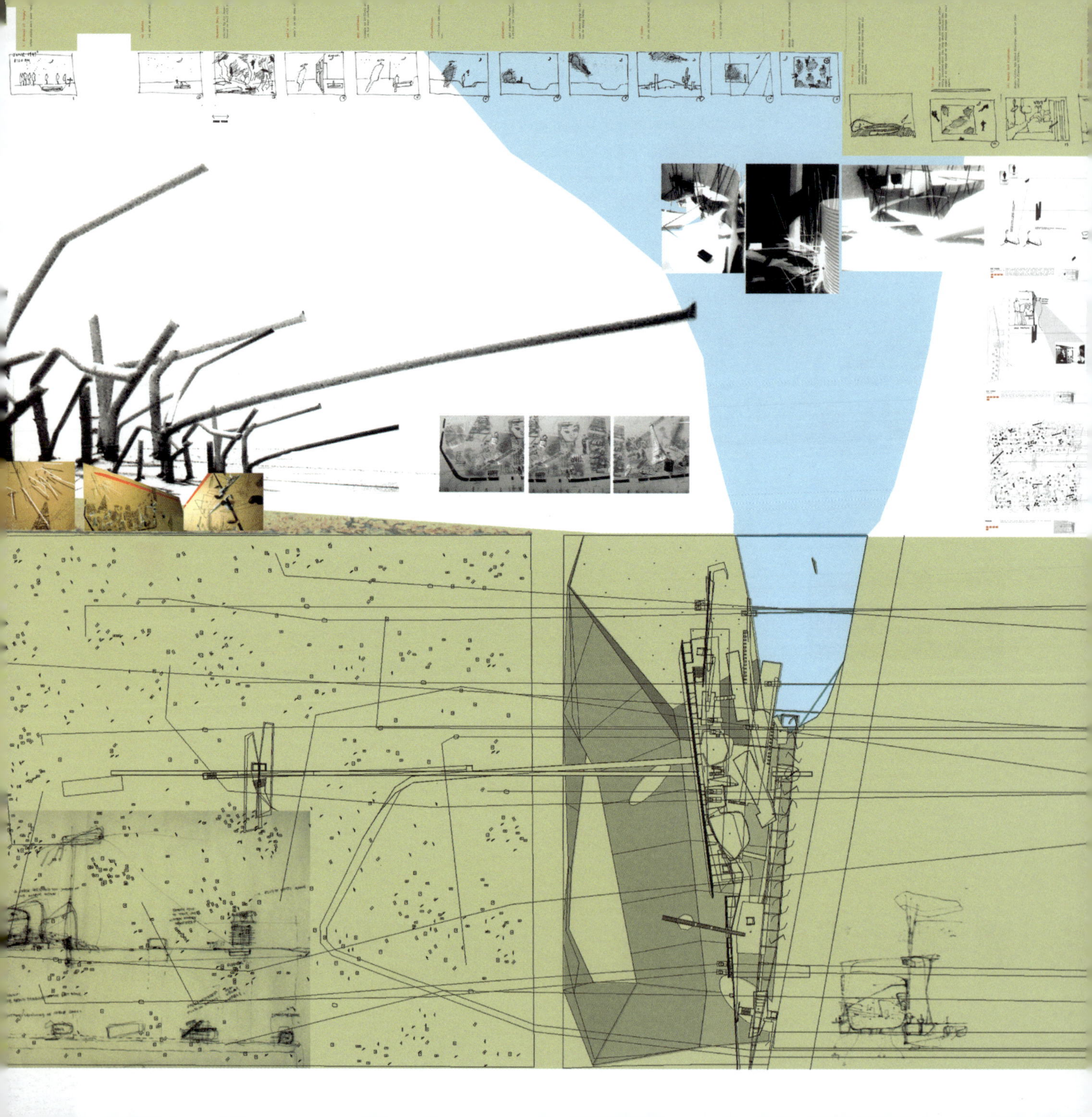

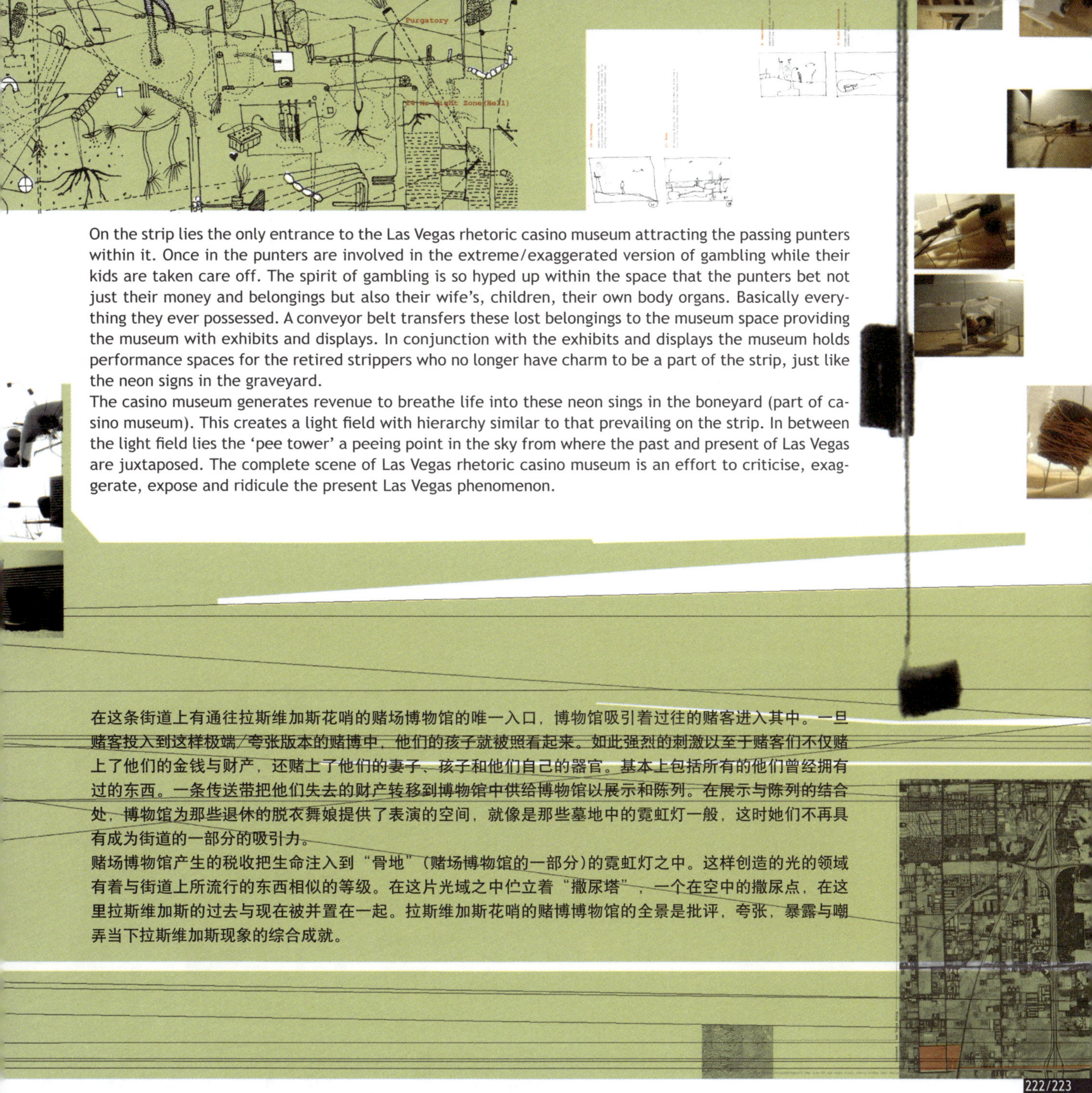

On the strip lies the only entrance to the Las Vegas rhetoric casino museum attracting the passing punters within it. Once in the punters are involved in the extreme/exaggerated version of gambling while their kids are taken care off. The spirit of gambling is so hyped up within the space that the punters bet not just their money and belongings but also their wife's, children, their own body organs. Basically everything they ever possessed. A conveyor belt transfers these lost belongings to the museum space providing the museum with exhibits and displays. In conjunction with the exhibits and displays the museum holds performance spaces for the retired strippers who no longer have charm to be a part of the strip, just like the neon signs in the graveyard.

The casino museum generates revenue to breathe life into these neon sings in the boneyard (part of casino museum). This creates a light field with hierarchy similar to that prevailing on the strip. In between the light field lies the 'pee tower' a peeing point in the sky from where the past and present of Las Vegas are juxtaposed. The complete scene of Las Vegas rhetoric casino museum is an effort to criticise, exaggerate, expose and ridicule the present Las Vegas phenomenon.

在这条街道上有通往拉斯维加斯花哨的赌场博物馆的唯一入口，博物馆吸引着过往的赌客进入其中。一旦赌客投入到这样极端/夸张版本的赌博中，他们的孩子就被照看起来。如此强烈的刺激以至于赌客们不仅赌上了他们的金钱与财产，还赌上了他们的妻子、孩子和他们自己的器官。基本上包括所有的他们曾经拥有过的东西。一条传送带把他们失去的财产转移到博物馆中供给博物馆以展示和陈列。在展示与陈列的结合处，博物馆为那些退休的脱衣舞娘提供了表演的空间，就像是那些墓地中的霓虹灯一般，这时她们不再具有成为街道的一部分的吸引力。

赌场博物馆产生的税收把生命注入到“骨地”(赌场博物馆的一部分)的霓虹灯之中。这样创造的光的领域有着与街道上所流行的东西相似的等级。在这片光域之中伫立着“撒尿塔”，一个在空中的撒尿点，在这里拉斯维加斯的过去与现在被并置在一起。拉斯维加斯花哨的赌博博物馆的全景是批评，夸张，暴露与嘲弄当下拉斯维加斯现象的综合成就。

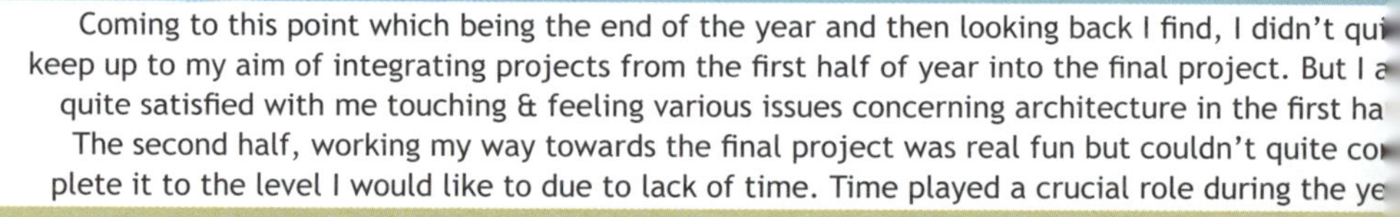

Coming to this point which being the end of the year and then looking back I find, I didn't qui
keep up to my aim of integrating projects from the first half of year into the final project. But I a
quite satisfied with me touching & feeling various issues concerning architecture in the first ha
The second half, working my way towards the final project was real fun but couldn't quite co
plete it to the level I would like to due to lack of time. Time played a crucial role during the ye

当到了这一年的末尾，当我回头再看时发现，我并没坚持自己把这一年前半部分项目结合到最终项目之中的目标。但是我非常满意自己在前半部分接触和感受了各种各样与建筑相关的事物。后半部分，用我的工作方式达成最终的项目是真正的乐趣，但是由于缺乏足够的时间，没能完全达到我所希望的程度。在这一年中，时间扮演了至关重要的角色，我花去了大量的时间在目标周围打转，然后击中目标。总体上，这是一段从我自身，以及从我朋友的经验中学习的愉快时光。通过这些课程，思想和与朋友们的协作，我将自己投入到“移动回应小组”之中，希望进行试验和在建筑中取得乐趣。

nd I spent a lot of time running around the target then hitting it. Overall had great time learning om my own, as well as experiences of my fellow friends. With these lessons, ideas and collaboraon with my friends I launched myself into practice of architecture with ‘Mobile Response Unit’, oping to experiment and have fun with architecture.

Central London, Stratford
Chelmsford, Romford
N. Circular A406
Ilford, Barking
Docklands (A13)
Huntleigh
Healthcare...it's part of our name

大卫
塔基曼
David
Tajchman

david tajchman architect dla, m.arch
大卫•塔基曼 建筑师，建筑学硕士

1977 birth the second of january, Brussels, Belgium
1994 secondary school certificate
1994/99 architecture studies at Institut Supérieur d'Architecture Victor Horta, Brussels, Belgium
1996 housing and theater project in Dubrovnik, Croatia
1999 architecture diploma, distinction
registered as a professional architect at Architecture Council, Brussels
video-installation set in the basement of the South Train Station, Brussels
1999/02 working for several architects in Brussels on competitions and private orders competition for a cultural center, Merelbeek, Belgium
Set design and cooperation in the direction of a play : 'Fragments du discours amoureux' from Roland Barthes
own initiative projects: evolutive lighting installation for a railway station, Brussels
transformation of a place in Brussels
member of juries: Institut Supérieur d'Architecture Victor Horta
2002/03 M.Arch(Master in Architectural Design) supervised by Pr. Peter Cook, the Bartlett School of Architecture, UCL, London, U.K.
2004 Opening own practice in Brussels
Working at Agence Dominique Perrault, Paris, France
2005/06 Working at Jacques Ferrier as project leading architect
2007 Working with Stéphane Maupin, Beckmann-n'Thepe
2008 Working with Patrick Jouin

1977 生于比利时布鲁塞尔
1994 中学文凭
1994/99 维克多宏特建筑高等学院学习建筑，比利时，布鲁塞尔
1996 住宅和剧院工程，杜布鲁尼克，克罗地亚
1999 获建筑执业文凭，优秀毕业
取得职业建筑师资格，布鲁塞尔
影像装置，布鲁塞尔火车南站
1999/02 参与多位建筑师主持的在布鲁塞尔的竞标和私人项目
布鲁塞尔梅鲁比克文化中心竞标
为源自罗兰•巴特的《恋人絮语》戏剧提供设计，进行
自发项目：火车站革命性照明装置，布鲁塞尔
布鲁塞尔一地的转型
维克多宏特建筑高等学院评委
2002/03 建筑设计硕士，伦敦巴特雷特建筑学院，在彼得•库克
的督导之下
2004 在布鲁塞尔独立开业
在法国巴黎为多米尼克•佩罗事务所工作
2005/06 作为领衔项目建筑师为雅克•费拉里工作
2007 先后为斯蒂芬•末彬，贝克曼—N' 特皮工作
2008 为帕特里克•基容工作

e irony of architecture
筑之反讽

ew up in Brussels, in the middle of three typologies: on the left ypical sixties modernist ensemble influenced by Le Corbusier as, on the right one of the only architectural examples exist- in Belgium: the Atomium dating from the international expo of 8.

studies in Brussels have given me the faculty to think, to judge, analyze and to program the project, in fact a way to approach job in pragmatic and theoretical views. In my former school, the portant thing was not so much the answer, but more the ques- n. At the Bartlett, it is exactly the opposite: people are more cerned by how is it going to look like?

m both methods of learning or making, I try to keep both in nd. I define myself as an intellectual making. I'm a product of h schools, influenced by what I see in my everyday life. I keep in d the optimism of the Bartlett, and the sensitivity of my teach- back in Brussels.

en you say you're from Belgium, foreign people will automatically er to surrealism, as if Belgium was a concentration of Magritte's nes. I'm not feeling surrealist at all, I'm more concerned by archi- ture as a tool to innovate without any filiation to any style

我成长于比利时，成长于三种建筑类型学之中：左边是典型的受勒·柯布西耶影响的1960年代现代主义的集体演出；右边是比利时仅有的具有建筑学意义的实例之一：始自1958年世界博览会的“原子球”。

我在布鲁塞尔的学习赋予我思考、决定、分析和制定方案的能力，这实际上是一种建立在实效性与理论性观点上的工作方法。在我以前的学校里，答案不那么重要，问题才重要。在巴特雷特学院，这一点完全相反：人们对它看起来会怎么样更感兴趣。

记取两种学习工作的模式，我试图兼收于心。我把自己定义为知性的产物。我受到日常见闻的影响，是两个学校共同的作品。我将巴特雷特式的乐观和布鲁塞尔师辈们的敏感同时留记心中。

当你说你来自比利时，外国人会自然地联想起超现实主义，好像比利时就是马格利特克隆人的集中地。我倒一点也不超现实，而更多地把建筑作为创造的工具，无关风格渊源

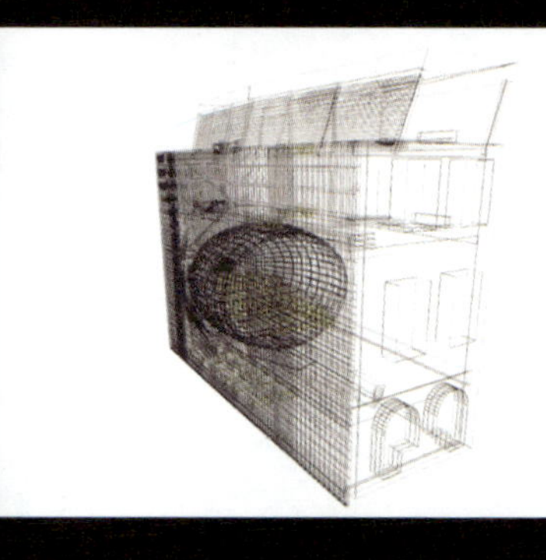
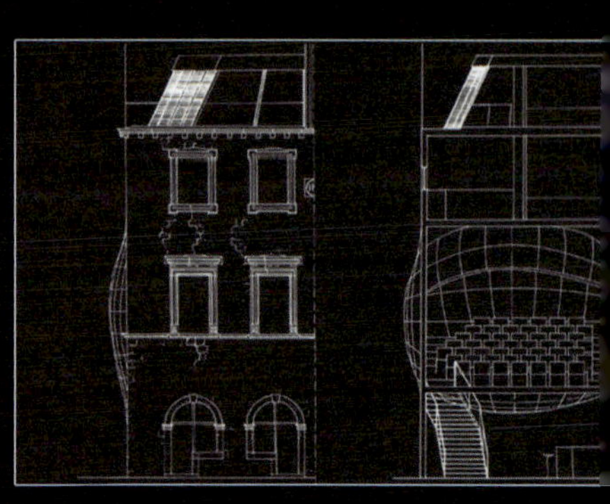
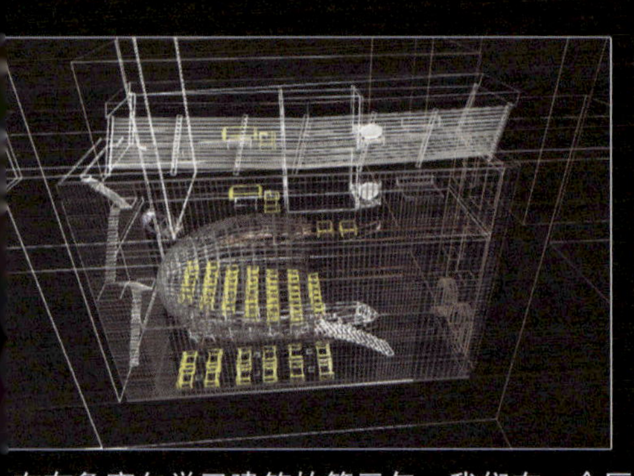
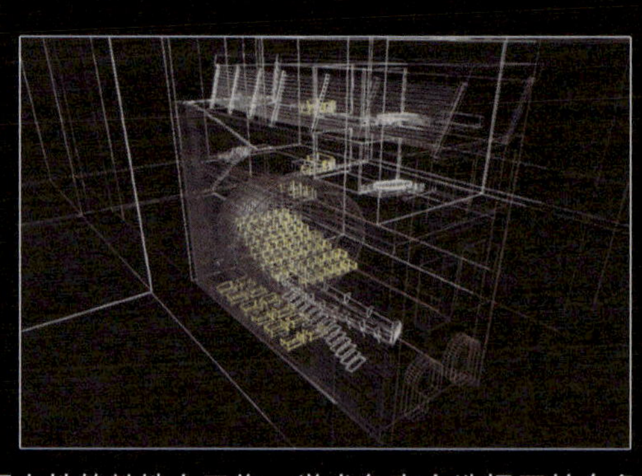
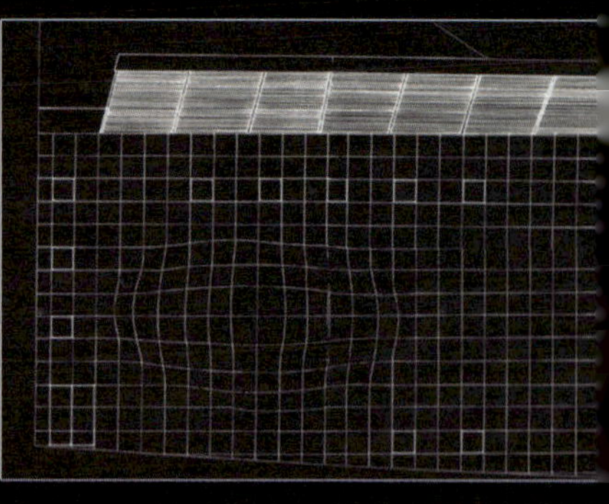

在布鲁塞尔学习建筑的第三年，我们在一个历史性的基地上工作。学术负责人选择了杜布鲁尼克，一个位于海山之间的克罗地亚城市被城垒所环绕，老城这样被保护起来，几近隔绝，像一座活着的博物馆。作为克罗地亚族和塞尔维亚族之间战争的受害者，这座城市一个旅游景点，带着它丰富的文化，作为一个历史性标记……然而，与我们来杜布鲁尼克时经过的景观相反，这里类似于一个无人带：被毁坏的房屋、废弃的海边度假地……杜布鲁尼克，虽历经轰炸，仍保有旅游城市的所有特征：它的活跃、它的密度、它的餐馆它的酒吧，所有这些造就了这座城市。我们可以选择在四块基地上工作。它们都在老城墙之内。它们都曾经被作为轰炸的目标。它们保留有当地石料制成的巨大墙体。四个基地中的一个位于老城的主街上，这条主街将现代城市联向港口。在这个基地上，没有楼板、有屋顶，只有巨大的墙体留存下来。我选择探讨内部建筑的外在表象。因为杜布鲁尼克是大多由遮挡阳光影响的窄巷构成，我对赋予个项目的主墙以街巷的感觉着迷，希望以此来暗示人们一些事情发生在其中，而不采用诸如图形或霓虹灯等其他的标识。

rrowness_ dubrovnik 狭窄 杜布鲁尼克 1996

96 During the third year of the architecture studies in Brussels, we worked on a historical site. The academic staff had chosen Dubrovnik, a city in oatia located between the sea and the mountains. Surrounded by ramparts, the old city is protected, nearly isolated, like a living museum. Victim o of the war between the Croats and the Serbs: the city looks like a tourist attraction, with its cultural richness, being a historical symbol... How-er, contrary to the landscapes we have crossed to get to Dubrovnik, resembling to a no-man's-land: demolished houses, deserted beach resorts... brovnik, although it had been bombed, still had all the features of a tourist city: its activity, its density, its restaurants, its bars, all these things de it a city. We had the choice to work on four different sites. All of them are within the walls of the old city. All of them have been the target bombs. All of them have kept their large walls made with the local stone. One of those four sites was located on the main street of the old city, ich links the modern city to the port. On this site, there were no floors, no roof, only the large walls remained. I chose to work with the external rception of what's inside a building. As Dubrovnik is made most of the time of narrow lanes to avoid sun effects, I found intriguing to give the feeling t the main wall of the project being in a lane, could suggest to people without any other signs like graphics or neon lights, that there is something ppening inside.

perveranda_ brussels 2004 超级阳台 布鲁塞尔 2004

ouple of psychologists came to me saying they get old and want to move from the first floor of their house to the basement and the ground floor, order to minimize climbing the staircase and enjoy more their large garden. The back facade of this house has been typically extended with years, d looks charming but not so well built. My proposal was first to bring natural light in the basement in order to fit for habitation and, by drawing a pe from the garden level to the basement level, create a relation between bathroom + bedroom and garden. This relation is visual and allow to the lars to be considered as a real floor. Then I remembered when I visited for the first time their house, that the view from this façade was amazing.

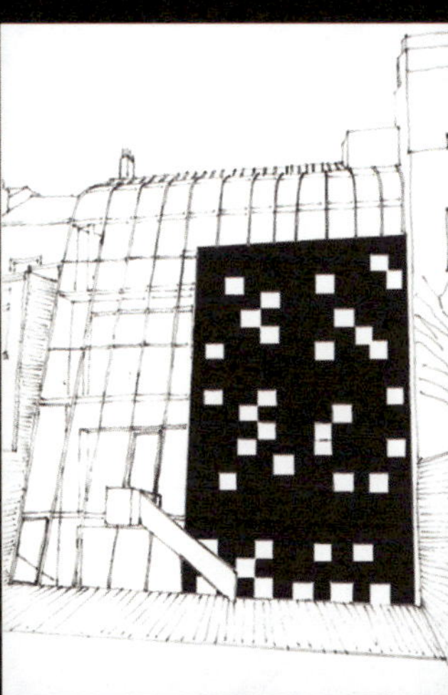

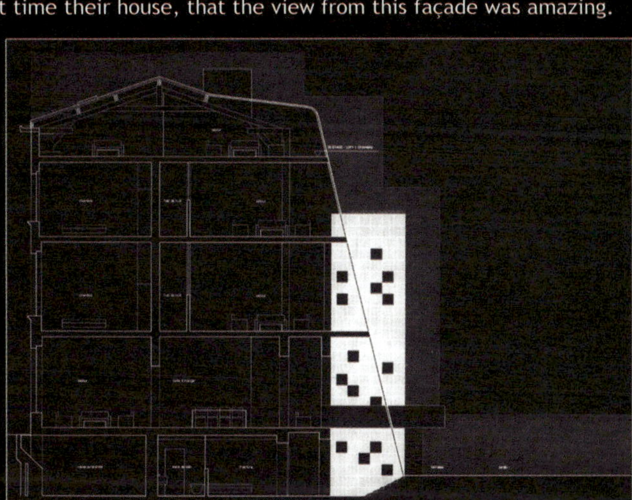

:ated on the top of a hill in the center of Brussels, you can enjoy the panorama with significant towers of churchs or roofs of old buildings. The cur-t occupation of the upper levels was not taking advantage of this situation. That's why I decided to continue the glass facade up to the roof level noving a part of it and replaced by glass as well. This new glass-skin wraps the existing facade and its extensions, giving it more homogeneity. The ck volume with square windows, containing kitchens or bathroom, diversifies and frames view points on the panorama. The last floor, formerly gar-, is known opened to the view and convertible.

对心理学家夫妇找到我说他们老了，想从二楼搬到一楼和地下室居住，来减少爬楼梯并更好地享受他们的大花园。这座房子的后立面型地随着岁月而拓展，看起来吸引人但做工不那么精良。我的意图首先是从花园到地下层建立一个坡道，为地下层带来适合起居的阳，并且制造出卫生间+卧室和花园的一种关系。这种关系是视觉的并且允许地下酒窖被当作真正的楼层。接下来，我记起当我第一次造他们的房子时来自里面的景观让人惊叹。它坐落于布鲁塞尔中心区的小山上，你可以欣赏到教堂尖塔和老建筑屋顶的全景画。现有的边的楼层没有利用这样的有利状况。这就是我为什么决定将玻璃立面延伸至屋顶，并将部分屋顶以玻璃取代。这个新的玻璃皮肤包裹现存立面和它的延伸，赋予它更多同质性。黑色的带方窗的体量包含有厨房和卫生间，丰富和框定全景画中的视点。最后一层，以前顶楼，现在可以变成敞篷的对景观开放。

endless park_ brussels 1999 diploma project At a time when the limits of a city are more and more difficult to determine and in a consciousn of the increasing density of construction in a given territory, "empty spaces" remain and will probably become the object of longing: railway slop highway borders, airports zones... are transit places, no-man's lands created to link points and to reduce the time between them. Given inaccessi voids within a city, the horizontal sweep and the density of which are increasing, the proposal accepts their current state while rendering them

无尽的公园　布鲁塞尔 1999 执业文凭项目 在这一刻，城市的界限越来越难以确定并且在给定的界限中有意识的增加建造的密度，"空地"留有来并且可能成为期盼中的物体：铁路坡道，公路边界，机场地区……成为转运地带，无人之地被创造出来以联结各点并且缩短其间的时间。在给定的城的不可到达的空地，在水平性的伸展和增加的密度下，设计意图在接受现有状况时渲染他们的可到达性。项目依赖于它自身变化的可能性：实际上大多

asphalt bethlehem_ brussels 2001 (with Théo Sarantoglou arch.) This project conserves the existing topographic givens, it questions the curr pavement by putting a single, supple material, an extension of the street. The ground becomes a support of different activities: social, play, comm

沥青伯利恒　布鲁塞尔2001 （与建筑师提奥•萨隆托格鲁合作）本设计保存现有的地形，以使用单一软质铺地的方式延伸街道了，质疑了现有的人道。 地面成为多种不同活动的支持体：社会的，玩乐的，商业的，文化的，教育的，事件和电影式。场地成为一种扎根的方式：寻求给与人们对场所的

fla(t)nders_ merelbeek 2000 (competition; with Renaud Menestret & Jean-Marc Sterno arc This cultural center is built on a boggy ground. Its linear configuration allow to cross the entire

佛兰德斯_梅鲁比克2000 （竞标：与建筑师雷诺•梅奈斯崔德和让-马克•斯德诺合作）这个文化中建在一块多沼泽的场地上。它的线性外形允许它横跨整个基地俯瞰佛莱芒人（译者：比利时佛兰德

sible. The project relies on the possibility of its own mutation: in fact most of the plants live on the site come from flows of trains. The treatment he axis in the project is different from the rest of the slope. The major element integrated into the conception of the park is the passage of trains, siderated as events, in this way, a work on the thickness of the ground by folds, hollows, crevasses... create the conditions to observe and be sur-sed by trains.

被生长在由火车流带来的基地之上。项目中轴线的处理与坡地的其他部分不同。主要结合进公园概念的元素是火车的驶过，被作为事件来考虑，以这样的式，一项以折叠、空洞、碎裂等手段在场地的深度上的操作创造出观察火车并为之震惊的条件。

, cultural, educational, for events and cinematic. The ground as a way of taking root: seeking to give roots and a sense of ownership of the space, ular replanting of new trees (Mediterranean pines) and other plants, leading to a more singular reading.

感和根植感。规则的重植以新树（地中海松）和其他植被，带来更加特定的解读。

king the flemish landscape. The roof of the building is used for external activities and looks like an agricultural landscape: carpets of stone, grass or vers cover the roofs but also the parking and the footpaths.

使用佛莱芒语的民族）的景观。建筑的屋顶被用作外部活动并且看起来像是农业的景观：石头、草和花的地毯覆盖屋顶以及停车场和步行道。

collecting

the first step of the project is to rapidly sketch ideas and illustrate with diagrams more complex phenomenon appearing in the project. As the project was worked on one year development, many drawings and diagrams were done, I show here samples of different steps of my process to start a project.

收集

这个项目的第一步是快速草绘出构思和对显现在项目中的更加复杂的现象给以图解。作为耗时一年来发展的项目，有很多的图纸和图解，在此我仅展示一些开始一个设计不同阶段的过程示例。

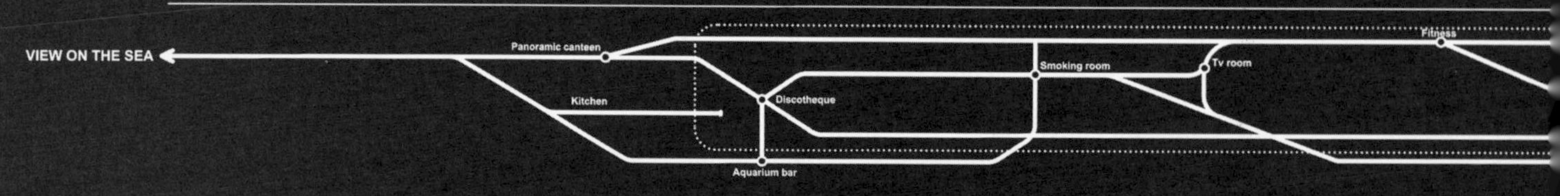

LABYRINTH
Plunge pool
PIER AND CELLS LEVEL
Surveillance path
Scuba diving
Visitors entrance
BLACKPOOL
Swimming pool
Whirlpool
Bowling
Milk-bar_Parlour
Free entrance
Sauna
Showers
Wc
Cloakroom
Check-in
Hall
Surveillance path
Check-out
Way out
Free entrance
BEACH LEVEL

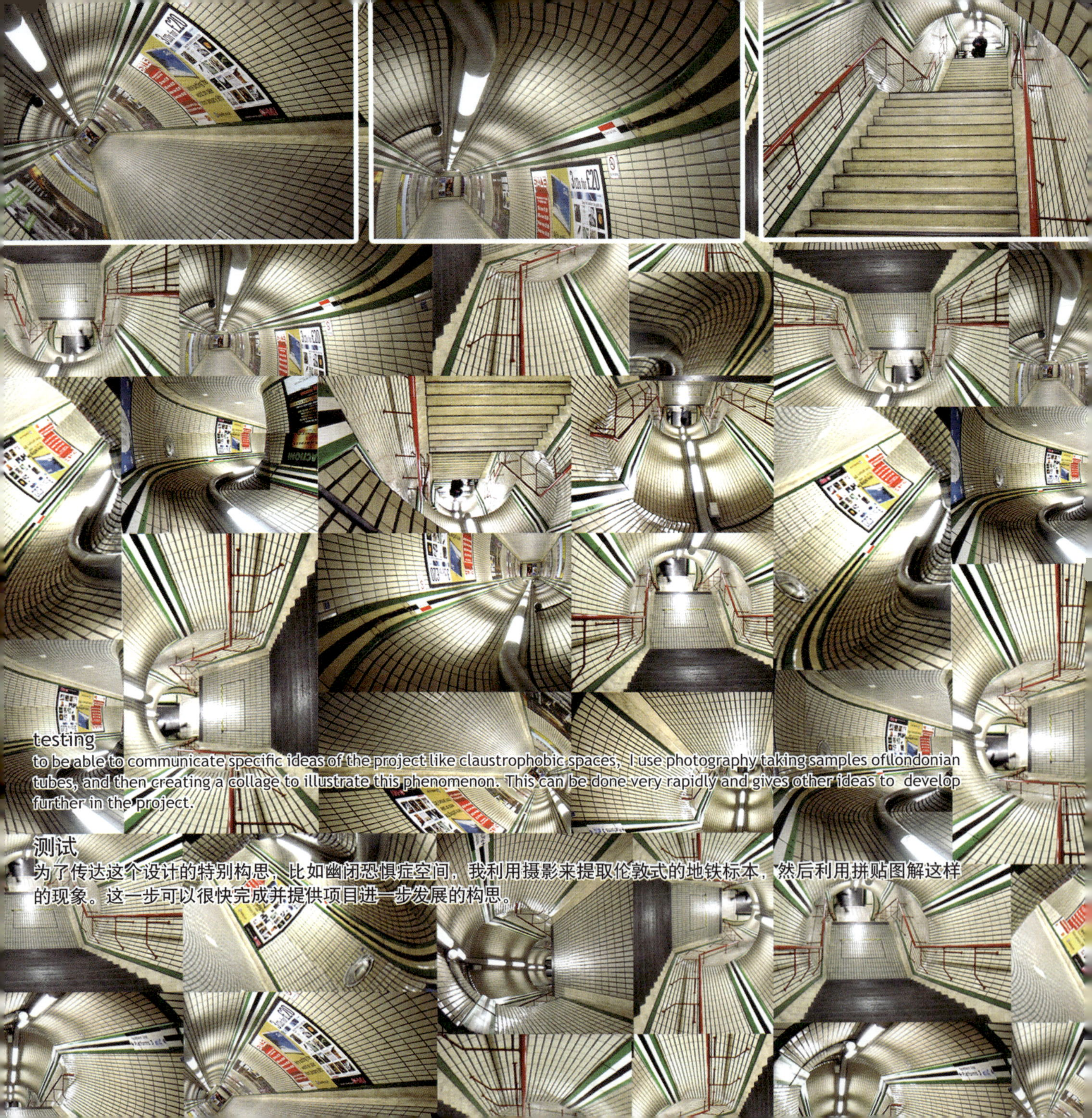

testing

to be able to communicate specific ideas of the project like claustrophobic spaces, I use photography taking samples oflondonian tubes, and then creating a collage to illustrate this phenomenon. This can be done very rapidly and gives other ideas to develop further in the project.

测试

为了传达这个设计的特别构思，比如幽闭恐惧症空间，我利用摄影来提取伦敦式的地铁标本，然后利用拼贴图解这样的现象。这一步可以很快完成并提供项目进一步发展的构思。

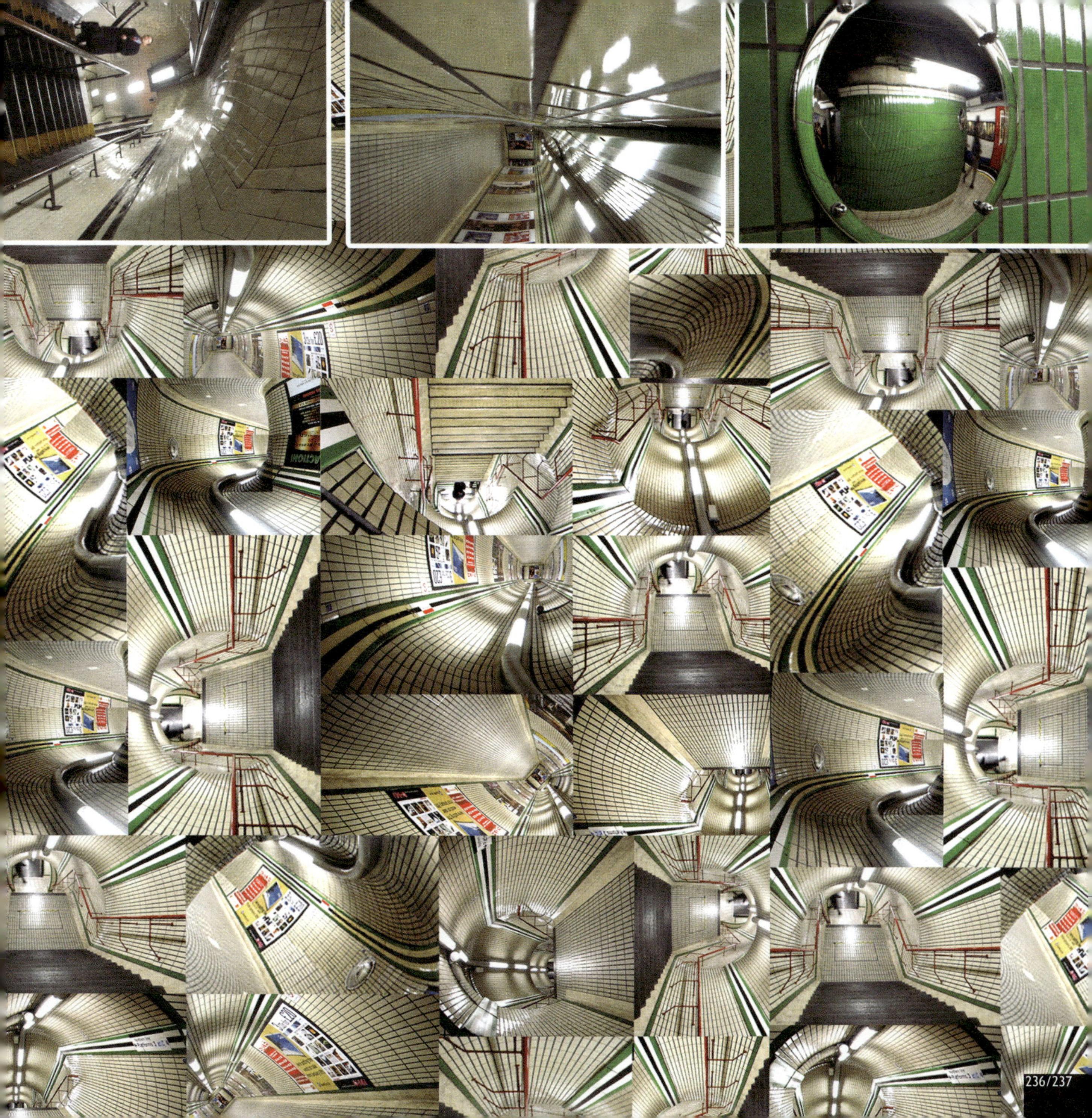

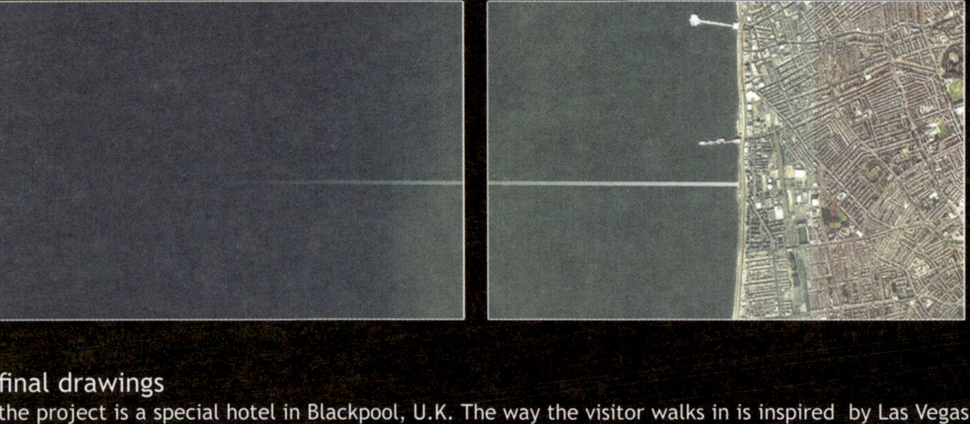

final drawings

the project is a special hotel in Blackpool, U.K. The way the visitor walks in is inspired by Las Vegas Resorts, where everything is done to maintain the visitor the longest period of time inside. Located along the beach, as a pier, this project criticises our consumer society. As people are always looking for new attractions, or entertainment, this hotel might be an answer to their requests.

最终绘图

本项目为英国布赖克浦的特殊旅馆。旅客在其中的行进方式受到拉斯维加斯度假地的启发，在那里，所作的一切都是为了将访客最长时间地留在其中。坐落于海边，就像一个栈桥，这个旅馆批判了我们的消费社会。既然人们总是寻求新的刺激和娱乐，这个旅馆可能会为他们的需求提供解答。

导言

伦敦一直以来为其欢迎和吸收世界性建筑图景的理念而感到骄傲：尽管近距离的体察会发觉“英国建筑”的价值与习俗依然在你莅临乡间时闪耀。

大量国际化的所谓“锋锐”建筑成为近来几年的特质。你得努力观察福斯特、哈迪德事务所的产品才能追寻到其中德国建筑师的份额（相当高的）。你能体悟到齐普菲尔德事务所瑞士和拉丁的成分。你能在巴特雷特和AA分辨出大量希腊、德国、中国、南美或伊比利亚的元素。

已然，层叠集体文化的英国传统中的某些东西体现在这本书中。优雅的混合技术、社会学、感性、文脉于一个对话中。避免论战，或者（我们英国人会这么说）在欧洲其他传统中作出乡土姿态。

看着巴特雷特的建筑学硕士课程在批判性价值的巅峰发展——就我来说，这种价值源自于我文化中开放的自由主义，由我从英格兰传统和在AA建筑联盟的一段特定时期继承和汲取而来——就是观察外国学生美妙的混合。我曾经有一次和FOA的阿利桑德拉·扎耳·保罗对此进行过讨论，他攻击这种他称之为“盎格鲁-撒克逊自由艺术传统”。很坦率地说，我相信这一“传统”发展于精神层面上的理解多于固守。后者常常被认为有法西斯主义和傲慢自大的弦外之音。

思及以往，我仍然可以回想起那个来自东方的学生说：“我要学巴特雷特风格”（我毫不怀疑马路下段他的女朋友会要求“学习AA风格”）（译者注：AA建筑学校在巴特雷特学院同一条街道的南段。）

经过一段时间，观察到最有意思的毕业生发展出他们自己的“风格”是令人鼓舞的——或者至少，进行观察后将观察变作回应，然后将回应变作可以被发展的格调。

一路顺风！

彼德·库克

INTRODUCTION

London has always prided itself on the idea of welcoming and absorbing a cosmopolitan architectural scene: though closer inspection will reveal that the values and mores of 'British Architecture' have remained somewhat blinkered when you get up country.

What is a characteristic of these last years is the massive internationalism of so-called 'cutting edge' architecture. You have to look hard at the output from the Foster or Hadid offices to trace the proportion (rather high) of German architects within them. You can discern the Swiss and Latin component of the Chipperfield office. You can ponder upon the plethora of Greek, German, Chinese, South American or Iberian tutors at the Bartlett and the AA.

Yet there is something in the English critical tradition that overlays the collective culture represented in this book. With a way of gently combining the technical, the sociological, the demonstrative, the contextual - all within one conversation. Avoiding the polemic or, (as we English would see it) the posturing endemic in other European traditions.

Watching the Master of Architecture course at the Bartlett develop on top of critical values - coming (in my case), out of the open-ended liberalism of my culture (picked-up in provincial England and at a certain period at the Architectural Association) - is watching a fascinating mix with the foreign students. I once had an argument with Allesandro Zair Polo (of FOA) who attacked what he called the 'Anglo-Saxon Liberal Arts Tradition. Quite frankly, I believe that this 'tradition' develops more out of a psychology of understanding rather than insisting. The alternative often has overtones of Fascism or arrogance.

Mind you, I can still recall the occasional student from the East saying: "I want to learn Bartlett style"(and I have no doubt that his girlfriend down the road would be asking to "learn AA style").

It is encouraging to observe that after a while, the most interesting Graduates are developing their own 'style ' - or at least, approach to observing and then turning the observation into response and then turning the response into Mannerisms that can be developed.

Bon Voyage !

Peter Cook

P194-

丽莎
修
Lis
Silv

P138-151
迈克
米切尔
Michael
Mitchell

P052-065
阿曼多
赫南德兹
Armando
Hernandez

P122-137
里奥纳多
拉塔瓦
Leonardo
Lattavor

P102-121
洛乌
昆资
Raoul
Kunz

P066-085
娜奈德
亚科斯基
Nannete
Jackowski

P002-017
卡提娅
阿柯曼
Katja
Ackermann

P180-193
提奥
…隆托格鲁
Theo
…arantoglou

P086-101
玛若
卡利马尼
Maro
Kallimani

P166-179
蒂诺斯
帕帕蒂米卓
Dinos
Papadimitriou

P018-033
沃肯
奥卡纳格鲁
Volkan
Alkanoglu

…226-239
大卫
…塔基曼
…avid
…ajchman

P034-051
丹尼尔
丹卓
Daniel
Dendra

P152-165
潘岩
Yan
Pan

P210-225
祖贝尔
苏提
Zubair
Surty

伦敦成为当下国际设计潮流的节点，头脑风暴呼啸的山谷。我们需要理解，在微观层面，什么样的事情，以怎样的方式在这里发生。

本书收录了15名来自世界各地的新生代建筑师，他们在不同的教育体制下，开始成为建筑师的人生之旅，建立各自对建筑的观点后，被伦敦所吸引，来到这里学习。对很多人来说，这段经历是他们建筑道路上的一个转折点。

本书总体分为两个部分，第一部分中，每位建筑师独立编为一个章节，我们邀请他们以一个小故事开始，谈谈关于自己和伦敦。建筑师们精心选取了自己最有代表性的作品，包括在伦敦建筑设计的中心之一——巴特雷特建筑学院学习时的，和到伦敦之前，之后的。在我们的要求下，除了精美的图片之外，每个人用文字详细地介绍了设计过程和思想方法，试图为国内读者提供最为真实的设计思维的研究资料。在可能的情况下甚至提供一份个人简历，以便我们看清一个独立的追询自我的建筑师的成长之路。

在这样的框架下，我们给予每个人最大的自由度，以保证充分的多样性。继以希望不仅提供作为设计项目层面的研究资料，更回归到人。不研究每个人经历，性格的特殊性就无法理解这些年，国际设计舞台纷繁芜杂的状态。对建筑师个人的研究显然可以帮助每个建筑从业者，建筑系、设计系学生关注自身，发现自己的兴趣，建立自己的建筑学。书中每个人独立在规定框架下设计了自己的单元，这给我们进一步提供了窥探不同思维路径的机会。不同的背景，不同国家建筑教育体制的影响在其中显现，这样，本书作为一扇窗口，不仅开向伦敦，而且引申向四大洲的多个国家。

本书的后一部分展示了部分参与者为北京设计的一个小装置，奥运对于北京的影响应该是可持续的，重要的不仅是奥运本身，而更是奥运之后。以此向中国读者致敬。这部分的初衷是为国内的读者提供一个即时的小范例，为对同一给定题目下的设计思维的探讨提供更多的资料。

感谢本书的每一位参与者，并感谢他们无条件的友谊。没有他们的支持，本书将不可能出版。特别感谢为本书的出版编辑作出贡献的诸位朋友。并感谢彼德库克爵士，他对建筑理解和对学生的热情使他不时浮现在本书之中。

London: the junction of design where trends intertwine and the brainstorm swirls. We need to understand at a micro level what and how things happen here.

This book documents 15 architects of the new generation from all over the world. They were educated in different education systems. To start building their ideas about architecture, they were all attracted by London and continued their further education in this vibrant metropolis. For most of them, their experiences in London were the turning points of their architectural careers.

There are two parts to this book. In the first part, each architect receives a separate chapter. They start with a short narrative, talking about themselves and London. Then, representative works are carefully selected by themselves, including projects completed during their study at the Bartlett School, which is one of the main architectural design centres in London and the world, and presented alongside projects before and after the Bartlett. Associated with the intention of the book, everyone offers fine images and texts about their design, to illustrate their strategies and approaches. Some CVs are given, which help the readers trace these self-reflective architects' career paths.

In this framework, every architect has maximum freedom to edit and design their own parts to create variety. This book is both about the detailed research materials of design projects, and the stories of individual human beings. Without knowing the unique features of each architect's experience and personality, we could hardly understand the complex and colourful international design trends. The studies of these architects can help the reader to discover their own interests, and to develop their own architecture. The influences of different cultural backgrounds and education systems in different nations are embodied here. Thus, the book is a window which not only opens towards London but towards many countries in four continents.

In the second part of the book, there are several schemes for "Pavilions in Beijing", which are associated with the 2008 Beijing Olympic Games, designed by some participants as a present for the event and for Chinese readers. The intention in this part is to offer some samples to illustrate different design strategies and approaches to the same topic.

Thanks to every participant, and thanks for their unconditional friendship. Without their support, this book would never have been possible. Special thanks to the friends who contributed to the editing and publication of the book, especially to Sir Peter Cook, whose notion of architecture and generosity to his students make his presence felt throughout the book.

Part2 第二部分

卡提娅·阿柯曼，娜奈德·亚科斯基，洛乌·昆资，里奥纳多·拉塔瓦，里卡多·奥斯图斯
Katja Ackermann, Nannette Jackowski, Raoul Kunz, Leonardo Lattavo, Ricardo Ostos

丽莎·修娃
Lisa Silver

迈克·米切尔
Michael Mitchell

玛若·卡利马尼
Maro Kallimani

阿曼多·赫南德兹
Armando Hernandez

沃肯·奥卡纳格鲁
Volkan Alkanoglu

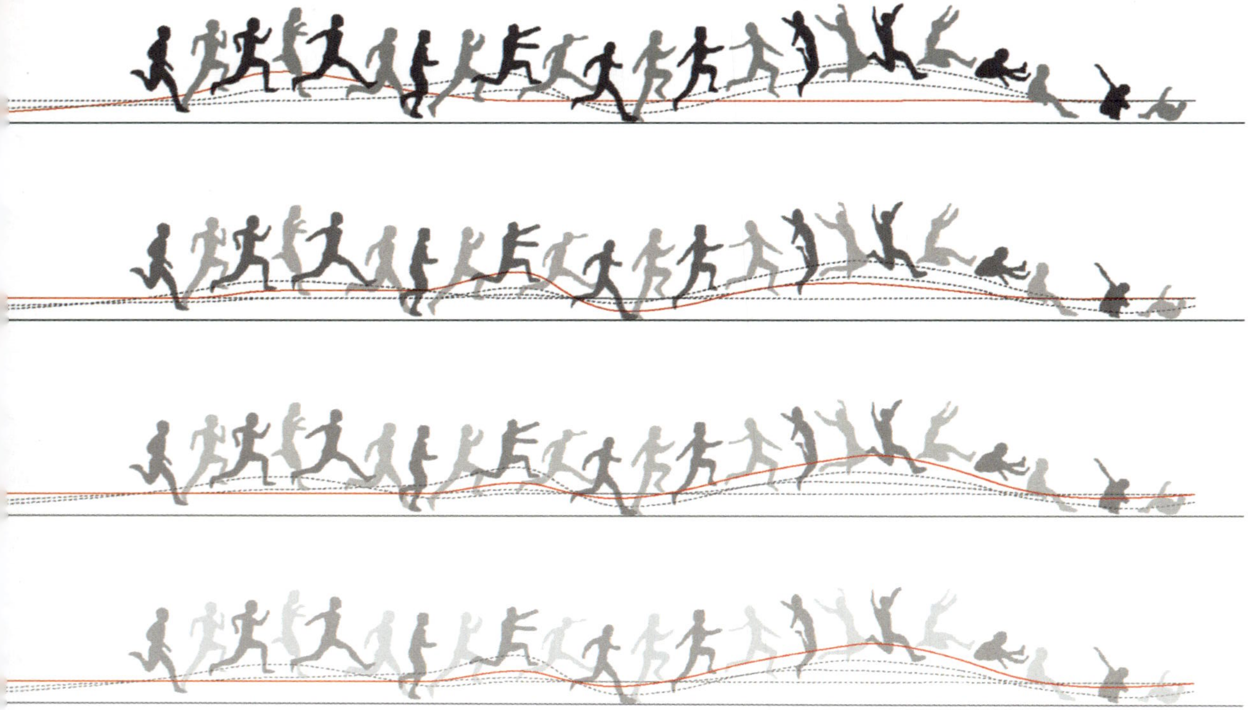

奥运会是一个精英的盛事。全世界最优秀优秀的运动员相聚，竞技他们的力量，意志和耐力。在过去的100多年里，奥运会已经到过大多数最令人向往的城市，它们为世界和平作出了贡献，促进和繁荣了跨文化的沟通与交流。对中国来说，2008年是一个特殊的年份，其首都北京进入了承办奥运会的世界级城市的行列。回首这个国家过去四千年的丰饶多产的历史，中国是一个伟大的国家。中国在科技和社会发展方面的进步表现为许多开创性的发明，这个国家在许多日常生活领域都是先行者。其独特的书写方式、哲学、艺术以及政治组织形式都有助于跨文化交流的繁荣。

在这种强烈的文化背景下，特别是其在艺术和手工艺方面的传统启发我们思考一个与中国传统紧密联系的设计。设计的目标是强化这种文化信心，并从文化遗产中汲取灵感，创造出新精神，结合传统与创新，以庆祝中国文化在奥林匹克精神王国取得的成就。这个项目的特殊挑战在于设计一个宣传使者，来提前宣传奥运精神，以营造兴奋的气氛，为未来的赛事增添魅力，同时也提供这项赛事过去所取得的成就的信息。人们能够清楚地看到，这个设计的任务不是要挑战发布在网路上的巨大信息量，或由大型电视屏幕提供的清晰细节，而是让游客们亲身体验与参与到同样尺度的活动中来。

Olympic Games are an event of the superlative. The best sportsmen of the world meet to compete and compare their strength, will power and endurance. For over 100 years the games of the new world have traveled to the most aspiring cities, which became contributors to peace, progress and cross-cultural fertilization. 2008 marks a special moment for China when its capital Beijing is to join the row of these world class cities honored to conduct the games. Looking back on over 4000 years of rich and fruitful history China is a country of the superlative. With magnificent technological and social progress manifested in many groundbreaking inventions it became a forerunner in many fields of daily life. Its distinctive system of writing, philosophy, art, and political organization helped the cross-fertilization of cultures.

The intense cultural background, particularly in the arts and crafts led us to consider a design closely linked to these traditions. It is the goal to strengthen cultural confidence and inspire a new spirit towards heritage, by presenting an opportunity to merge tradition and innovation and celebrate the achievements of Chinese culture within the realm of the Olympic spirit. The particular challenge lay in designing a herald that described this spirit of the event in advance -to create an atmosphere of excitement and fascination for the future as well as to inform of the achievements of the past. It became clear that the task was not to challenge the amount of information delivered by the Internet, or the level of close-up detail offered by large TV screens, but rather the opportunity for the visitor to physically experience and interact 1:1 with an event of this scale.

Team: Katja Ackermann, Nannette Jackowski, Raoul Kunz, Leonardo Lattavo, Ricardo Ostos.

BEIJING 2008

of the Olympic symbol each prototype is formed from one light core and extensions.

原型研究——展示五个初步可能性中的两个。展示的灵感源于中国的纸张工艺传统。每个原型都是由一个发光核心和其延展部分构成的，拥有奥林匹克标志的五种颜色。

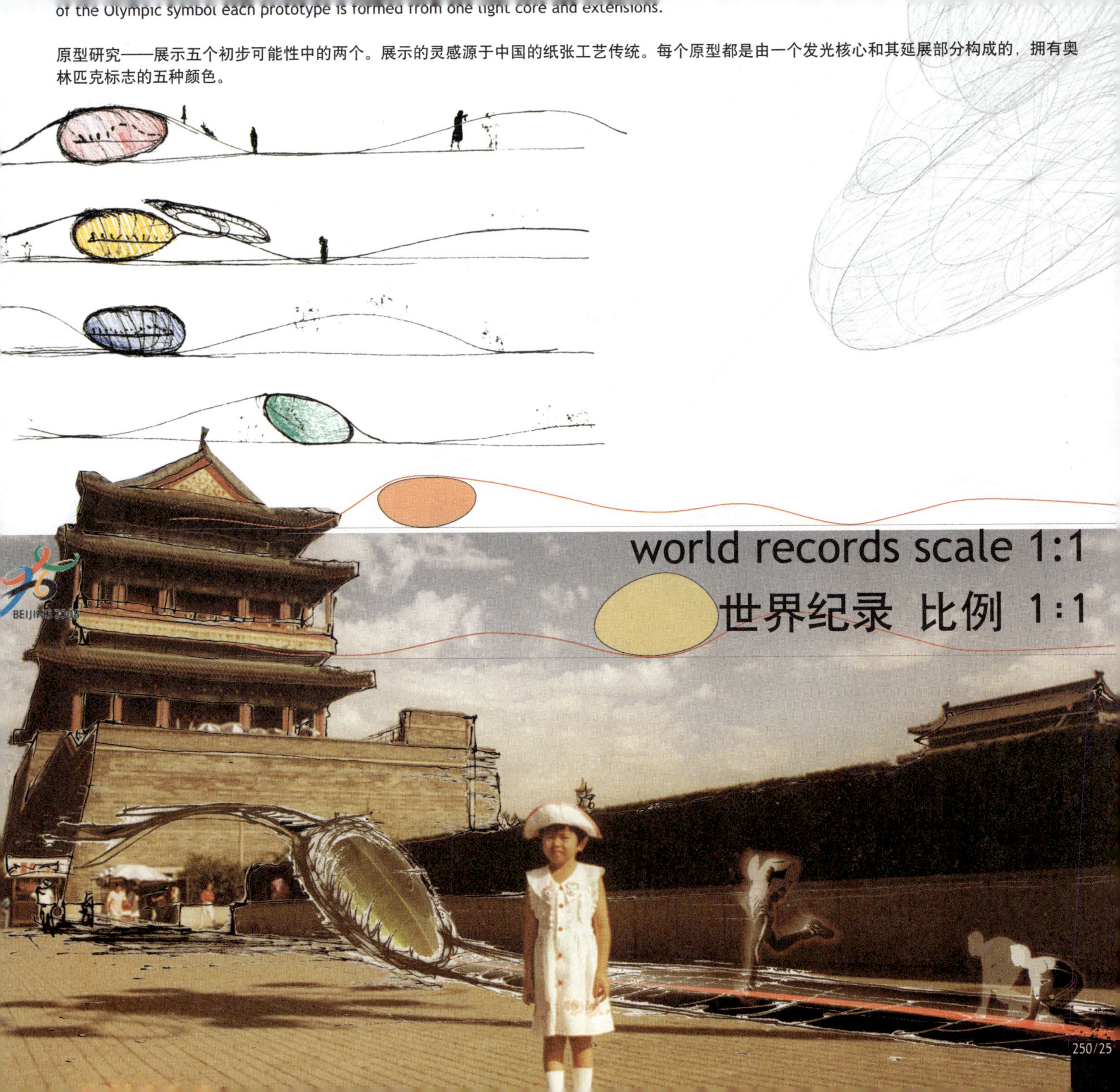

structure_at_rest
spacial__arrangement
area_arrangement
line_arrangement

world records scale 1:1

世界纪录 比例 1:1

The resulting form change of the object allows moving patterns in speed and height along a single or a multidirectional matrix. The installation is thought to occupy vast spaces wall, square, rice fields emphasize the relationship of an individual object to the powers a mass. The structure turns a common object into a device of intriguing flexibility.

结果来自于物体在单方向上或多方向上的不同速度和高度的矩阵运动，该装置被设计成用来占据广阔的空间、墙壁、广场、稻田，强调了单个物体与群体组合之间的关系。这种结构把一个平常的物体变成了一个具有耐人寻味灵活性的装置。

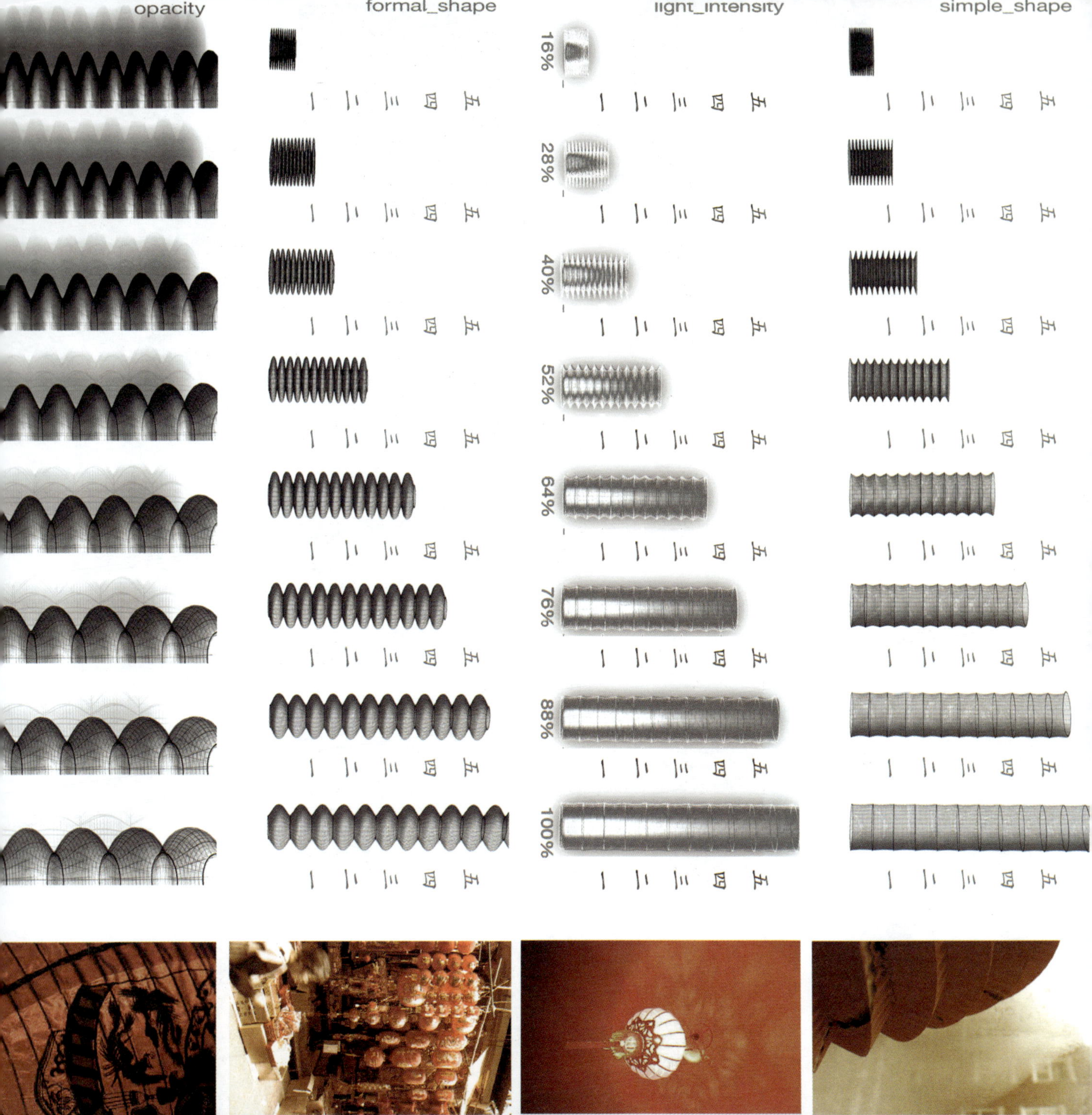
opacity
formal_shape
light_intensity
simple_shape
16%
28%
40%
52%
64%
76%
88%
100%
一 二 三 四 五

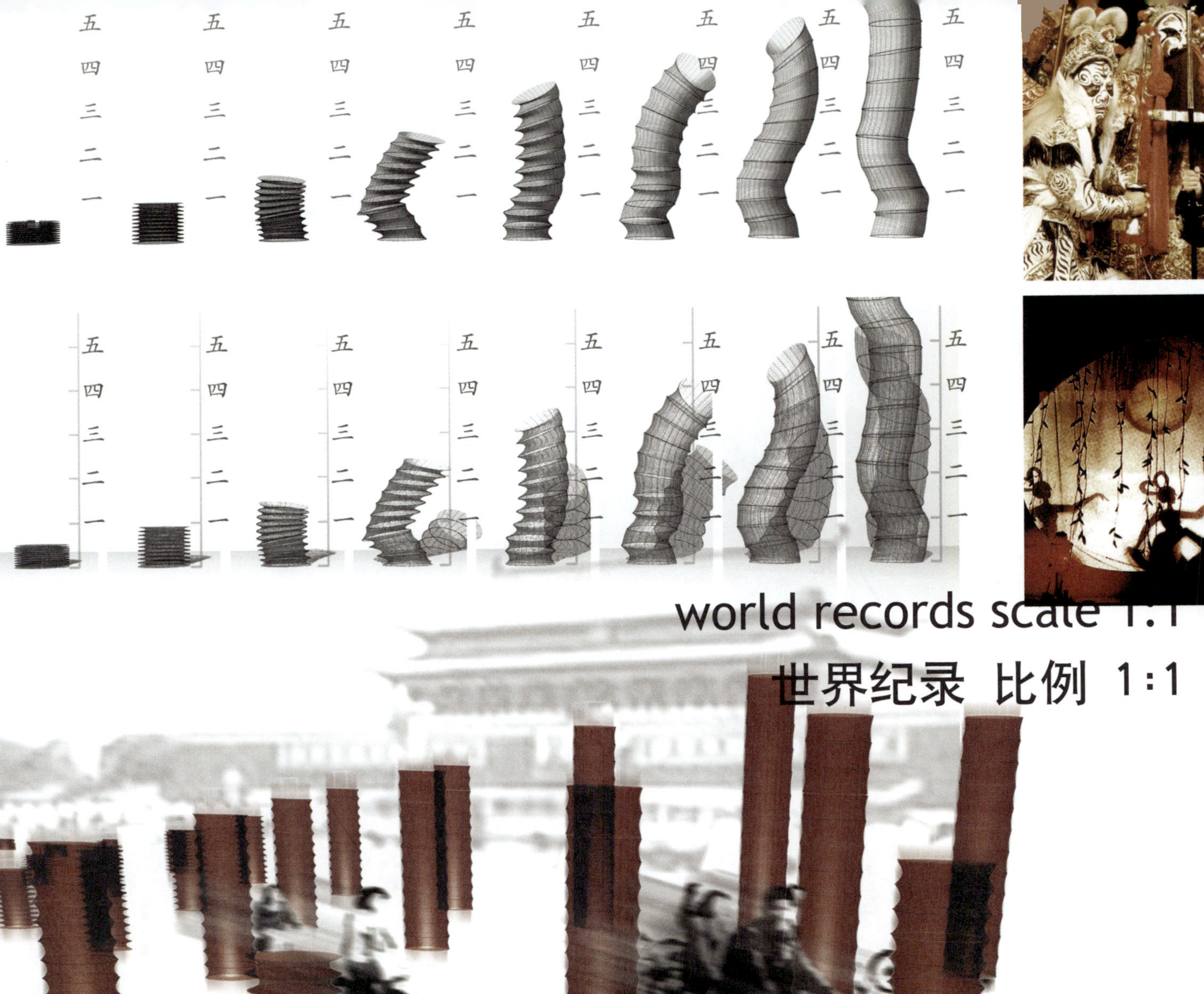

world records scale 1:1

世界纪录 比例 1:1

Studies of light, shadow and transparency explore the materiality of the medi[illegible] its physical context they create an atmosphere of mythical power, imagination and anticipation. Each device is equipped w[illegible]egulating the alternating over and under pressure boosts, which let the shade speedily expand and contract.

对光线，阴影和透明性的研究，探索媒介的材料特性，并把它放在一个实体的背景下[illegible]出一种神秘力量、想像与预言的气氛。每台设备上都配备阀门调节控制气压，以让阴影迅速地扩大和缩小。

world records scale 1:1

世界纪录 比例 1:1

The lantern made of one of the most noticeable of the ancient Chinese inventions and a distinct cultural symbol served as main inspiration since it combines innovation and tradition - like the festive spirit at the lantern festival Yuan Xiao, marks the past and the future at the end of the Chinese New Year.

灯笼作为中国古代最引人注目的发明创造中的一个和独特的文化象征，作为主要的灵感，因为它结合了创新与传统——就像元宵节的节日气氛，标志着过去和将来在年底对中国新年的开始。

OLYMPIC PAVILLION 奥林匹克亭

A MOBILE INFORMATION BOX FOR THE 2008 BEIJING OLYMPIC GAMES

一个为2008年北京奥运会设计的移动信息箱

MAPS AND PLANS OF OLYMPIC AREA 奥运场地的地图与平面图

PHOTOGRAPHIC IMAGES 摄影图像

ARCHIVE OF MEDALISTS /RECOD HOLDERS 奖牌得主/纪录保持者的档案

A MULTITUDE OF PRECIOUS TIME CAPSULES, EACH CONTAINING A CONSEALED ARCHIVE OF INFORMATION, WILL FORM THE BUILDING BLOCKS FOR THE PAVILLION. EACH CAPSULE, HEAVY AND DAZZELING LIKE A GOLD BAR, IS IDENTICAL TO THE NEXT IN ITS EXTERNAL FORM, ALLOWING THE CAPSULES TO INTERLOCK OR HINGE ABOUT EACH OTHER TO FORM MANY DIFFERENT SPACIAL CONFIGURATIONS. IN EACH NEW LOCATION, THOSE THAT CONFIGURE THE CAPSULES WILL DECIDE THE OVERALL FORM, ALLOWING THE LOCAL CONTEXT TO DETERMINE THE DESIGN. BY VIRTUE OF THEIR VERSITILITY THE CAPSULES MAY FORM AN ENCLOSURE, A ROUTE, A SCULPTURE OR INTERVENTION PROVOKING THE USERS TO INTERACT DIFFERENTLY WITH THE PAVILLION IN EACH NEW SITE. EACH CAPSULE REFERS TO ONE YEAR OF THE OLYMIPIC GAMES CONTAINING INFORMATION ON MEDALISTS, WORLD RECORDS AND LOCAL MAPS AND PLANS OF THE OLYMPIC AREA AND STADIA. IN THIS WAY THE PHYSICAL FABRIC OF THE PAVILLION WILL EMBODY THE INFORMATION INSTEAD OF MERELY CONTAINING IT.

奥林匹克亭由许多精致的舱体堆积建造而成，每个舱体内都封存着档案资料。每个舱体都很重、闪闪发光，像金条似的，它们的内部结构完全相同，彼此相互之间可以链接或铰接，整体上形成许多不同的空间配置。每当奥林匹克亭来到一个新的地方，对这些舱体不同的空间配置将会改变亭子的整体形式，以便让基地的特征来决定设计方案。凭借着这种舱体组合“多才多艺”的特性，奥林匹克亭在每个新的站点都是不同的，或许是一种围合、一条路径、一件雕塑，激发使用者以不同方式与之互动。每个舱体里都装有某一年奥运会的相关资料信息，其中包括奖牌得主的信息，世界纪录，奥运场地和体育场馆的地图与平面图。这样，奥林匹克亭将会以实体的形态和结构来体现信息，而不仅只是一个容纳信息的装置。

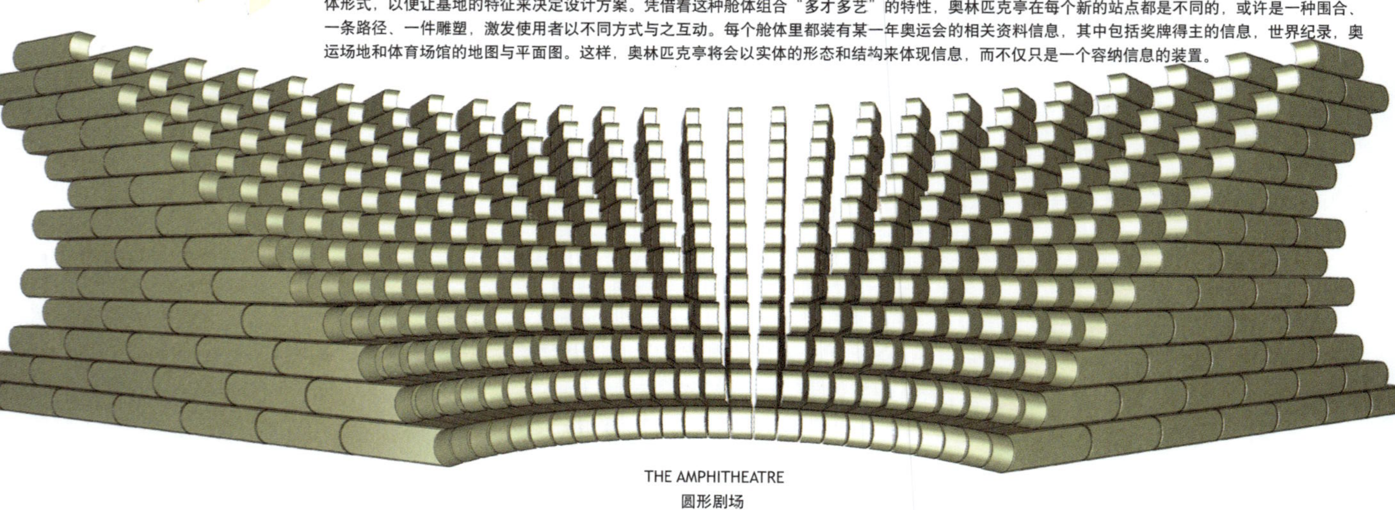

THE AMPHITHEATRE

圆形剧场

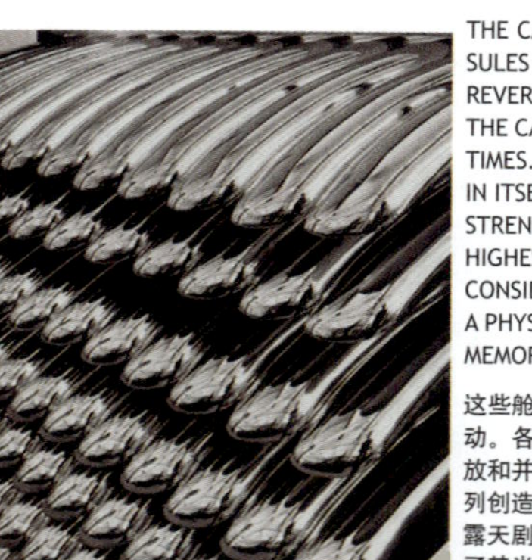

THE CAPSULES ARE SMOOTH AND TACTILE AND PROVOKE DIRECT INTERACTION. VARIOUS SURFACES ARE ESTABLISHED ACCORDING TO HOW THE CAPSULES ARE ORIENTATED AND JUXTOPOSED WITH EACH OTHER. A STEPPED ARRANGEMENT ON ONE SIDE CREATES A CURVED RIPPLE EFFECT ON THE REVERSE. THIS AMPHITHEATRE ARRANGEMENT PROMOTES ITS SEMI ENCLOSED HEART AS A SPACE FOR SOCIAL ACTIVITY. WHILE PEOPLE WANDER AMONG THE CAPSULES, MEETING, PERFORMING AND WAITING WILL TAKE PLACE SIMULTANEOUSLY, AND A CONNECTION IS MADE TO THE COLLUSEUM OF ANCIENT TIMES. GATHERING INFORMATION BECOMES COMPETITIVE IN ITSELF WHEREBY ONE MUST USE THEIR STRENGTH AND ENERGYTO CLIMBP TO THE HIGHEST CAPSULES ANDDISCOVER INFORMATION CONSIDERED MOST NOTEWORTHY. USERS TAKE AWAY A PHYSICAL SENSEOF THE GAMES WHICH MAY HELP TO MEMORISE THE GLEANED INFORMATION.

这些舱体光滑而富于触感，能够激发人们与之直接互动。各种不同形式的表面是由这些舱体相互之间摆放和并置的方式确定的。一个阶梯式的排列创造了一个反向涟漪效果的形态。这个露天剧场作为一个社会活动的空间，发挥了其半封闭的聚合效应，当人们徜徉在这些舱体之间，聚会

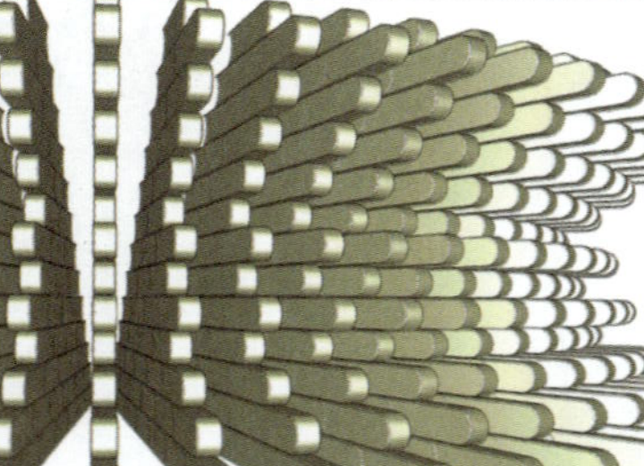

THE CAPSULES MAY FORM A ROUTE, PAVEMENT OR BRIDGE THROUGH OR LEADING TO PART OF THE TOWN OR NEIGHBOURHOOD. A YELLOW BRICK ROAD OF DISCOVEREY SUBTLY INTERTWINES WITH THE LOCAL FABRIC, REMAINING LOW ON THE GROUND PLANE AS A PATH OR RAISED UP AS A THIN SLITHER IN THE SKYLINE. CAPSULES COULD BE HUNG STACKED OR SUPPORTED BY OTHER BUILDINGS OR STRUCTURES ALLOWING THE PAVILLION TO TAKE ON NEW MEANING ACCORDING TO ITS LOCATION. THE BRIDGE FORMATION SUGGETS A LOGICAL ORDERING OF INFORMATION ALONG THE ROUTE WHETHER IT BE CHRONOLOGICAL, GEOGRAPGICAL, ALPHABETICAL OR OTHER. PHYSICAL INTERACTION IS AGAIN NECESSARY; USERS MAY HURDLE FROM ONE SIDE TO THE OTHER AND USE THE CAPSULES AS A PLATFORM, SEAT OR STEPPING STONE. TO GATHER INFORMATION ONE MUST CAREFULLY BALANCE AND MANOUVER THEIR BODIES USING SKILL AND CONTROL TO HARNESS THE CONTAINED INFORMATION.

桥

这些舱体可能会形成一种路径，通往城市或居民区的某一部分的步行道或桥。一条黄色的“发现之路”与当地的城市肌理相微妙地交织纠缠，在地面的部分作为一条路径，在空中升起的部分则构成一条窄细的飘带出现在天际线中。舱体也可以洪堆放或支持其他建筑物或建筑物允许亭采纳了新的内涵，根据其位置。桥的形式暗示了信息排列的逻辑顺序，这种顺序可能是时间的，地理的，字母的或其他的顺序。身体互动也是必要的；使用者也可以从一边跨到另一边，把舱体作为一个平台，一排座椅或者一级台阶。为了搜集信息，人们必须小心地平衡和调动他们的身体，并使用技巧和控制来得到舱体里包含的信息。

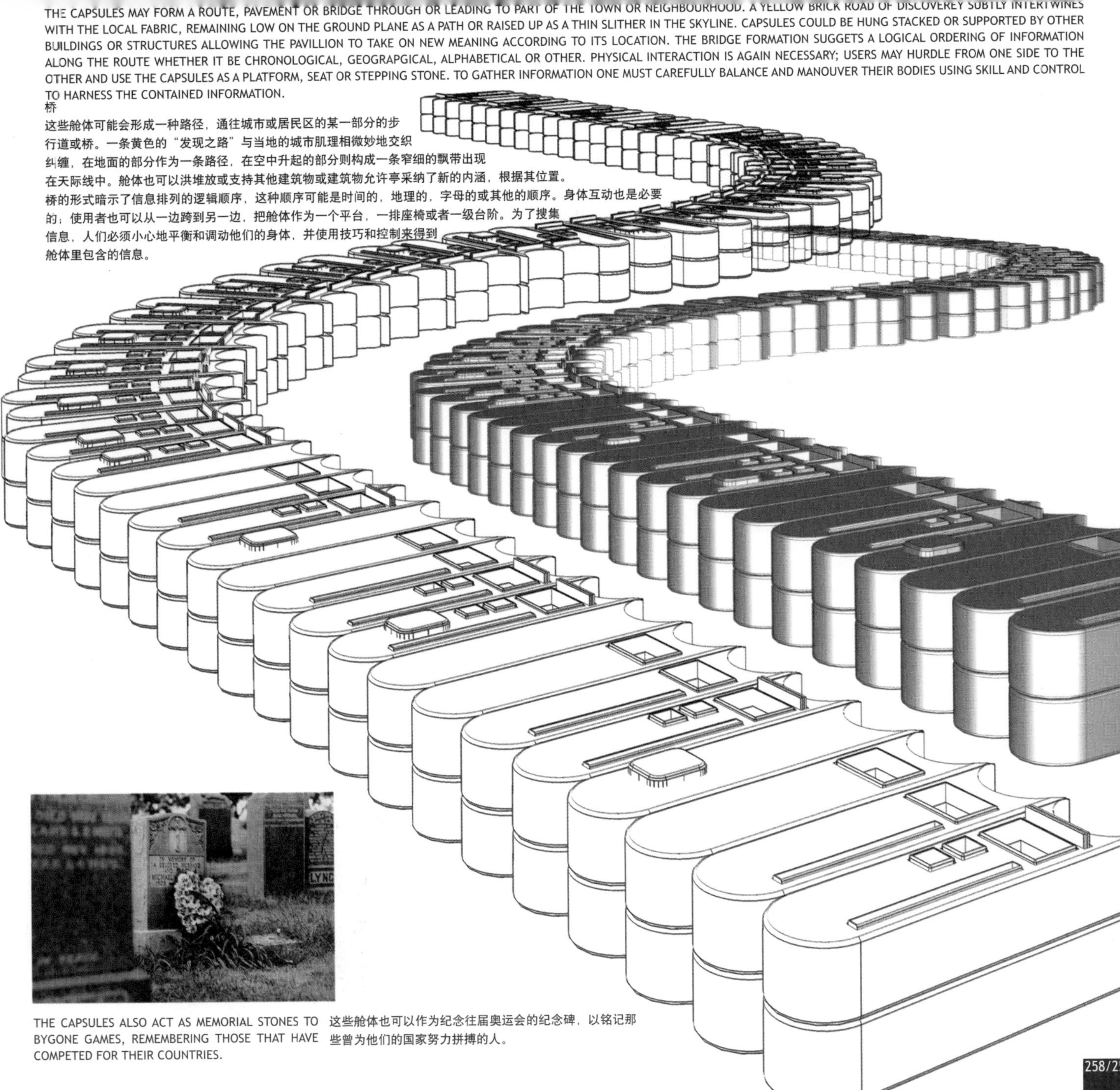

THE CAPSULES ALSO ACT AS MEMORIAL STONES TO BYGONE GAMES, REMEMBERING THOSE THAT HAVE COMPETED FOR THEIR COUNTRIES.

这些舱体也可以作为纪念往届奥运会的纪念碑，以铭记那些曾为他们的国家努力拼搏的人。

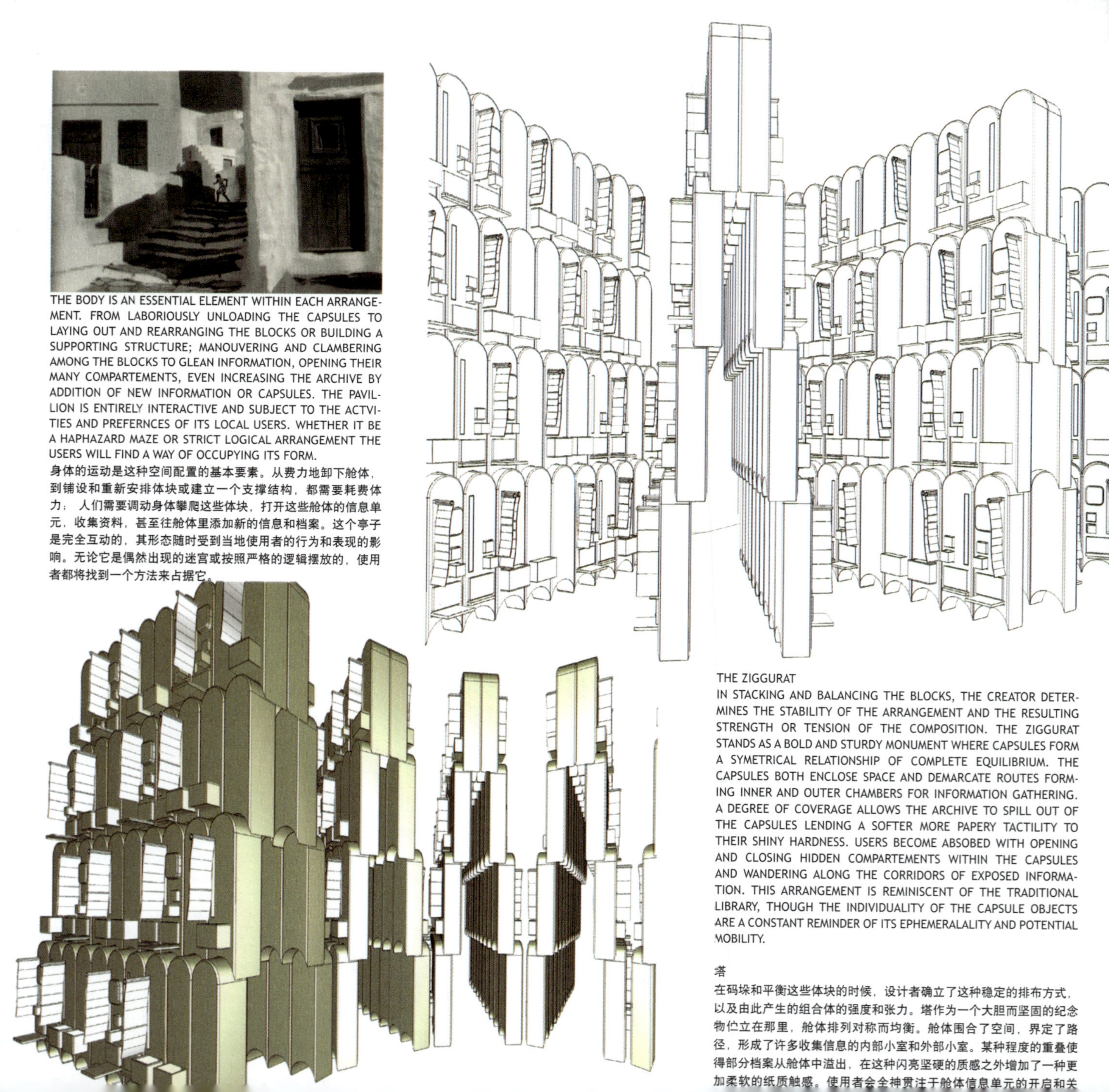

THE BODY IS AN ESSENTIAL ELEMENT WITHIN EACH ARRANGEMENT. FROM LABORIOUSLY UNLOADING THE CAPSULES TO LAYING OUT AND REARRANGING THE BLOCKS OR BUILDING A SUPPORTING STRUCTURE; MANOUVERING AND CLAMBERING AMONG THE BLOCKS TO GLEAN INFORMATION, OPENING THEIR MANY COMPARTEMENTS, EVEN INCREASING THE ARCHIVE BY ADDITION OF NEW INFORMATION OR CAPSULES. THE PAVILLION IS ENTIRELY INTERACTIVE AND SUBJECT TO THE ACTVITIES AND PREFERNCES OF ITS LOCAL USERS. WHETHER IT BE A HAPHAZARD MAZE OR STRICT LOGICAL ARRANGEMENT THE USERS WILL FIND A WAY OF OCCUPYING ITS FORM.

身体的运动是这种空间配置的基本要素。从费力地卸下舱体，到铺设和重新安排体块或建立一个支撑结构，都需要耗费体力；人们需要调动身体攀爬这些体块，打开这些舱体的信息单元，收集资料，甚至往舱体里添加新的信息和档案。这个亭子是完全互动的，其形态随时受到当地使用者的行为和表现的影响。无论它是偶然出现的迷宫或按照严格的逻辑摆放的，使用者都将找到一个方法来占据它。

THE ZIGGURAT

IN STACKING AND BALANCING THE BLOCKS, THE CREATOR DETERMINES THE STABILITY OF THE ARRANGEMENT AND THE RESULTING STRENGTH OR TENSION OF THE COMPOSITION. THE ZIGGURAT STANDS AS A BOLD AND STURDY MONUMENT WHERE CAPSULES FORM A SYMETRICAL RELATIONSHIP OF COMPLETE EQUILIBRIUM. THE CAPSULES BOTH ENCLOSE SPACE AND DEMARCATE ROUTES FORMING INNER AND OUTER CHAMBERS FOR INFORMATION GATHERING. A DEGREE OF COVERAGE ALLOWS THE ARCHIVE TO SPILL OUT OF THE CAPSULES LENDING A SOFTER MORE PAPERY TACTILITY TO THEIR SHINY HARDNESS. USERS BECOME ABSOBED WITH OPENING AND CLOSING HIDDEN COMPARTEMENTS WITHIN THE CAPSULES AND WANDERING ALONG THE CORRIDORS OF EXPOSED INFORMATION. THIS ARRANGEMENT IS REMINISCENT OF THE TRADITIONAL LIBRARY, THOUGH THE INDIVIDUALITY OF THE CAPSULE OBJECTS ARE A CONSTANT REMINDER OF ITS EPHEMERALALITY AND POTENTIAL MOBILITY.

塔

在码垛和平衡这些体块的时候，设计者确立了这种稳定的排布方式，以及由此产生的组合体的强度和张力。塔作为一个大胆而坚固的纪念物伫立在那里，舱体排列对称而均衡。舱体围合了空间，界定了路径，形成了许多收集信息的内部小室和外部小室。某种程度的重叠使得部分档案从舱体中溢出，在这种闪亮坚硬的质感之外增加了一种更加柔软的纸质触感。使用者会全神贯注于舱体信息单元的开启和关

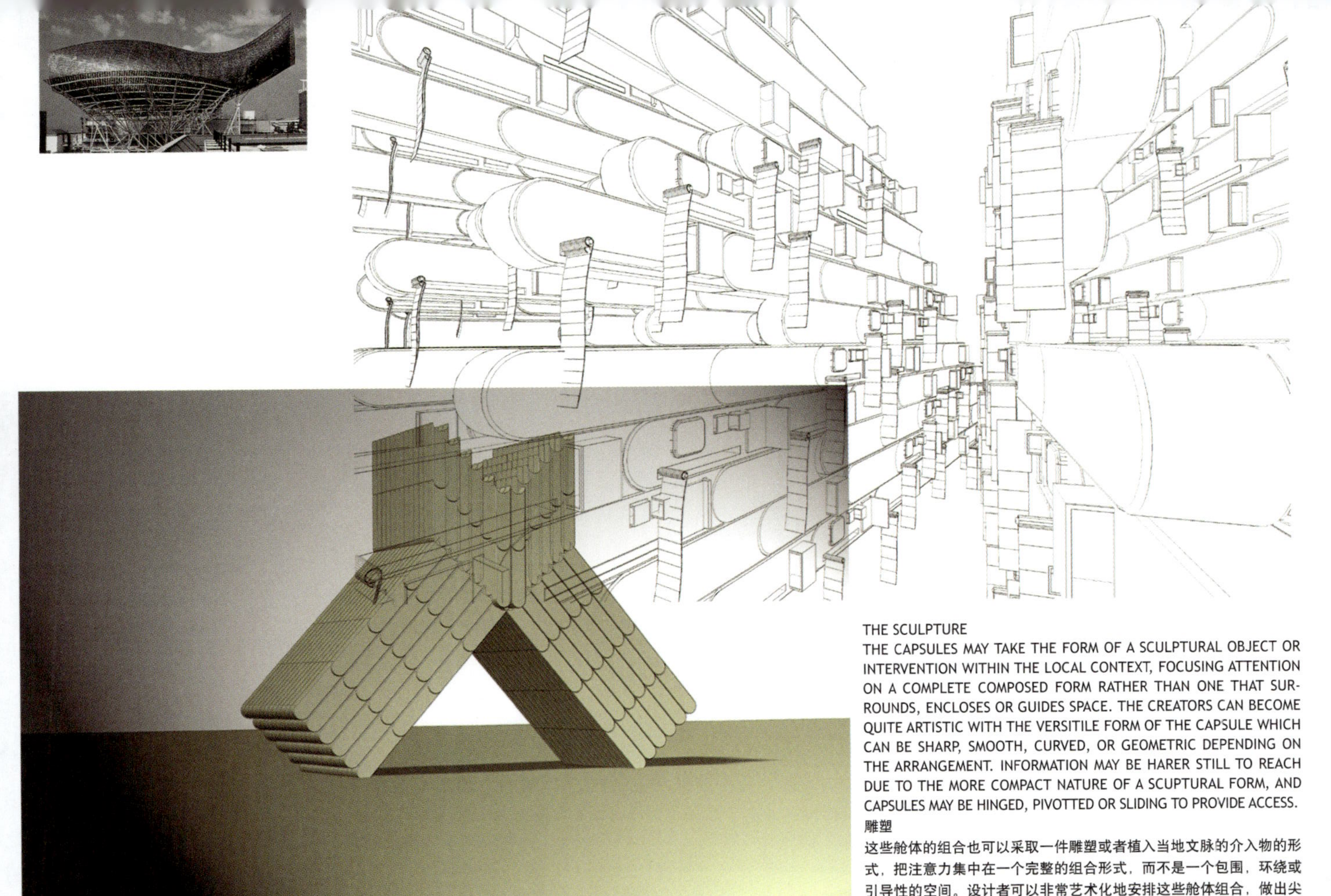

THE SCULPTURE

THE CAPSULES MAY TAKE THE FORM OF A SCULPTURAL OBJECT OR INTERVENTION WITHIN THE LOCAL CONTEXT, FOCUSING ATTENTION ON A COMPLETE COMPOSED FORM RATHER THAN ONE THAT SURROUNDS, ENCLOSES OR GUIDES SPACE. THE CREATORS CAN BECOME QUITE ARTISTIC WITH THE VERSITILE FORM OF THE CAPSULE WHICH CAN BE SHARP, SMOOTH, CURVED, OR GEOMETRIC DEPENDING ON THE ARRANGEMENT. INFORMATION MAY BE HARER STILL TO REACH DUE TO THE MORE COMPACT NATURE OF A SCUPTURAL FORM, AND CAPSULES MAY BE HINGED, PIVOTTED OR SLIDING TO PROVIDE ACCESS.

雕塑

这些舱体的组合也可以采取一件雕塑或者植入当地文脉的介入物的形式，把注意力集中在一个完整的组合形式，而不是一个包围，环绕或引导性的空间。设计者可以非常艺术化地安排这些舱体组合，做出尖锐的、光滑的、弯曲的、或者是依据空间配置方式形成的几何形式。由于雕塑更加紧凑的特点，信息可以被更容易的获得，这些舱体可以是铰接的、轴动的或滑动以提供一个通道。

TRANSPORTATION

WHEN READY TO MOVE ON, THE PAVILLION CAN BE DISASSEMBLED INTO ITS COMPONENT CAPSULES AND COMPACTED INTO A TIGHT BLOCK FOR TRANSPORTATION. THE INFORMATION IS SAFELY HELD INSIDE THE COMPARTEMENTS OF THE CAPSULES AND EACH CAPSULE IS SECURED TO THE NEXT BY ITS INHERENT FORM. THE CAPSULES ARE ENTIRELY WATERPROOF WITH A HARDY AND ROBUST OUTER SKIN. DURING TRANSPORTATION, THEIR PHYSICAL FORM IS ONE OF A GLIMMERING HOMOGENOUS BLOCK, WHICH PROMOTES THE GAMES EVEN ON ROUTE. THEREFORE, THIS OLYMPIC PAVILLION IN ALL ITS FORMS ACTS AS A CONTINUAL REMINDER OF THE REWARDS FOR PHYSICAL EFFORT, COMPETION AND PATRIOTISM.

运输

当准备前往下一个地点时，这个亭子可以被拆解成若干个舱体并被压缩成一个紧密结合的块体以便运输。信息被安置在每个舱体内部独立的单元内以便安全地携带，由于其内在的结构，每个舱体相对旁边的部分都独立安全。这些舱体拥有坚固强韧的外表面，能够完全防水。在运送过程中，他们的物理形态是一个闪着微光的块体，甚至在运输的路途上也可以用来宣传推广奥运会。因此，这个奥运体亭在任何形式下，都会不断提示人们体育锻炼的好处、宣扬竞技精神和进行爱国主义教育。

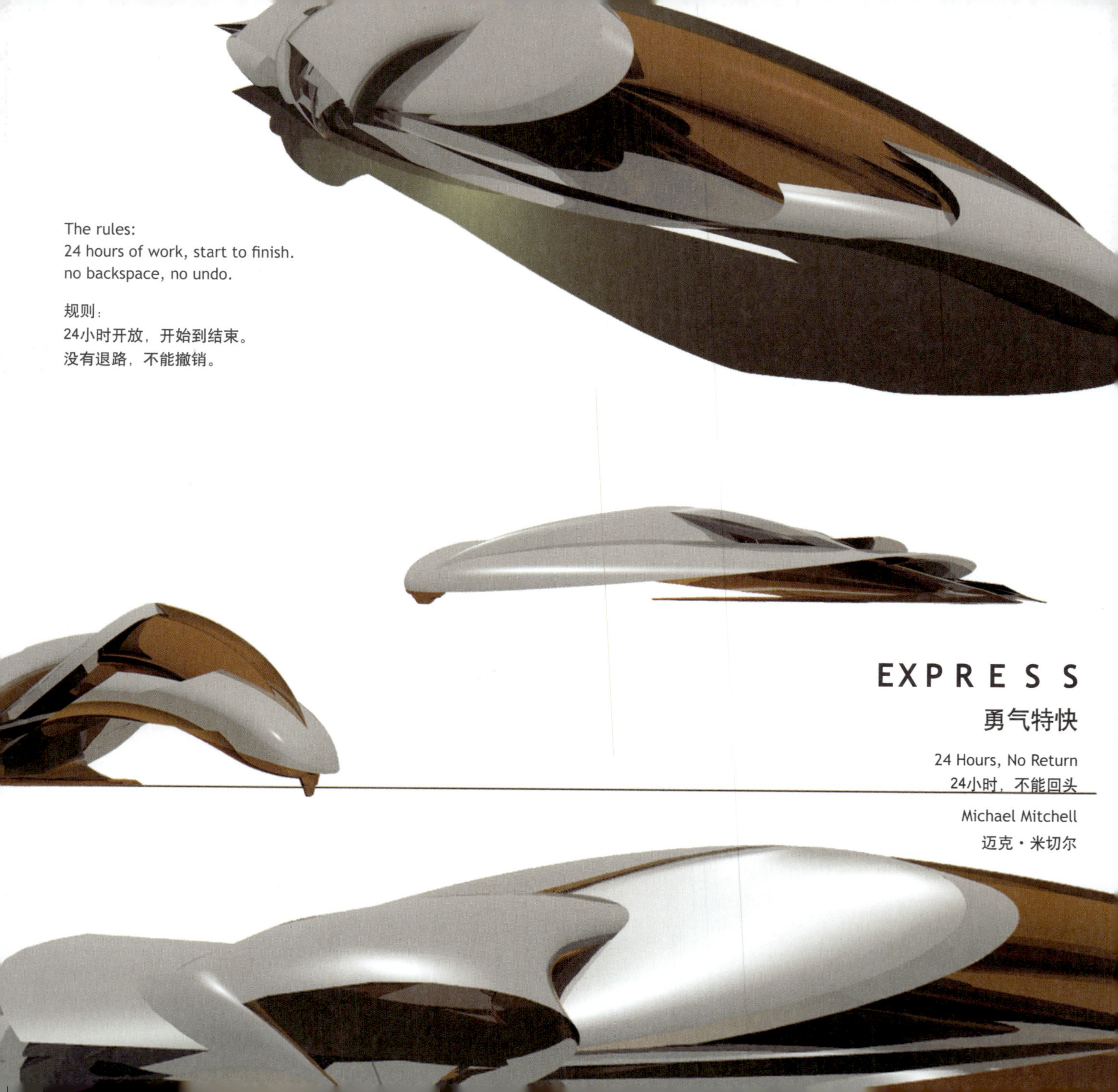

The rules:
24 hours of work, start to finish.
no backspace, no undo.

规则：
24小时开放，开始到结束。
没有退路，不能撤销。

EXPRESS
勇气特快

24 Hours, No Return
24小时，不能回头

Michael Mitchell
迈克・米切尔

The pavilion encourages the one to stand for the first time before the few. The form is provocative yet protective, and contains a small stage and room for about 30 seats. There are theatrical lighting and speakers embedded in the building and the only other program is a matchbox sized bar above to provide 'liquid courage' for those going on stage.

这个亭子鼓励人们在少数人面前首次走上舞台。亭子的形式既能激发勇气也能保护勇气，包括一个小型舞台和大约容纳30个座席的空间。建筑物里有内置的剧场灯光和喇叭，唯一的其他项目是一个火柴盒大小的酒吧，为那些准备上舞台人提供“带来勇气的液体”。

Open modestly to the left and to the right, the pavilion invites people outside to observe the performance at arm's length without being directly involved.

亭子适度地向左侧与右侧开放，使得亭子外面的人们在一尺之遥不用直接参与，也能观看表演。

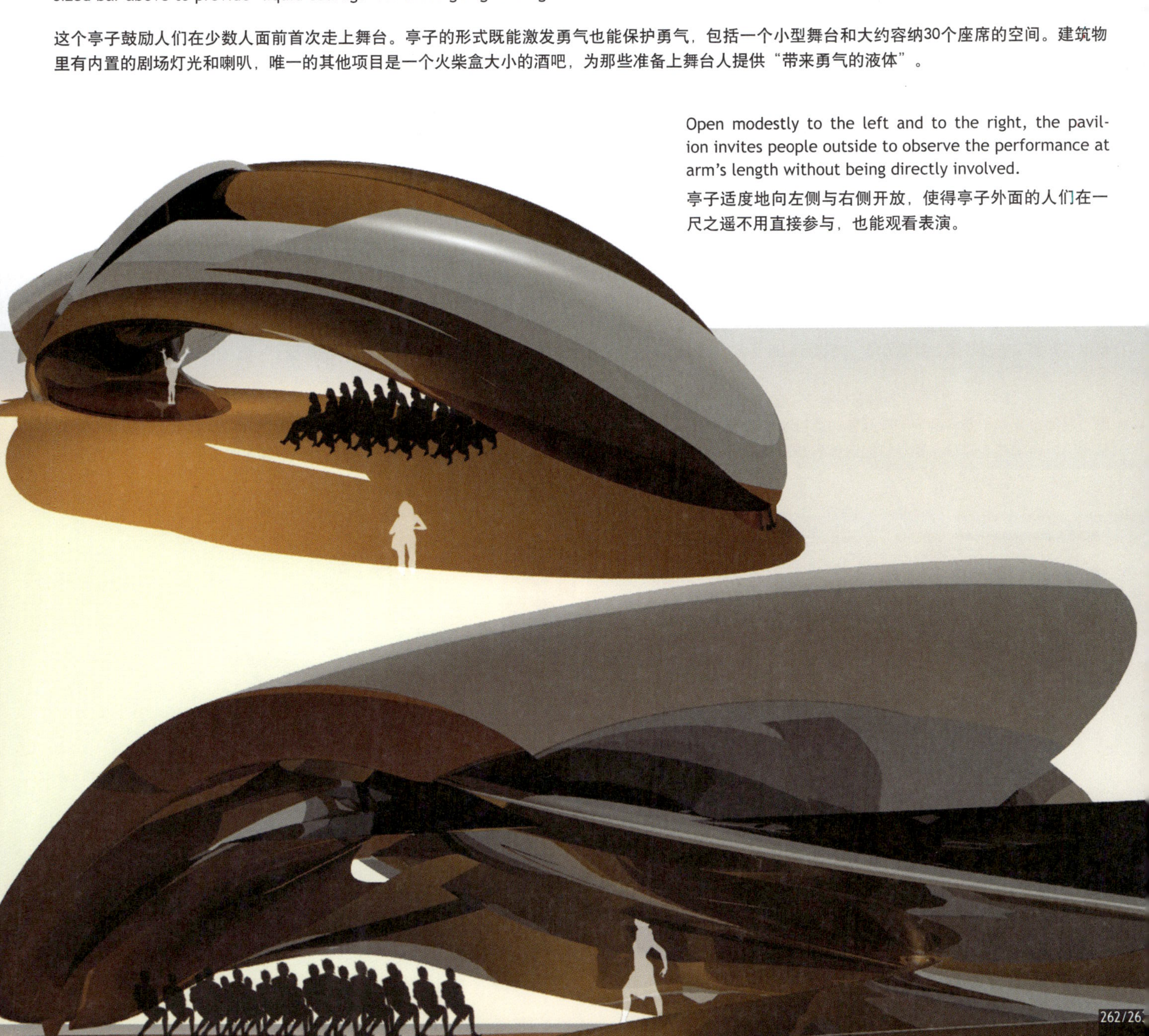

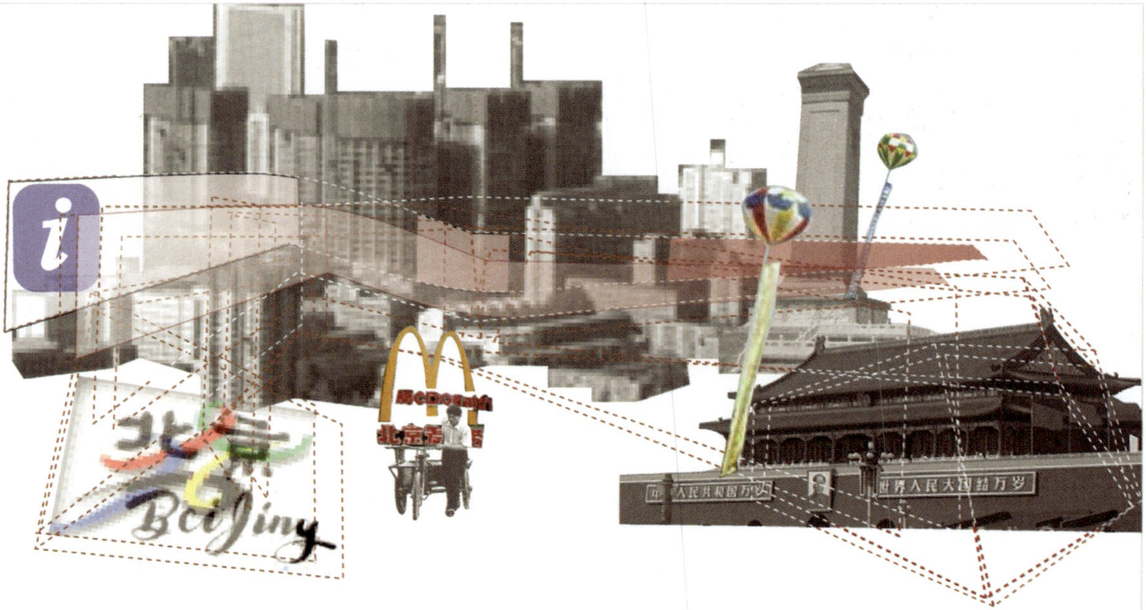

信息在变化、在游走、在没完没了地运动。这是一种当其成为历史时能加以控制的能量。信息在北京保持着相同的特点，在这个城市中，旧的传统的与新的当代的相对峙。过去的深厚教养与当前的西方化相对立；尺度很小的胡同和园林建筑与大尺度的混凝土街区和总部大楼形成对比。这个亭子，描绘了信息的性质，并考虑到北京的多元性格，是一个拉长的姿态散布在这个城市中，并统一了其多重特征。

信息亭是由”皮肤”和”塔” 组成的。 “皮肤”是一个可以灵活展开的”丝带”，创造出空心和过渡效果。它包括两个高分辨率的屏幕，用许多单独的预制面板组合在一起形成一个大的图像，与城市巨大的尺度相联系。发送的数据是由一个中心站控制的，这里每天24小时一周7天的全天侯不间断播出奥运会的信息。在钢架结构上，有一部分被选择出来，覆盖上可以控制的信息接入的互动监视器。它的片段特性给予了形成不同的空间形态的自由，根据在城市各个不同地点的特征。 “塔”是一个静态体量，作为一个地标——一种观光景点的信号。这个玻璃塔般的建筑占据了最小的空间，是一个由人管理的提供印刷资料的信息处。其发光的顶部，根据奥林匹克标志的颜色不断变换色彩，表明了该信息点在城市里的位置。

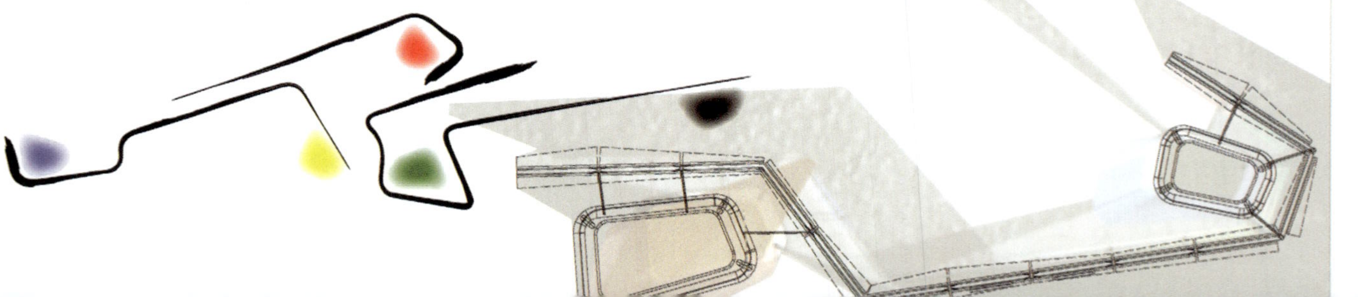

Information is altering, is travelling and perpetually moving. It is the energy, which can only be controlled when it becomes history. Information maintains the same characteristics in Beijing, a city, where the old and traditional is opposed to the new and contemporary. The sophistication of the past is opposed to present westernisation. The small scale of the hutongs and garden architecture is in contrast to the big scale of the concrete blocks and the headquarters. The pavilion, which depicts information's nature and takes into account Beijing's diverse character, is a drawing gesture spreading in the city and unifying its multiple characteristics.

The info pavilion is composed of the "skin" and the "tower". The "skin" is a flexible unfolding "ribbon" creating hollows and transitions. It consists of two high-resolution screens made of prefabricated separate panels that work together and form a large image related to city's big scale. The sent data are controlled from a central station, which broadcasts information for the Olympic games in 24/7 modes. Selected parts of its structural steel frame are covered with interactive monitors for controllable info access. Its fragment able nature gives also the freedom for different space formations, according to the urban characteristics of various locations in the city. The "tower" is a static volume functioning as a landmark - a sign of attraction. This glass tower-like construction occupies the minimum space for a manned info bureau providing printed information. Its luminous top, coloured according to the Olympic sign colours, indicates the position of the information point in the city.

玛若 • 卡利玛尼

Maro Kalimani

olympic information kiosk

taking into consideration the direction of moden technology, this project is meant to reflect what can be donme in the future with new technology and materials.

project

translucent kiosk with expandable screens to accomodate an increasing population of users. screans are transparent LCD displays that can be used from both sides creating an interaction between two users and the information being sought. navigation is through touch screen systems. main structure is composed of translucent GRP that can have any possible color of illumination.

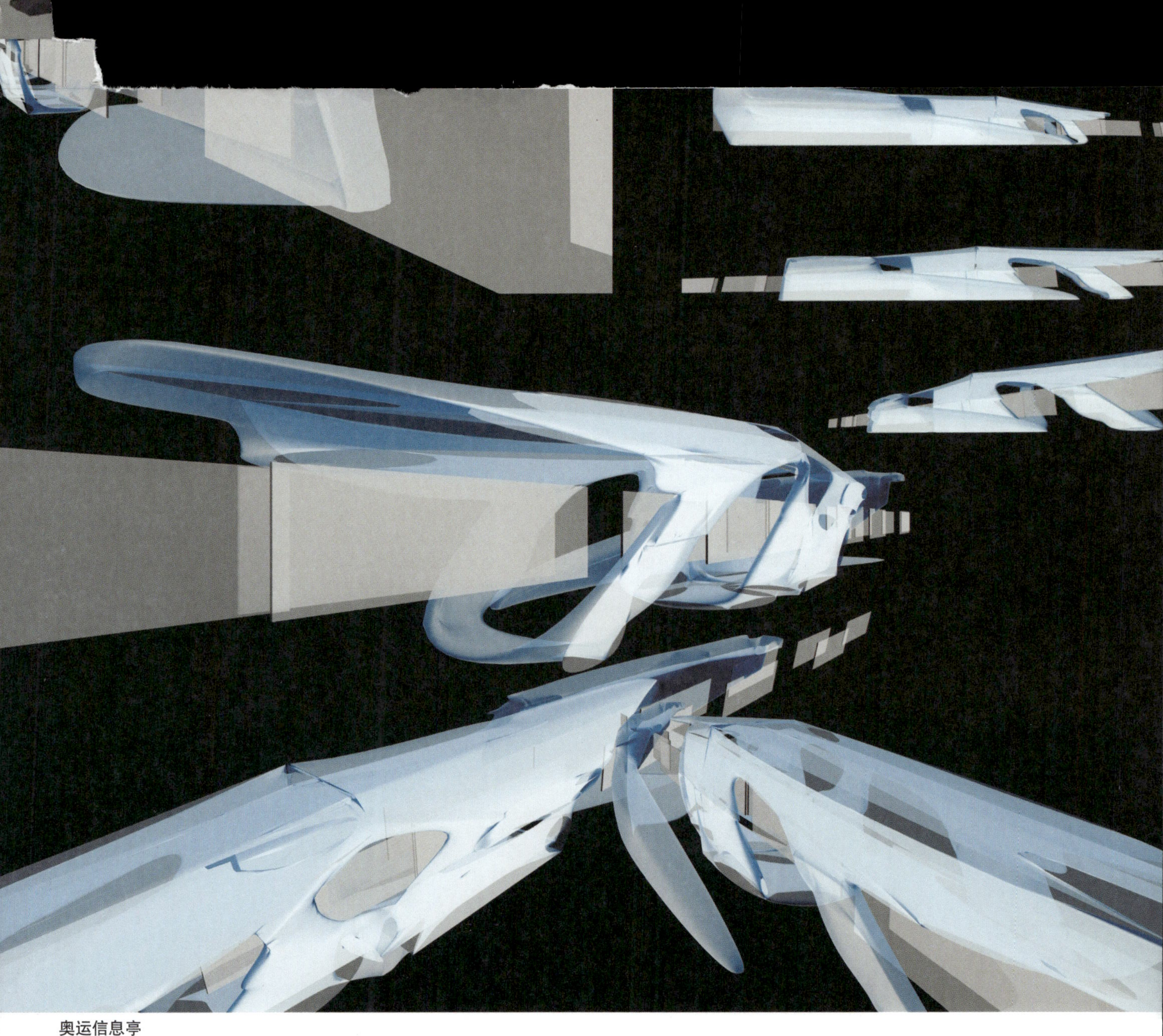

奥运信息亭

考虑到现代科技发展的方向，这个项目是想展示用新的技术和材料在未来可以做些什么。

项目：

有可扩展屏幕的半透明亭子可以容纳不断增加的用户。屏幕是透明的、两面都可以显示的液晶显示器，可以创造出两个用户与正在搜寻的信息之间的互动。导航是通过触摸屏系统进行的。主体结构是由半透明的玻璃钢组成的，可以用任何可能的颜色照明。

阿曼多 • 赫南德兹
Armando Hernandez

... AND THE OLYMPIC GAMES 2008 WILL BE IN......BEIJING
...2008年奥运会将在......北京

Artificial narrator puppet in glazed blob sphere
人工叙述木偶
在光滑的泡表面

First design sketches of pavilion
亭子设计素描第一稿

>Conceptual idea The pavilion for the Olympic Games in Beijing 2008 will be placed in several cities around china to inform the population about the outstanding event.It will also refer to the history and spirit of the Olympic Games since their first appearance in ancient Greece. The conceptual idea of this pavilion is to place a object in a local public square and occupy the surrounding space by unfolding a flexible banner. The object can easily be transported with a truck. The surface of the fabric banner will provide prints and pictures of former Olympic events and the new sport facilities and developments in Beijing. The banner can be unrolled and set up in certain positions to create flexible space arrangements. The object itself is divided in two functional parts. One is housing the rolled banner and the other part is a glazed sphere. Within this sphere will be a mechanical puppet with the look of a human being. It will tell the story of the Olympic Games and act like a narrator. The sound will be distributed by speakers flush mounted on side of the housing. So people can gather in front of the object and listen to the narrator and watch his artificial body movement and gestures. The technique and use of these puppets was popular and common within lunaparks in Europe in the early 80`s. The form of the object is round and blobby. The main orientation of the object is the sphere but due to the shape still maintaining as a hole.

>概念性想法
为2008年北京奥运会设计的亭子，将被放置在几个中国的城市，向大众宣传这项盛事。它也将回溯自从第一届古希腊雅典奥运会以来的历史和奥运精神。这个亭子概念性的想法，是要把一个物体放置在一个当地的公共广场上，并且通过展开一个可活动的织物横幅，来围合占据周边的空间。这个物体可以很容易被装进一辆卡车以便运输。织物横幅的表面有历届奥运会的印刷品和照片，也有北京奥运会新建体育设施和发展的信息资料。横幅可以被打开，并立在某些位置来创造灵活的空间安排。该物体本身被分成两个功能部分。一部分可以容纳卷起的横幅，另一部分则是一个光滑的球体。在这球体部分，有一个人类模样的机械木偶。它会作为一个叙述者讲述奥运会的故事。声音将从屋子两侧的扬声器里发出，以使人们聚集在这个物体的前面，倾听木偶的讲述，观看它人工身体的动作和姿态。在1980年代初的欧洲，这种技术和木偶的使用在月亮公园已经非常普遍。这个物体的形状是圆的，水滴状的。由于这个物体其主要方向上是一个球体，其形状仍保持为一个洞。

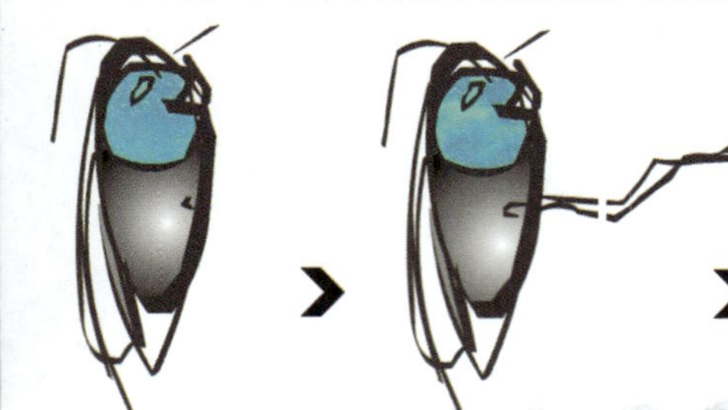

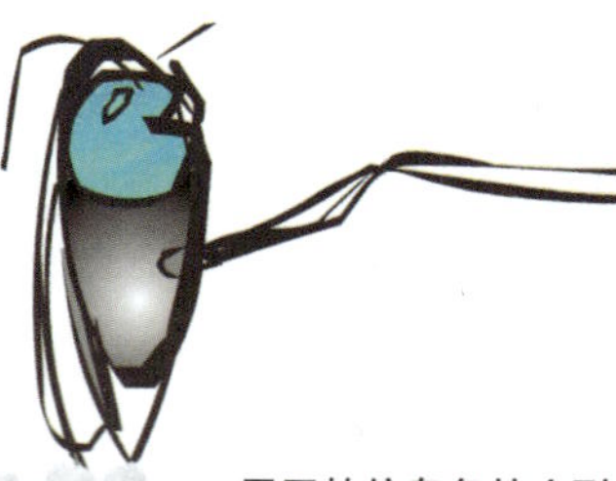

seque
展开的信息条的序列

top view of infoblob in extended position
信息泡在延长外置的顶视图

perspective of infoblob in use
信息泡在使用中的视图

图书在版编目(CIP)数据

伦敦开题：15位国际新生代建筑师的伦敦建筑教育体验及实践/潘岩等编著. -北京：中国建筑工业出版社，2009
ISBN 978-7-112-10360-7

I.伦… II.潘… III.建筑设计-作品集-世界-现代 IV.TU206

中国版本图书馆CIP数据核字（2008）第155810号

责任编辑：徐 纺

伦敦开题：
15位国际新生代建筑师的伦敦建筑教育体验及实践
潘岩等编著
*
中国建筑工业出版社出版、发行（北京西郊百万庄）
各地新华书店、建筑书店经销
恒美印务（广州）有限公司制版、印刷
*
开本：889×1194毫米 1/20 印张：13.8 字数：322千字
2009年3月第一版 2009年3月第一次印刷
定价：110.00元
ISBN 978-7-112-10360-7
（17163）